Synergetics

A Workshop

Proceedings of the International Workshop on Synergetics at Schloss Elmau, Bavaria, May 2–7, 1977

Edited by H. Haken

With 136 Figures

Springer-Verlag Berlin Heidelberg New York 1977

Professor Dr. Hermann Haken

Institut für Theoretische Physik der Universität Stuttgart
Pfaffenwaldring 57/IV, D-7000 Stuttgart 80, Fed. Rep. of Germany

ISBN 3-540-08483-5 Springer-Verlag Berlin Heidelberg New York
ISBN 0-387-08483-5 Springer-Verlag New York Heidelberg Berlin

Offset printing and bookbinding: Zechnersche Buchdruckerei, Speyer

2153/3130-543210

Preface

This volume contains most of the invited papers presented at the International Workshop on Synergetics, Schloss Elmau, Bavaria, May 2 to May 7, 1977. This workshop followed an International Symposium on Synergetics at Schloss Elmau, 1972, and an International Summerschool at Erice, Sicily, 1974.

Synergetics is a rather new field of interdisciplinary research which studies the self-organized behavior of systems leading to the formation of structures and functionings. Indeed the whole universe seems to be organized, with pronounced structures starting from spiral galaxies down to living cells. Furthermore, very many of the most interesting phenomena occur in systems which are far from thermal equilibrium. Synergetics in its present form focusses its attention on those phenomena where dramatic changes occur on a macroscopic scale. Here indeed Synergetics was able to reveal profound analogies between systems in different disciplines ranging from physics to sociology. This volume contains contributions from various fields but the reader will easily discover their common goal. Not only in the natural sciences but also in ecology, sociology, and economy, man is confronted with the problems of complex systems. The principles and analogies unearthed by Synergetics will certainly be very helpful to cope with such difficult problems.

I use this opportunity to thank the Volkswagenwerk Foundation for its support of the project Synergetics and in particular for sponsoring the International Workshop on Synergetics.

Finally, I thank my secretary, Mrs. U. Funke, for her great help in the organization of this workshop and her assistance in the preparation of these proceedings.

Stuttgart, May 1977 H. Haken

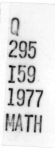

Contents

CHEMICAL WAVES AND TURBULENCE

MORPHOGENESIS

BIOLOGICAL SYSTEMS

GENERAL SYSTEMS

List of Contributors

Buckley, Prof. Dr. Walter, University of New Hampshire, College of Liberal Arts, Social Science Center, Durham, NH 03824 USA

Bullough, Prof. Dr. Robin K., University of Manchester, Institute of Science and Technology, Dept. of Math. P.O.B.88, Manchester M601QD, England

Cowan, Prof. Dr. Jack, Max-Planck-Institut für biophysikalische Chemie, 34 Göttingen, Fed. Rep. of Germany

Czajkowski, Dr. G., Institute of Mathematics and Physics, Technical and Agricultural Academy, Hanki Sawickiej 28, 85-084 Bydgoszcz, Poland

Dodd, Dr. R.K., Department of Mathematics, University of Manchester, P.O.B.88, Manchester M601QD, England

Fenstermacher, Dr. P.R., Physics Department, The City College of the City University of New York, New York, NY 10031 USA

Fröhlich, Prof. Dr. Herbert, University of Liverpool, Dept. of Physics, P.O. Box 147, Liverpool L69 3BX, England

Fröhlich, Mrs. Fanchon, Liverpool, England

Gollub, Dr. J.P., Physics Department, Haverford College, Haverford, PA 19041 USA

Haken, Prof. Dr. Hermann, Institut für theoretische Physik der Universität Stuttgart, Pfaffenwaldring 57, 7000 Stuttgart 80, Fed. Rep. of Germany

Joseph, Prof. Dr. Daniel D., University of Minnesota, Dept. of Aerospace Engineering and Mechanics, 110 Union Street S.E., Minneapolis, MN 55455 USA

Kastler, Prof. Dr. Alfred, Université Paris VI, Laboratoire de Spectroscopie Hertzienne de l'Ecole normale superieure, 24, Rue Lhomond 75231 Paris, France

Kippenhahn, Prof. Dr. Rudolf, Max-Planck-Institut für Physik und Astrophysik, Institut für Astrophysik, Föhringer Ring 6, 8000 München 40, Fed. Rep. of Germany

Kirchgässner, Prof. Dr. Klaus, Mathemat. Institut A der Universität Stuttgart, Pfaffenwaldring 57, 7000 Stuttgart 80, Fed. Rep. of Germany

Koschmieder, Prof. Dr. E.L., College of Engineering, The University of Texas at Austin, Austin, TX 78712, USA

Kuhn, Prof. Dr. Hans, Max-Planck-Institut für biophysikalische Chemie, Postfach 968, 3400 Göttingen-Nikolausberg, Fed. Rep. of Germany

Kuramoto, Prof. Dr. Yoshiki, Institut für theoretische Physik der Universität Stuttgart, Pfaffenwaldring 57, 7000 Stuttgart 80, Fed. Rep. of Germany (Department of Physics, Kyoto University, Kyoto, Japan)

Maresquelle, Prof. Dr. H.J., Institut de Botanique, 28 rue Goethe, 67000 Strasbourg, France

Meinhardt, Dr. Hans, Max-Planck-Institut für Virusforschung, Molekularbiologische Abteilung, Spemannstr. 35, 7400 Tübingen, Fed. Rep. of Germany

Nitzan, Prof. Dr. Abraham, Tel-Aviv University, Institute of Chemistry, 61390 Ramat-Aviv, Tel-Aviv, Israel

Pacault, Prof. Dr. A., Centre de Recherche Paul Pascal, Domaine Universitaire, 33405 Talence, France

Rössler, Dr. Otto E., Universität Tübingen, Institut für physikal. und theoretische Chemie, Auf der Morgenstelle 8, 74 Tübingen, Fed. Rep. of Germany and Institute for Theoretical Physics, University of Stuttgart, Fed. Rep. of Germany

Sattinger, Prof. David H., University of Minnesota, School of Mathematics, 206 Church Street S.E., Minneapolis, MN 55455 USA

Swinney, Prof. Dr. Harry L., The City College, The University of New York, Dept. of Physics, New York, NY 10031 USA

Thom, Prof. Dr. René, Institut des Hautes Etudes Scientifiques, 91440 Bures-sur-Yvette, France

Welge, Dr. Martin K., Universität Köln, Seminar für Allgemeine Betriebswirtschaftslehre und Organisationslehre, Albertus-Magnus-Platz, 5000 Köln 41, Fed. Rep. of Germany

General Concepts

Some Aspects of Synergetics

H. Haken

With 14 Figures

The speakers and participants of this International Workshop on Synergetics come from different disciplines, namely mathematics, physics, astrophysics, chemistry, biology and sociology. Accordingly, their topics seem to be quite different. Therefore I feel I should give a short introduction why I believe that such a workshop is meaningful and moreover, why "synergetics" is a promising new field of interdisciplinary research. The word "synergetics" is composed of two Greek words and means "working together". In many disciplines, ranging from astrophysics over biology to sociology, we observe that very often the cooperation of many individual parts of a system leads to macroscopic structures or functionings. In its present state, synergetics focusses its attention to those situations, where the structures or functionings of the systems undergo dramatic changes on a macroscopic scale. In particular, synergetics investigates how the subsystems (parts) produce these changes in an entirely self-organized manner. The subsystems are usually discrete, e.g. atoms, cells, or humans. In a number of cases it is advantageous to treat such subsystems, for example molecules, as continuously distributed. Instead of going on with abstract definitions I rather present some explicit examples. Some of them will be discussed in much more detail during this workshop and many further examples will be treated by other speakers.

1. Some Typical Phenomena

1.1 Spatial Structures

The first class of examples concerns the formation of spatial structures (fig. 1). Let me start with the well known example of a liquid layer heated from below. For small temperature gradients, heat is transported by conduction from the lower to the upper surface and no macroscopic motion shows up. When the temperature gradient exceeds a certain critical value, suddenly a macroscopic motion occurs in the form of rolls. A related phenomenon is the Taylor instability. Here the liquid is enclosed between two coaxial cylinders, the outer one kept fixed, the inner one rotating. Beyond a critical rotation speed of the inner cylinder, a macroscopic pattern of fluid motion in the form of Taylor vortices occurs. Another class of problems reported at this meeting refers to chemical reactions. When certain chemicals are put together one finds spatial pattern of different chemical products as indicated in fig. 1. Finally we give an example from biology. There exist nowadays models of morphogenesis where by means of chemical agents a homogeneous tissue is transformed into a spatial pattern (fig. 1 lower part). In all these cases by change of external parameters (heat flux, rotation speed, concentration of chemicals) suddenly a new pattern on a macroscopic scale appears. Such phenomena are wide spread. For instance, the Bénard instability occurs in many examples of practical interest: in cloud formation (glider pilots are well aware of the vertical motion of air between these clouds and use it for being lifted), spatial structures of stars, in crystal

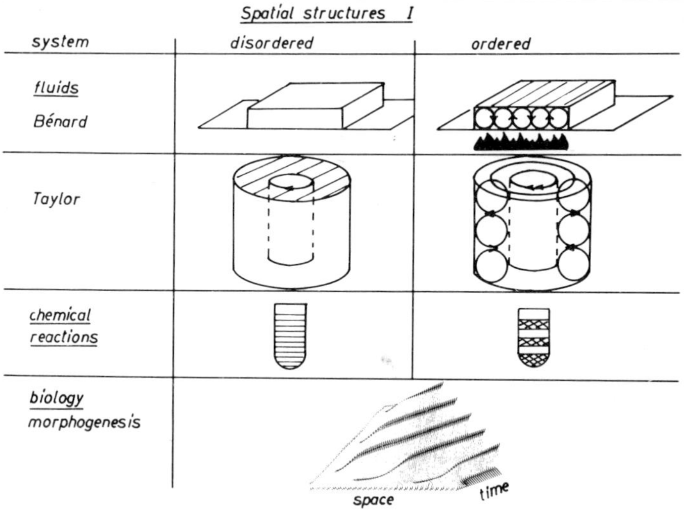

Fig. 1 *Example of spatial structures I*

growth processes etc. The shape of spatial structures is not limited to those of fig. 1. For instance under different conditions one finds convection with hexagonal cells where the fluid rises in the center and falls down at the borders (or vice versa)(fig. 2). Exactly the same pattern shows up in mechanics, when thin plates are deformed in the so-called post-buckling region. Also chemical reaction models may

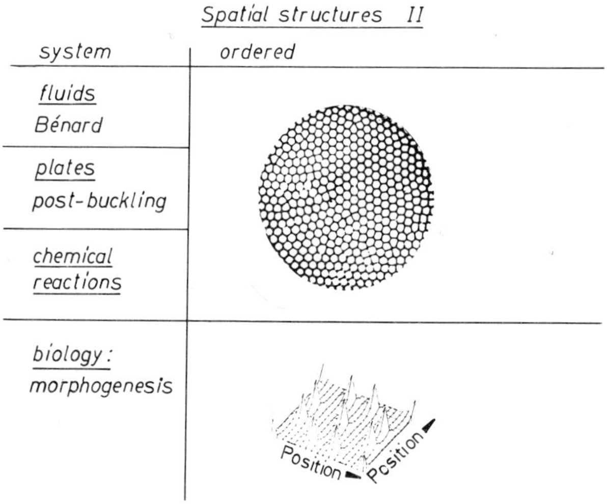

Fig. 2 *Example of spatial structures II*

exhibit such patterns. While computer calculations of morphogenetic models by Meinhardt and Gierer have yielded patterns as indicated in fig. 2 (lower part) analytical calculations by Olbrich and myself show under certain conditions again hexagons.

1.2 Temporal Structures (Oscillations)

The next group of phenomena refers to oscillations which occur on a macroscopic scale in a self-organized manner (fig. 3). A well known example is nowadays the laser. Here a rod of laser active material with two mirrors at its endfaces is pumped energetically from the outside and the atoms emit light. When the energy flux is small the electric field strength of the emitted light consists of random wave tracks. Beyond a critical value of the pump power the emitted light is nearly perfectly sinusoidal, i.e. the whole set of the laser atoms oscillates in a macroscopic manner and so does the emitted light.

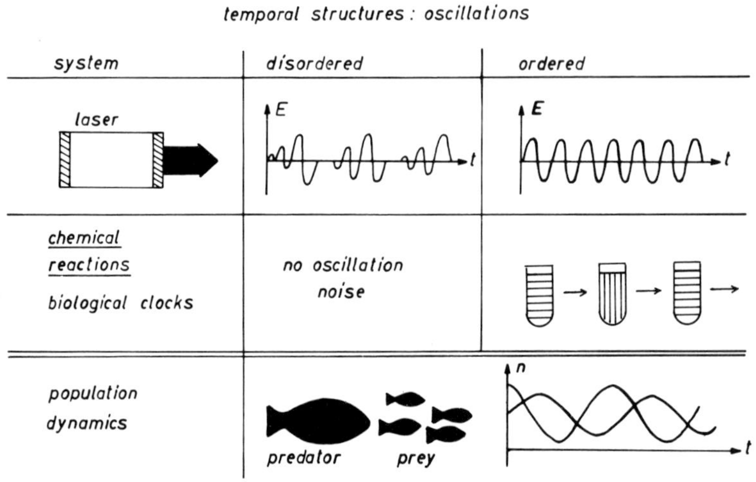

Fig. 3 Example of oscillations

A number of chemical reactions show oscillating patterns in which the colour changes, for instance from red to blue. Our third example is well known in population dynamics, or ecology, where, for instance, two kinds of fishes, the predator and the prey interact with one another, giving rise to oscillations of their individual numbers.

1.3 Pulses

A further group of phenomena is related to the formation of pulses (fig. 4) which occur in quite different disciplines: When we increase the temperature gradient of the convection instability, suddenly the rolls start a wavy motion along their axis. A related phenomenon is found in the Taylor instability where the Taylor vortices start a wavy motion when the rotation speed is increased. The reader will find more about these and more complex instabilities in the contribution by Swinney. When the laser is pumped still more it ceases emitting a constant wave and starts emitting ultrashort pulses. A similar phenomenon occurs in the Gunn effect, where in certain semiconductors

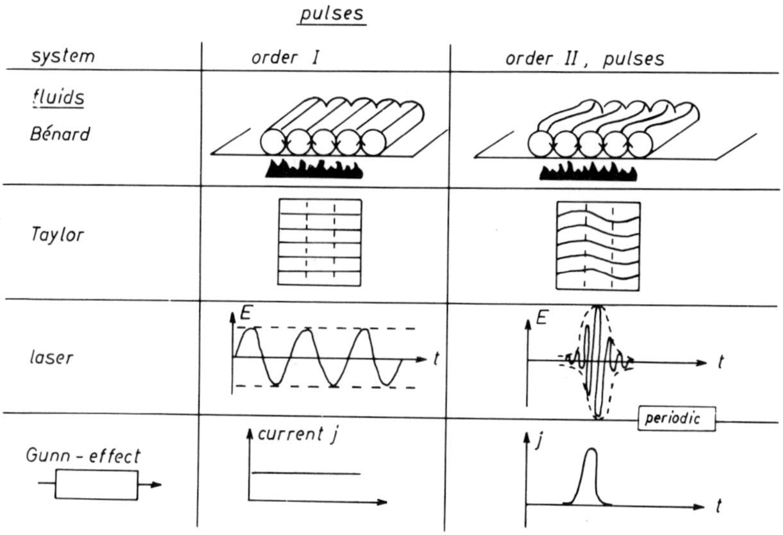

Fig. 4 Examples of pulses

the current j suddenly breaks into pulses. Finally nerve pulses are well known in biology.

1.4 Solitons

While these pulses are periodic in time, an interesting further class of phenomena occurs with respect to solitons. Solitons, which are solutions of nonlinear wave equations, retain their shape and show no dispersion in contrast to usual behavior of waves. In particular, two or more solitons may collide without loosing their original shape after the collision. Solitons or solitory waves have been discovered or are discussed in many fields of physics ranging from water waves over plasma and solid state physics to elementary particle physics. An example of two solitons is shown in fig. 5, where two solitons meet each other. Details about these phenomena are reported in the paper by Bullough.

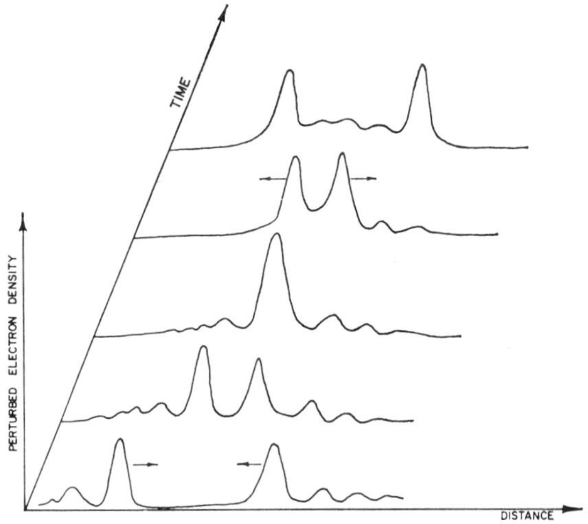

Fig. 5 Two solitons hitting each other

1.5 Spirals and concentric waves

Still more complicated patterns are found in certain chemical reactions showing spirals (fig. 6).
Entirely the same patterns occur in biology in the aggregation of cells of a slime mold (fig. 7). The essential steps of growth of slime mold are indicated in fig. 8.

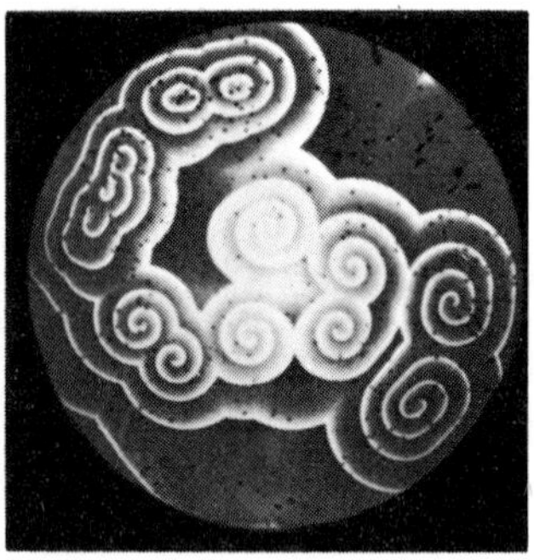

Fig. 6
Spirals of chemical activity in a shallow dish (Photograph taken by A.T.Winfree with Polaroid SX 7o)

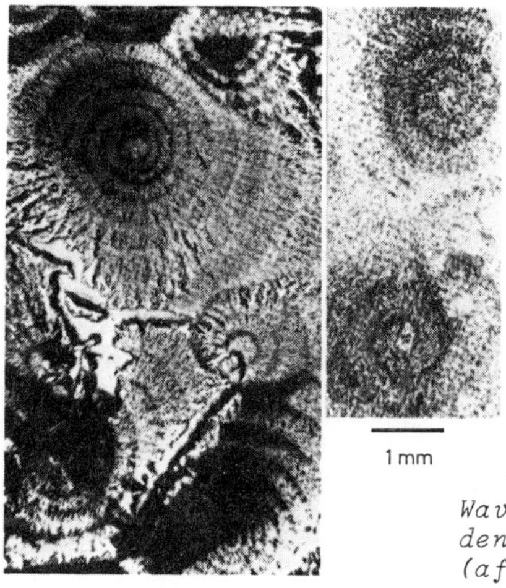

Fig. 7
Wave pattern of chemotactic activity in dense cell layers of slime mold (after Gerisch et. al.)

First a number of cells is present on a substrate and is able to emit a certain chemical, namely cAMP. When cells are hit by this chemical they are stimulated to emit this chemical so that we have the effect of spontaneous and stimulated emission of cAMP. The cells are sensitive to gradients of the chemical concentration and can move by means of pseudo pods against the gradient of the concentration field. In this way they assemble and form a mushroom by cell differentiation. In the present context the form of the chemical

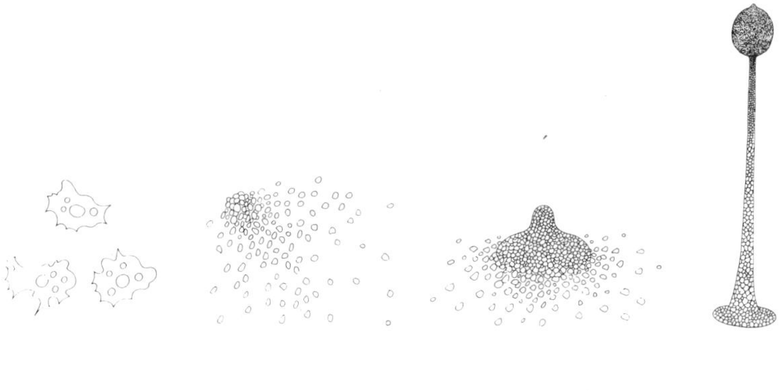

Fig. 8 The different stages of growth of slime mold

waves is extremely interesting. They are either concentric waves or spirals very much the same as in the chemical reaction I have mentioned above.

1.6 Chaos

So far I have presented examples where regular spatial or temporal patterns occur. During the past years a new kind of behavior has attracted the attention of mathematicians, physicists and biologists, namely "completely random" motion "caused" by deterministic equations (fig.9). Again it appears that such behavior is found in different disciplines i.e. in models of turbulence of fluids, in the laser, in models of the earth magnetic field, in chemical reaction models, and in population dynamics. This topic is treated by Rössler and by Kuramoto in this volume.

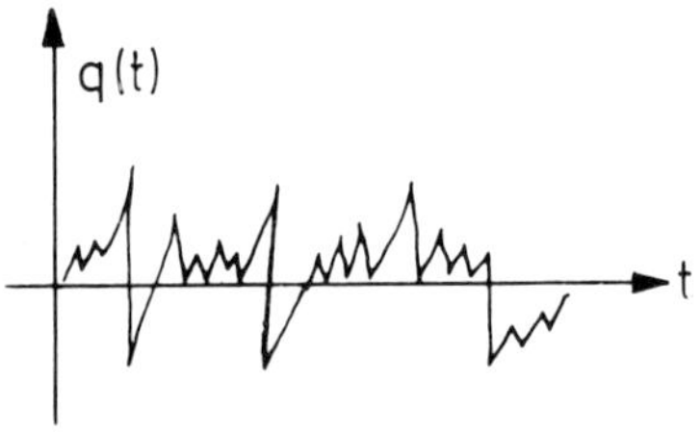

Fig. 9 Chaotic motion (schematic)

1.7 Summary. Other disciplines

So far I have been talking about phenomena which can be found in quite different disciplines mainly in the natural sciences. Of course, my list is by no means complete. Let me summarize these phenomena (fig. 1o) . We find spatial structures, temporal structures (oscillations), regular pulses, solitons, and we find more complicated forms, for instance spirals and finally very strange kinds of motion called chaos. Processes leading to structures occur also in other domains: in ecology, society, development of cities, industrial companies languages etc. Some of the contributions to this volume deal with these aspects (Buckley, Welge, F.Fröhlich), further examples have been presented in previous publications on synergetics (cf. list of references).

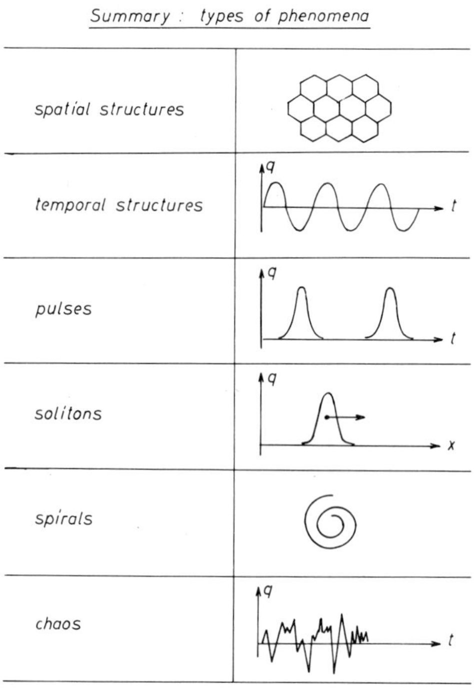

Fig. 1o
Different types of phenomena as discussed in the text

2. Selforganization and Slaving

2.1 Nonlinear equations and local stability

Let me now follow up the question if these seemingly quite different phenomena have something in common. First of all these phenomena are produced by systems which are composed of very many subsystems. Furthermore, these phenomena emerge mostly out of homogeneous phases or out of phases of lesser"structure" when external parameters (temperature gradient, energy flux, concentration of chemicals, etc.) are changed. In the rest of my talk I want to show how many of these phenomena can be traced back to a common root. While this is so, it would be unwise to base the future of synergetics on this line of thought alone. It might be desirable to develop still more general or even entirely different concepts to cope with the problems of complex systems. In this sense, the reader should interpret the following discussion only as an example of how synergetics may proceed. Now let me do a little bit of mathematics. I describe a system by a set of variables q_j which, in general change in the course of time. This change is determined by equations of the form

$$\dot{q}_j = N_j \, (q_1, \ldots, q_n; \sigma) + F_j \, (t) \tag{1}$$

where N_j are nonlinear functions of the variables and σ symbolizes external "control" parameters. Because in many cases the variables q depend on space, N_j may also contain derivatives with respect to spatial coordinates, $\underset{\sim}{x}$. It is important to note that systems are usually subject to fluctuations of the surrounding or the internal fluctuations. These are taken care of by time dependent fluctuating forces, $F_j(t)$. Let me first neglect these fluctuations and let me consider a rather simple situation which, however, reveals a good deal of the essential facts. I assume that I have found a stationary solution q_j^0. The solution may, for instance, represent a fluid at rest or a stationary state of a chemical reaction. Since under certain circumstances new patterns may occur, we check the stability of a given configuration. Therefore I consider small deviations $u_j = q_j - q_j^0$ which will depend on space and time. I then, in a first step, follow up the usual stability analysis by inserting $q_j = q_j^0 + u_j$ into the nonlinear equations and linearizing them with respect to the u's. This leaves me with equations of the type

$$\dot{u}_j = \sum_{j'} L_{jj'} u_{j'} \tag{2}$$

In their simplest form the equations read $\dot{u} = \lambda u$, where a negative eigenvalue λ indicates a stable solution and a positive eigenvalue an unstable solution. Of course, in general I shall find a whole set of such solutions which can represent, for instance, waves or dynamic configurations of the system.

2.2 Slaving

To determine the variables q_j without restriction to a linear approach I represent q_j as superposition of the modes $u_j^{(\lambda)}$ with still unknown amplitudes ξ_λ. Again I confine my analysis to a typical and most simple case taking into account only two modes, one stable, $u_j^{(s)}(\underset{\sim}{x})$, the other one unstable, $u_j^{(u)}(\underset{\sim}{x})$.

$$q_j(\underset{\sim}{x},t) = q_j^0 + \xi_u(t)\, u_j^{(u)}(\underset{\sim}{x}) + \xi_s(t)\, u_j^{(s)}(\underset{\sim}{x}) \tag{3}$$

Skipping all the details, I eventually find two equations, one for the amplitude ξ_u of the unstable and another one, ξ_s, for the stable mode. A typical example is

$$\dot{\xi}_u = \lambda_u \xi_u - \xi_u \xi_s \tag{4}$$

$$\dot{\xi}_s = -|\lambda_s| \xi_s + \xi_u^2 \tag{5}$$

Now a very simple but equally important point follows. When I change the external parameters, σ, in such a way that the first mode with the index u passes from the stable to the unstable regime, λ_u or at least its real part must pass through zero. That means, however, that this mode varies very slowly. Now let me assume that the behavior of the total system is eventually governed by ξ_u. According to eq.(5) this means that also ξ_s changes very slowly. Therefore I can neglect the temporal derivative, $\dot{\xi}_s$, compared to $|\lambda_s|\xi_s$. This allows me to solve this equation immediately and to express ξ_s by ξ_u

$$\xi_s(t) = |\lambda_s|^{-1}\, \xi_u^2(t) \tag{6}$$

I shall say in the following that ξ_s is slaved by ξ_u, i.e. ξ_s adapts itself immediately. Such phenomena are not limited to physics or chemistry but are found in quite different disciplines. Consider for instance humans. After their birth they quickly learn the language so that they are slaved by the language. We see in a minute that these slaved entities then carry on the macroscopic features. Inserting (6) into equation (4) I obtain the equation

$$\dot{\xi} = \lambda\xi - |\lambda_s|^{-1}\xi^3 + F(t) \tag{7}$$

where I have dropped the index u. I have now again included the fluctuating force, F(t). In the general case with many modes all stable modes can be eliminated. A systematic elimination technique including all higher orders of ξ_u, komplex λ_u's, "small band" excitations and fluctuating forces has been given in my previous papers and are described in my 1977 text book on synergetics (see references). In practice it often turns out that only a few modes are unstable so that an enormous reduction of the numbers of degrees of freedom results. This procedure allows us to study most of the phenomena I have described in the first part of my talk in an adequate way, if a number of refinements are taken into account.

2.3 Order parameters

The slaving principle has a very important consequence: once ξ_u is known, all ξ_s are also known. Therefore the ξ_u's determine the macroscopic behavior or, in other words, the order. I shall call them order parameters. They play a similar role in physical systems far from thermal equilibrium or in nonphysical systems, as do the order parameters of conventional phase transition theory of systems in thermal equilibrium.

To interpret eq.(7) I add an acceleration term $m\ddot{\xi}$ and interpret ξ for the moment being as the coordinate of a particle under the action of a deterministic force, $K(\xi) = \lambda\xi - \xi^3$, and a fluctuating force, F. Both in physics and in mathematics it is known that we can write the right hand side as derivative of a potential function $V(\xi)$, $K(\xi) = -\partial V/\partial\xi$. Depending on the size of the eigenvalue λ we find two configurations indicated by a dashed and a solid line in fig. 11. This allows me to establish links to phase transition theory in statistical physics. For a detailed discussion of these phase-transition analogies see for instance my 1977 text book on synergetics.
In the context of this workshop I want to stress another aspect. When I plot the equilibrium position as a function of the parameter λ I find the following situation (fig. 11 r.h.s.).
As long as λ is negative the equilibrium position is $\xi = 0$. However, for positive values of λ I obtain two stable configurations. Since the resulting curve has the form of a fork this phenomenon is called bifurcation. I have presented here the simplest case one can presumably think of which represents, however, already some typical features of certain classes of nonlinear differential equations dealt with by a branch of mathematics called bifurcation theory. The contributions to this volume by Kirchgässner, Joseph and Sattinger represent recent results of these authors out of this field. Since my paper is intented to give only a brief survey I cannot go into more details but just want to point out that there are close connections between the order parameter concept and the slaving principle on one side and concepts of bifurcation theory on the other. In particular in a number of cases the space spanned by the order

parameters coincides with the "null space" (Ljapunov and Schmidt) of bifurcation theory.

ξ_u order parameters
one degree of freedom :

$(m\ddot{\xi}+)$ $\dot{\xi} = \lambda\xi - b\xi^3 + F(t)$

$V(\xi) = -\frac{\lambda}{2}\xi^2 + \frac{b}{4}\xi^4$

$\xi = 0$ $\xi = \pm\sqrt{|\lambda|/b}$

critical slowing down, soft mode
critical fluctuations
broken symmetry

bifurcation

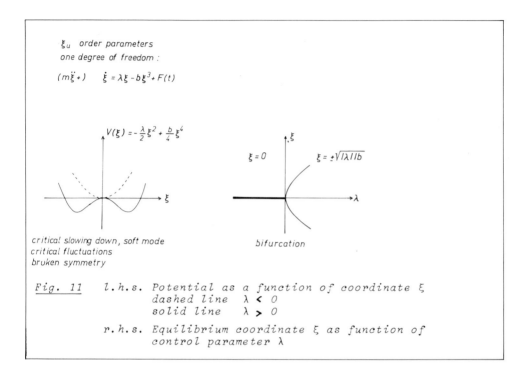

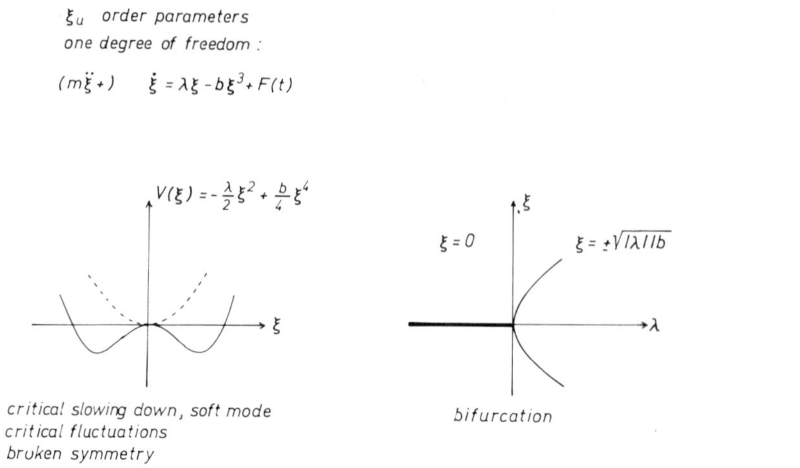

Fig. 11   l.h.s. Potential as a function of coordinate ξ
                   dashed line $\lambda < 0$
                   solid line $\lambda > 0$

            r.h.s. Equilibrium coordinate ξ as function of
                   control parameter λ

2.4 Formation of structures by competition or cooperation of order parameters. An example.

To show how pattern formation can now be understood I consider the case of the Bénard instability. Here the unstable modes have the form of plane waves in the horizontal plane, $u_j(x) \propto \exp(ikx)$, while $|k|$ is fixed, the horizontal directions of k are still arbitrary. With each direction k an order parameter ξ_k is associated, so that a whole set of order parameters is present. Their equations are still looking fairly complicated. However, one may show that they can be derived from a potential. In a certain range of λ the minimum of the potential is realized when three ξ_k's with $k_1 + k_2 + k_3 = 0$ are equal and unequal 0. Remember that the variable, q, for example the z-component of the velocity field, is a superposition of the individual modes. Because $\xi_{k_1} = \xi_{k_2} = \xi_{k_3} = \xi$, whose vectors k form a triangel.

$$q = \xi_{\underset{\sim}{k}_1} \exp(i\underset{\sim}{k}_1\underset{\sim}{x}) + \xi_{\underset{\sim}{k}_2} \exp(i\underset{\sim}{k}_2\underset{\sim}{x}) + \xi_{\underset{\sim}{k}_3} \exp(i\underset{\sim}{k}_3\underset{\sim}{x}) + \text{c.c.} \quad (8)$$

the maximum of q forms hexagons. Thus by the cooperation of three order parameters a new spatial configuration is created. However, if we change λ which is done experimentally by increasing the heat flux the modes start a competition and the deepest minimum is realized for only one order parameter unequal zero. In this case the velocity field has the form of rolls. Thus the order parameters are responsible which spatial or temporal or pulse-like structures are eventually realized. More recently, Olbrich and I have found a similar cooperation of order parameters when patterns are formed by chemical reactions or when solving models of morphogenesis e.g. that of Meinhardt and Gierer.

2.5 The potential case

Talking of potentials reminds us strongly of Thom's theory of catastrophes and the reader will find much more in his contribution. Here I want to make only a few comments. As you have seen the general equations I consider here are of the form

$$\dot{q}_j = N_j (q_1, \ldots, q_n) + F_j(t) \quad (9)$$

Thom's theory assumes that the N_j's can be derived from a potential

$$N_j = - \partial V / \partial q_j \quad , \quad (1o)$$

and that the random forces F_j vanish. In numerous cases I have treated explicitly I found that the original equations do not allow for a potential. This must be even so because a potential implies detailed balance which is absent in open systems. However, my above elimination procedure of slaved modes leads to effective equations for the order parameters which in many cases have a potential. By change of external parameters or applying external fields we obtain Thom's unfoldings. It appears to me, however, that Thom's ideas about pattern formation are rather different from those I have outlined here. I just want to mention one further aspect which seems to be promising because one can transfer catastrophe theory to cases in which random forces are included. When I write the solution of the corresponding stochastic equation i.e. the probability distribution in the form

$$P(q_1, \ldots, q_n, t) = \exp\left(-\tilde{V}(q_1, \ldots, q_N, t)\right) ,$$

$\tilde{V}$ can now be interpreted in very much the same way as V in Thom's theory.

2.6 Fluctuations

Fluctuations (represented above by $F_j(t)$) play a vital role in disorder-order transitions. To demonstrate this let us consider fig.12. A macroscopic state of the system is here represented by a point q. Fluctuations may drive the system from one state to another one.

Rôle of fluctuations

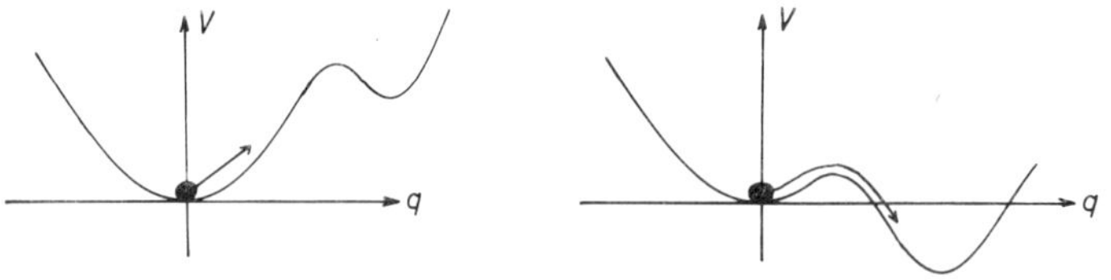

q : order parameter

rôle of surroundings : change of parameters

Fig. 12 Some typical examples for diffusion of order parameter q in a potential

One sees immediately that without fluctuations the system will never discover that there exists a state of lower potential energy, or. in other situations, a state of higher efficiency. These phenomena are essential for switching processes which was discussed in detail by Landauer.

3. Summary

3.1 Mechanism of formation of new structures

Above I have considered physical, chemical, biological or other systems. When we change external parameters (often called control parameters), the old structure may become unstable and a new spatial, temporal or functional structure arises. When the old structure becomes unstable, in general only a few "modes" become unstable while the other modes remain stable. The unstable modes slave the stable modes. This allows us to do away all the superfluous modes and we are left with much fewer modes which serve as order parameters describing and determining the macroscopic configurations and also the dynamics of the total system. The order parameters themselves obey certain (differential) equations. In many cases their structures can be classified using rather general arguments such as symmetries and smallness of order parameters. We are thus led to the definition of universality classes. This explains the surprising fact that quite different systems are governed by the same types of equations for the order parameters. This in turn explains why seemingly quite different systems behave in such a similar manner. This is especially so when we consider the dynamics i.e. the way the new structures

The general scheme

old structure $\longrightarrow$ instability $\longrightarrow$ new structure
change of external parameters

old structure $\begin{array}{l}\nearrow \\ \searrow\end{array}$ **unstable modes** $\left.\begin{array}{l} \\ slave \\ \textbf{stable modes}\end{array}\right\} \rightarrow$

$\rightarrow \begin{cases} \textbf{unstable modes} \\ \textbf{order parameters} \end{cases} : \begin{array}{l} same\ type\ of\ equations \\ for\ different\ systems \end{array}$

$\rightarrow$ universality classes ; new structure

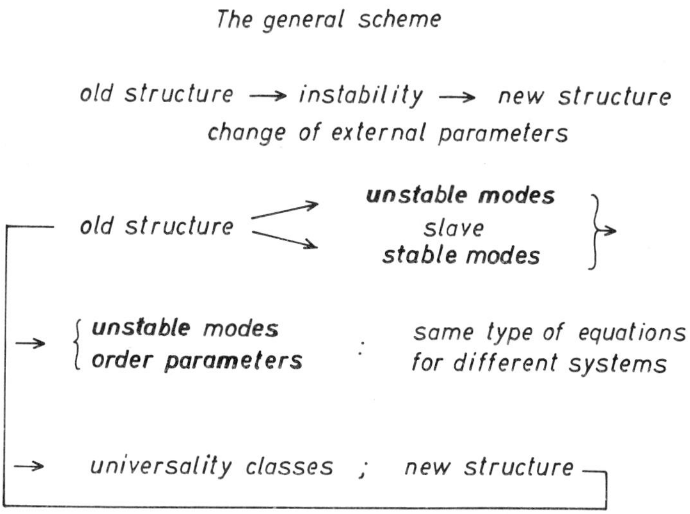

Fig. 13

arise. When we change the external parameters further the new
structures can again become unstable being replaced by another
structure and so on.

3.2 Time-evolution

When talking about the formation of structures, we may distinguish
between two approaches:
1) We make a Gedankenexperiment in which we change the external
("control") parameter σ and determine for each value of σ the
corresponding structure, especially close to,below and above the
bifurcation point. This procedure has been described above for
instance in connection with the Bénard cells.
2) In many practical cases the final structure is formed in the
course of time after the control parameter σ has been changed
suddenly, say from that of the dashed potential curve to the solid
potential curve of fig. 11. Another example: we adopt from the
very beginning the solid potential curve of the ordered state, but
let the system initially be in a certain disordered state. The
slaving principle allows us also to cope with these new problems:
We first solve the corresponding time-dependent problem for the
order parameters. For each time, we then can calculate the amplitudes of the slaved modes and thus the configuration of the total
system. In this way the temporal evolution of systems can be treated.
An example is exhibited in fig. 14 showing the build up of an ultrashort laser pulse obtained by this method.

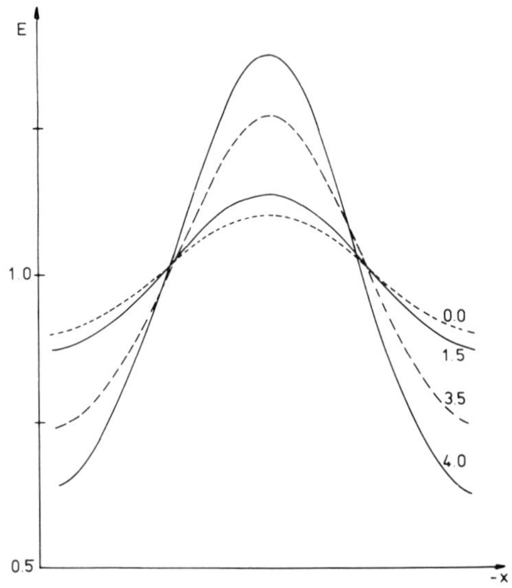

Fig. 14 *Build-up of ultrashort laser pulses as theoretically determined by means of slaving principle. (after H.Ohno and H. Haken)*

So far I have been speaking about sciences which can be formalized by sets of differential equations, a procedure well known in physics, chemistry, and mathematical models of biology. But I am sure that similar approaches of a more qualitative nature are applicable also to other sciences such as sociology and ecology, and indeed there are nowadays a number of models available. We have now gained some insight into the mechanisms which are at work when dramatic changes of systems happen on a macroscopic scale. Of course, still a lot of things has to be done and I hope that this workshop will contribute to develop new ideas on what's going on in more complicated transitions from one ordered state to another one.

To end up let me indicate how synergetics has ties to older already more established disciplines. We have seen that we have to deal with dynamic systems described by differential equations so that dynamic systems theory (especially bifurcation theory) with its analytical and topological branches comes in. Since the occurrence of structures can only be understood properly if we take fluctuations into account, the theory of random processes plays an important role. Near the transition points, the experience of physicists made with phase transitions are of utmost importance. In particular renormalization group techniques have to be further developed and applied to nonequilibrium phase-transitions. Finally the success of synergetics will certainly serve as a stimulus to researchers working on General Systems Theory.

Further Reading on Synergetics

H. Haken, R. Graham, Umschau 6, 191 (1971)

H. Haken (ed.): Synergetics (Proceedings of an International Symposium on Synergetics, Elmau 1972) B.G.Teubner, Stuttgart 1973

H. Haken (ed.) : Cooperative Effects, Progress in Synergetics (Proceedings of a Summerschool at Erice, Sicily 1974) North Holland, Amsterdam 1974

H. Haken: Rev. Mod.Phys. 47, 67 (1975)

H. Haken: Synergetics. An Introduction. Nonequilibrium Phase Transitions in Physics, Chemistry and Biology, Springer, Berlin, Heidelberg, New York 1977
This book is written in a more elementary fashion and has the character of a textbook.

For further references consult the contributions to the present volume.

Radiation and Entropy, Coherence and Negentropy

A. Kastler

With 2 Figures

The problem of entropy of radiation was first treated by Max Planck in the years 1895 - 19oo. He considered himself as a disciple of Clausius, the founder of thermodynamics, the man who had introduced the concept of entropy into physics. Planck's aim, when he began his research of black body radiation, was to establish an expression for the entropy of this radiation. He was convinced, when he began this work, that thermodynamics, combined with Maxwell's theory of electromagnetism, would give the answer to his question.

He started from the fundamental thermodynamic relation

$$\left(\frac{\partial S}{\partial U}\right)_V = T^{-1} \tag{1}$$

where S is the entropy, U the energy of a system and T the thermodynamic temperature.

Defining a spectral density of radiation u_ν and a spectral entropy s_ν he showed that the preceding relation could be specified:

$$\frac{ds_\nu}{du_\nu} = T^{-1} \tag{2}$$

His first step was to establish the relation between the spectral density of radiation u_ν and the mean value $\overline{u_1}$ of the energy+) of a resonator (') in equilibrium with this radiation. Kirchhoff had shown that, for a cavity, u_ν was independent of the shape of the cavity and the nature of the walls and that it was an universal function of ν and T. Using this result Planck imagined a wall of the simplest structure, containing for each frequency ν linear oscillators of this monochromatic frequency which he called "resonators". He showed by electromagnetism that the relation between u_ν and u_1 was of the form

$$u_\nu = g\, u_1 \tag{3}$$

where

$$g = \frac{8\pi\nu^2}{c^3} \tag{4}$$

being the frequency of the resonator and c the velocity of light.

To establish this relation it took Planck fifty pages of cumbersome calculations.

In 19oo Lord Rayleigh [2] showed that relations (3) and (4) can be established very easily but with a different physical meaning.

+ For simplicity we write u_1 instead of $\overline{u_1}$.

He showed that radiation imprisonned inside a cavity takes the form of standing waves. These standing waves are "quantized" and the expression g of (4) represents the number of standing waves or "modes" of radiation for a cavity of volume V = 1 and a frequency interval $\Delta\nu = 1$ between ν and $\nu + \Delta\nu$. This result can be easily established for a cavity of cubic shape and reflecting walls, and Kirchhoff's law indicates that the result must be valid for a cavity of any shape and nature. Lord Rayleigh's derivation shows that a "mode" of standing waves of radiation inside a cavity has properties identical to those of a linear oscillator. We know that this idea is the basic idea of quantum field theory.

Planck introduced also the "mean entropy" s_1 of a resonator (or a cavity mode) and he showed that

$$s_\nu = g\, s_1 \tag{5}$$

We must insist on the fact that this relation is not trivial. To be admitted as correct, Planck had to introduce the postulate of "natürliche Strahlung" which means that the vibrations of different resonators (or different modes) must be incoherent. Their phases must be distributed at random. This hypothesis is essential to guarantee the additivity of the entropies.

Planck showed further that relations (1) and (2) could be written also

$$\frac{ds_1}{du_1} = T^{-1} \tag{6}$$

To go further Planck used Wien's displacement law established in 1894 in the form

$$u_\nu = \nu^3 f\left(\frac{\nu}{T}\right) \tag{7}$$

and he showed that combined with the preceding relations s_1 could be brought into the form

$$s_1 = \phi\left(\frac{u_1}{\nu}\right) \tag{8}$$

which means that this expresion depends only on the unique variable u_1/ν.

We know that to represent the results of measurement of u_ν of black body radiation in the visible and the ultraviolet Wien proposed in 1896 the law

$$u_\nu = \alpha\nu^3 \exp\left(-\frac{\beta\nu}{T}\right) \tag{9}$$

which according to (3) and (4) is equivalent to

$$u_1 = h\nu \exp\left(-\frac{\beta\nu}{T}\right) \tag{1o}$$

The parameter h being related to Wien's parameter α by

$$\alpha = gh \qquad \text{or} \qquad \alpha = \frac{8\pi\nu^2}{c^3}\, h\,.$$

Planck showed that combining (1o) with (6) he could, by an integration, calculate an expression of s_1, and he showed also that by a derivation, he could obtain the expression

$$E = -\left(\frac{d^2 s_1}{d u_1^2}\right)^{-1} = \beta\nu\, u_1 \tag{11}$$

This expression is a linear function of u_1 in the range where Wien's law is valid.

In early 19oo the measurement made in the far infrared, especially those of Rubens, showed that in the range of small frequencies Wien's law was not valid and that in this range u_1 and u_ν were linear functions of the temperature T.

Later on Rayleigh and Jeans showed that this was in agreement with classical statistical mechanics according to whose equipartition principle we should have:

$$u_1 = kT \tag{12}$$

where k is the Boltzmann constant. Planck showed that in this frequency region the expression of the second derivative takes the form

$$E = -\left(\frac{d^2 s_1}{du_1^2}\right)^{-1} = \frac{1}{k}\, u_1^2 \tag{13}$$

E is a quadratic function of u_1.

In this situation it was obvious that Wien's law on one side and Rayleigh-Jeans law on the other side could only be valid as asymptotic laws, one being valid on the highfrequency side (or for small values of u_1), the other being valid on the low frequency side (or for large values of u_1). This gave Planck the idea to combine the expressions (11) and (13) and to write the "Ansatz":

$$E = -\left(\frac{d^2 s_1}{du_1^2}\right)^{-1} = \beta\nu u_1 + \frac{1}{k}\, u_1^2 \tag{14}$$

Combining this with (6) he obtained easily

$$u_1 = \frac{h\,\nu}{\exp\left(\frac{\beta\nu}{T}\right) - 1}$$

and applying this to the limiting case of small ν, he showed that $\beta = h/k$ so that finally

$$u_1 = \frac{h\,\nu}{\exp\left(\frac{h\nu}{kT}\right) - 1} \tag{15}$$

and

$$u_\nu = \frac{8\pi\nu^2}{c^3} \frac{h\,\nu}{\exp\left(\frac{h\nu}{kT}\right) - 1} \tag{15a}$$

which is Planck's famous radiation law first proposed by him on October 1o, 19oo.

He showed further that by integration, and taking (8) into account, the following expression for the entropy was obtained

$$s_1 = k \left[\left(1 + \frac{u_1}{h\nu}\right) \ln \left(1 + \frac{u_1}{h\nu}\right) - \frac{u_1}{h\nu} \ln \frac{u_1}{h\nu} \right] \tag{16}$$

His problem was solved, but he was not satisfied. He considered his "Ansatz" (14) as a happy guess. To find a theoretical foundation of his law he now turned to Boltzmann who had encouraged him to do so [3]. Boltzmann had founded statistical mechanics, dealing with the problem of distributing energy of translation among the molecules of a (perfect) gas. The gas had a discontinuous structure: A certain amount of gas was composed of N molecules. Boltzmann represented the translational energy of the molecules by a vector in velocity space (phase space introduced by Gibbs was not familiar to him). To solve his problem of distribution, to find a value for a probability of distribution, Boltzmann had been obliged to divide the velocity space into elements of extension, which he called the "cells".+) We may note that this was already a "quantum" procedure: Boltzmann had to give to the velocity space a discontinuous structure.

Planck's problem was to distribute radiation energy (u_ν by unit volume and unit frequency range) among resonators or "stationary modes" of a cavity. Resonators or "modes" have a discontinuous structure: there are g modes per unit volume and unit frequency range. But the radiation energy had a continuous structure. To do like Boltzmann, Planck had to give him a discontinuous structure. He divided it into "elements of energy" which he called ξ , so he could write

$$u_\nu = N\xi = g\, u_1 \tag{17}$$

He introduced also the mean numbers of energy elements per resonator

$$n = \frac{N}{g} = \frac{u_1}{\xi} \tag{18}$$

For the probability of distribution of N energy elements among g resonators Planck took from combinatory analysis the formula:

$$W = \Pi \frac{(N_i + g_i - 1)!}{N_i!\,(g_i - 1)!} \tag{19}$$

and not the expression used by Boltzmann

$$W = N!\, \Pi \frac{g_i^{N_i}}{N_i!} \tag{20}$$

+) The word "cells" does not appear in Boltzmann's original papers. He uses it for the first time in his book on "Gastheorie" edited in 189o.

Why? Planck does not explain this. Probably he had instinctively the intuition that the energy elements ξ had to be treated as "undiscussible" objects, but he did not mention explicitly this property. This question was only cleared up in 1911 by NATANSON [4] . Formula (19) which Planck used in 19oo is nothing else than the fundamental expression for Bose-Einstein statistics.

Starting from this formula, calculating $\ln W$ and using Stirling's approximation Planck obtained

$$\ln W = \sum_i \left[(N_i + g_i) \ln (N_i+g_i) - N_i \ln N_i - g_i \ln g_i\right]$$

Introducing $n = \frac{N}{g}$ this can be put into the form:

$$\ln W = \sum_i g_i \left[(1+n_i) \ln (1+n_i) - n_i \ln n_i\right]$$

Introducing now Boltzmann's postulate connecting entropy to probability

$$S = k \ln W \quad +) \qquad (21).$$

Planck obtained (we may replace suffix i by suffix ν):

$$S = k \sum_\nu S_\nu = k \sum g_\nu \left[(1+n_\nu) \ln (1+n_\nu) - n_\nu \ln n_\nu\right]$$

which, as $S_\nu = g_\nu S_1$ is equivalent to

$$S_1 = k \left[(1+n) \ln (1+n) - n \ln n\right] \qquad (22)$$

This according to (18) can be rewritten:

$$S_1 = k \left[(1+ \frac{u_1}{\xi}) \ln (1+ \frac{u_1}{\xi}) - \frac{u_1}{\xi} \ln \frac{u_1}{\xi}\right] \qquad (23)$$

Comparing (23) given by statistics to (15) given by thermodynamics and the Ansatz, Planck concluded that

$$\xi = h\nu \qquad (24)$$

was the expression for energy elements of radiation of frequency ν.

Planck showed also that for a progressive light beam in free space the number of quanta of radiation energy corresponding to a "cell" is given by

$$n = \frac{B_\nu \lambda^2}{h\nu} = \frac{B_\nu c^2}{h\nu^3} \qquad (25)$$

+) This formula was first written in this form by Planck

where B_ν is the spectral luminosity factor of the light beam.
We have seen that Planck's derivation is based essentially on two assumptions:

1) that the vibrations corresponding to different resonators or modes are incoherent in black body radiation,
2) that the different energy elements ξ are to be treated as indiscussible objects.

Another important point may be stated here, a point which was clearly expressed in Louis de Broglie's thesis [5] in 1924: "If a same resonator or a same "mode" of stationary vibrations or a same "cell" of phase space contains several elements n, the corresponding "waves" are always coherent. Their state can be represented by a same and unique wave-function".

This point is important for electromagnetic radiation, but also for the de Broglie-waves representing material particles: We know that this property is essential for the understanding of the superfluid and the superconducting states.

May I also mention that the expression E used by Planck (formula 14) has a physical meaning which was discovered by Einstein in 1910 [6] : It represents the intensity of fluctuations of boson particles, and can be written

$$\overline{\Delta n^2} = n + n^2 = n(1+n) \tag{26}$$

This formula has been derived for the case where N and g are large numbers. If not, if the number of phase-cells of the distribution is small, it must be replaced by the more general expression [7]:

$$\overline{\Delta n^2} = n(1+n)\,\frac{g-1}{g+2} \tag{27}$$

which for g = 1 leads to $\overline{\Delta n^2} = 0$ which means that in the case of Einstein-condensation of a Bose-gas or in the mono-mode operation of a laser, fluctuations are absent.

Let us now come to the fundamental problem indicated in our title: How must we change the expression of entropy s of radiation if light beams are coherent? Planck proposed to his research student Max von Laue to study this problem and von Laue did so in two fundamental papers published in "Annalen der Physik 1906 and 1907" [8].
Max von Laue started from the following simple case (Figure 1): A monochromatic light beam I_0, of intensity normalized to unity, falls on a glass plate G G' and splits into two beams: a reflected beam I_r of intensity r and a transmitted beam T_t of intensity t = 1-r.+) These both beams remain phase-coherent. Introducing two mirrors M_1 and M_2 we constitute a Michelson type interferometer, the two beams I_r and I_t joining again at B on the glas plate and splitting each again into a reflected and into a transmitted beam, so that we get a total of four outcoming beams whose intensities are indicated.

+) We suppose that the effects of multiple reflections on both sides of the glas plate are included in these coefficients.

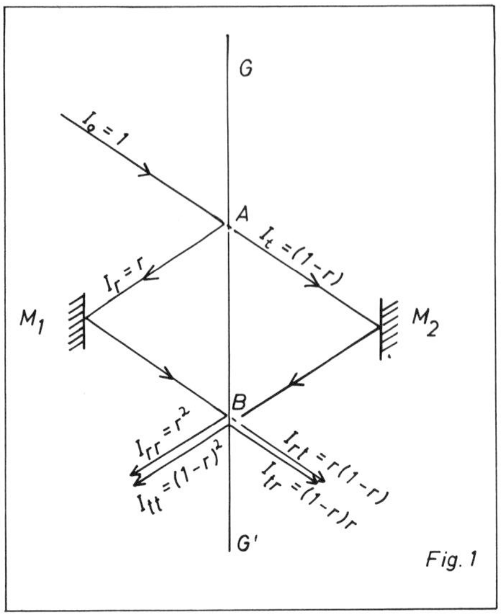

Fig. 1

As the two beams I_r and I_t are coherent, the emerging beams on each side will be coherent also and interfere. We can choose the optical path difference in such a way that there will be phase-agreement on the left, between I_{rr} and I_{tt} and phase opposition on the right, between I_{rt} and I_{tr} (the geometric path difference for this case must be equal to $\lambda/2$). In this case the outcoming intensity on the right hand side will be zero, and on the left hand side the amplitudes of I_{rr} and I_{tt} will add and give:

$$A_{rr} + A_{tt} = r + 1-r = 1.$$

The initial light beam I_o is restored. So the whole process of splitting and recombination must be a reversible adiabatic process, a process which conserves entropy.

Let us call n the number of photons per phase-cell in the incoming beam I_o. The number of corresponding photons for the two beams I_r and I_t will be

$$n_r = nr \qquad \text{and} \qquad n_t = (1-n)r.$$

If the two light beams would be incoherent, their total entropy would be

$$S_{inc} = S_1\,(n_r) + S_2\,(n_t) \tag{28}$$

inserting for S_1 and S_2 expressions like (22). But as the two light beams are coherent, their entropy must be identical to the entropy of the incoming beam:

$$S_{co} = S(n) = S(n_r + n_t) \tag{29}$$

So the coherence introduces the entropy-difference

$$\Delta S = S_{co} - S_{inc} = S(n_r + n_t) - S(n_r) - S(n_t) \quad (3o)$$

It can be shown that this expression is always negative. It is a "negentropy" corresponding to the coherence, and it expresses a certain degree of order.

Von Laue has also established formulae for partial coherence. We refer to his papers [8] .

Let us just mention that a "laser" is a device which produces coherent light. Figure 2 gives a rough scheme of its operation: It is an open thermodynamic system which rewaves an input of energy (containing free energy), and it has two outputs: the laser beam and heat production. This last property makes of it what Prigogine calls a "dissipative structure". But its principal interest is that it produces at the same time an ordered structure, a beam of highly coherent monochromatic light, whose frequency is determined by the internal structure of the device. So we may look on this device as a "self-organizing" structure.

It is a very simple and crude model of more complex devices of this type.

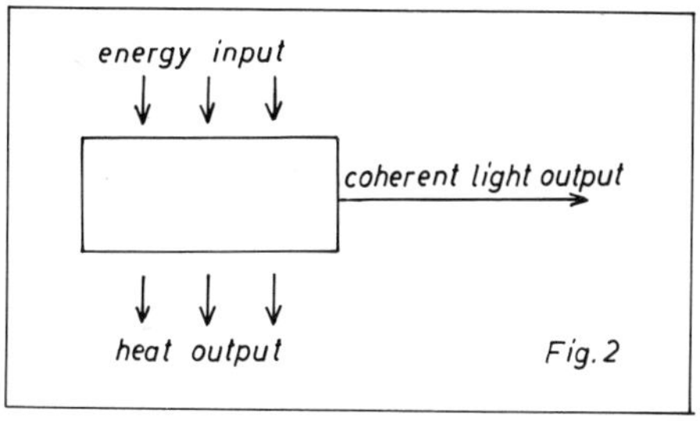

Fig. 2

References

1) M. Planck, Annalen der Physik 19o1, 4, 553; 6, 818
Theorie der Wärmestrahlung, Leipzig 19o6

2) Lord Rayleigh, Phil.Mag. 19oo, XLIX, 539

3) L. Boltzmann, Berliner Berichte 1897, 1o16 and Wissenschaftliche Abhandlungen, III.Band, 618

4) L. Natanson, Physikalische Zeitschrift 1911, 12, 659

5) L. de Broglie, thêse 1924, reéditée en 1963 chez Masson, Paris: "Recherches sur la theorie des quanta."

6) A. Einstein, Physikalische Zeitschrift, 19o9, 1o, 185

7) B. Decomps and A. Kastler, C.R.Ac.Sci. 1963, 256, 1o87

8) Max von Laue, Annalen der Physik 19o6, 2o, 365
and 19o7, 23, 1

What is Catastrophe Theory about?

R. Thom

When we start evaluating what is Catastrophe theory, it is very easy to be led into wrong conclusions just for terminological reasons. Of course, many people know already that in "Catastrophe theory", the word "Catastrophe" does not have the sinister meaning it has in every day language. But few people realize that the word "theory" itself has here to be understood in a very peculiar sense. It would be completely wrong to equate Catastrophe theory with one of the standard scientific theories, like the Newtonian theory of gravitation, or the Darwinian theory of evolution. In such cases, one has to expect that the theory has to get some experimental confirmation, it has to be founded (or at least may be "falsified" in Popper's sense) by experiment. The plain fact is that C. theory escapes this criterion : it cannot be "proved", nor "falsified" by experiment. I owe this fundamental remark to a British Biologist, L. Wolpert, who once - in a meeting - told me :"Your theory seems to be able to explain anything, but a theory which explains everything explains nothing". After some time of thinking, I had to accept this objection as perfectly valid. Being aware of this fact, we could try to find a way out by saying that Cat. Theory is a "mathematical theory". Here again, I am afraid this statement is not correct. Of course, Cat. theory deals with mathematical theories, and mathematical objects. It has led already to many interesting results in Differential Analysis, theory of smooth maps, in bifurcation theory, qualitative Dynamics, etc. But the fundamental fact remains that C.T. deals with "reality" (or claims to deal with), whereas mathematics, by itself, has nothing to say about reality. Hence Cat. theory is not part of mathematics - and many "pure" mathematicians look at it now with some misgivings as a somewhat unprecise and farfetched (if not delirious) speculation.

I think the correct answer is to say that C.T. is a method (or, more generally, a language). As ordinary language, it can depict (or describe) reality. But as for ordinary language, the description may be erroneous, or false. Catastrophe theory is being used ordinarily as follows : suppose we observe a given process during some time interval $t_o t_1$ and look at the corresponding morphological accidents. Find then the minimal dynamical model giving rise to these accidents. Extrapolate then this effect in time interval $(t_1 t_2)$ following (t_1) .

Such a prediction can never be said to have an absolute value. First, it is in general impossible to give precise quantitative thresholds for the time length $(t_1 t_2)$. It is possible to say : Knowing what happened between t_o and t_1 , then the most natural thing to expect between $(t_1 t_2)$ is such and such morphology. If the prediction turns out to be true, then this only justifies the validity of the simple minimal model. If, in the contrary, the prediction fails because of the appearance of a new, unexpected morphological accident, then this proves that the initial minimal model was too simple, and it has to be made more complex in order to account for the new fact. In that respect, C.T. is in general more interesting when it fails than when it is successful. But to be able to give criteria for deciding when a morphology is surprising or not is already something quite valuable, as in Biology, for instance, one of the most challenging questions is : what should we wonder about?

Let us now consider the basic ideas of C.T. One could say that the C.T. formalism aims at realizing a synthesis of two fundamental notions in Science, the idea of function (or map) on one hand, the idea of dynamical system on the other hand. This synthesis is realized by the notion of automaton, taken in the smooth category. This is a system where state s varies in a state space M (smooth manifold); there are inputs $u \in U$ and outputs $y \in Y$ $(U,Y$ smooth manifolds) and we suppose that there are differentiable laws :

$$\frac{ds}{dt} = F(s;u) \qquad Y = G(u;s)$$

giving the time evolution of s , and the output as a function of the input and the state of the system. Now the basic assumption is that the internal dynamic s is relatively fast with respect to the variations of u and y . If we suppose that this dynamic is in a stationary régime, that is, s is in a closed neighborhood of an attractor (like a closed trajectory) which is "structurally stable" and on which the flow $s' = F(s;u)$ is ergodic. If the dynamic is very fast, the output $Y = G(u;s)$ may be replaced by its time average for a time τ large enough for s to run over the whole attractor $\bar{y} = \frac{1}{\tau} \int G(u,s(t))dt$; hence we get a direct map $u \to \bar{y}$.

In the so-called "elementary catastrophe" theory one admits that for a given $u \in U$, the internal dynamic $s = F(s,u)$ has only a finite number of stationary régimes of the before-mentioned type. Hence the correspondance between u and $\bar{y}$ is such that its graph $\Gamma(u,\bar{y})$ meets each vertical $u = u_o$ in only a finite number of points. In most cases the stronger assumption is made that this graph is (a part of) a smooth manifold of $\dim k = \dim U$.

Remark : These assumptions may be difficult to justify, if one takes into account the pathology which was found recently in Qualitative Dynamics : Flows which have an infinite number of attractors on a dense subset of flows, existence of an infinite number of topological types of an attractor (the Lorenz attractor) for an arbitrarily small variation of the flow. The above assumptions are satisfied when the local dynamic is a gradient dynamic $s = -\operatorname{grad} V_s(s;u)$, then the stationary régimes are the minimizing points of $V(s;u)$ for $u \in U$; if h is compact, these points are generically finite in number. This is the subject of the so-called "elementary catastrophe theory", which mathematically boils down to the theory of singularities of functions.

Relation with morphogenesis.

Suppose a natural process (of arbitrary nature) takes place in some domain U of space-time. We then divide U into local cells U_α ; in each cell U_α the local process may be considered as a local automaton of the above type : the spatial coordinates in U_α being input functions for this automaton . Suppose now any observable of the process is an output of the local automaton. The points of U may be divided into two sets : regular points, for which the graph Γ is continuous (smooth) a function of the coordinates, the catastrophe points for which this is false.

For a part u to be regular, it suffices that the dominating stationary régime of $s' = F(s,u)$ be structurally stable (at least in the ergodicity point of view).

If we consider the output space as the space of colours, each local automaton may be considered as a black box with a window through which the local output colour is seen. The global process in U may be considered as a field of such coloured automata; Catastrophe points appear as points where one gets a jump in colour As such, they play a fundamental role in the visual morphology of the process.

<u>The Universal unfolding of a singularity</u>.

To classify the Catastrophe points according to local topological type, one uses the notion of universal unfolding of a singularity. The main idea of unfolding may be described as follows : Suppose dim Y = 1 ,then the graph Γ has an equation $G(y;u) = 0$; where $\frac{\partial G}{\partial y} \neq 0$, the equation $G(y,u) = 0$ may be solved with respect to y (Implicit function theorem), we obtain that way a local function $y = \varphi(u)$; hence the corresponding point u_o is regular (provided $\varphi(u)$ remains the dominating solution in a neighborhood of u_o) . If at a point (u,y) the derivative $\frac{\partial G}{\partial y}(u,y_1)$ is zero, the corresponding point is critical and in general, the presence of critical points does create catastrophes.

The mathematical theory of singularities tells us that critical points may be divided into two classes : First, critical points of so-called <u>finite codimension</u> for such points there exists at least one derivative $\frac{\partial^k G}{\partial y^k}$ which is not zero. Second, if all derivatives $\frac{\partial^r G}{\partial y^r}$ are zero, the critical point is said of <u>infinite codimension</u>.

For points of finite codimension, there is a local model given by the Weierstrass - Malgrange theorem

$$G(u,y) = y^k + \sum^{k-1} a_i(u) y^i \quad \text{if} \quad \frac{\partial^r G}{\partial y^r}(0) = 0 \quad \text{for} \quad r < k \qquad \frac{\partial^k G}{\partial y^k}(0) \neq 0 .$$

To study such a point, one considers the universal polynomial (P) $Y^k + \sum_{i=o}^{k-1} \alpha_i Y^i$. The space $\widetilde{U}$, space of coefficients (α_i) is called the universal unfolding space. It contains the discriminant variety (Δ) of (P) . It can be shown that all local properties of the system around $(u_o y_o)$ may be described through the local map

$a = U \to \tilde{U}$ and its position w,u to the discriminant variety (U). In other words, the local study of the map may be made by using only a local algebraic model defined above the universal space. As a classical (well known) case, the study of any potential point of type $V(y)/0 = y^4/4 + \varphi_3(y)$ may be reduced to the study of the canonical polynomial $x^3 + ux + v$ above the (u,v) space (thus defining the cusp surface).

If we admit that a critical point leads to some unstability, hence to some lack of determinism in the process, this point involves some virtuality. The totality of all actual situations arising out of this "virtual" situation is precisely described by the unfolding space.

The use of such local models can be said to create for the process a "soft theorization", as opposed to the "hard theorization" taking place in Physics by the use of physical laws. Such a theorization has the following character:

1) It has a local, qualitative character, and does not permit (in itself) quantitative prediction.

2) It deals with fundamentally irreversible phenomena, as the Aristotelian passage from virtual to actual is irreversible.

Of course, some instances are known of elementary Catastrophes in "exact" sciences. Let me quote them :

1) Lagrangian formalism (Work of I. Ekeland).
2) Hamilton Jacobi theory - wavefront propagation and Caustics.
3) Shockwaves in the Riemann equation $u_t = f(u, u_x)$ (in one dimension)
4) Hydrodynamics and breaking of waves (cf. Thom).
5) Applied Mechanics and bifurcation theory (Koiter's law).
6) Landau's theory for phase transitions.

To what I might also add, the Kléman-Toulouse principle of topological classification of defects in ordered media (which may be considered as a generalization of C.T. when local symmetries are taken into account).

But in practically all of these examples, the Catastrophe - discontinuity -

appears as an epiphenomenon - a secondary effect superimposed on a global continuous process, which has no effect (as for the Caustics), or very little (shock waves) on the propagation of the process. This explains why - up to now - C.T. has had very little importance in Physics.

In contrary Soft Theorizing has found illustrations in practically all known disciplines : from Biology (Embryology, Ethology) to Psychology, Sociology, History, Linguistics). The impact for this new way of seeing needs a proper assessment. In many of these cases, the use of C.T. leads to a geometrization of a verbally perfectly clear situation, and it is not clear which extra feature the mathematization may introduce. But as a general theory of analogy, Cat. Theory is certainly very effective. There is certainly - as a taxonomy of analogous situations - the main interest of C.T.

But more generally, we might ask whether C.T. may not be considered as a general theory of forms. Having an experimentally given morphology, we decompose any catastrophe set into local morphogenetical fields, and then try to construct a formal theory which will reconstruct the observed morphologies out of a finite system of fields, and concatenation (spatial aggregation)of such fields . This requires that the substrate space has itself metric properties, and a global metric equivalence group. This is the case for formal systems where the metric equivalence is used in the definition of the printing character : it would be nonsensical to allow all dilatations (or contractions) of a letter in a formal system. For then, we could write an infinite sequence of letters in a finite length. The elementary catastrophe theory in itself does not allow such an algebraization, because it is purely local in character. Moreover the semantic spaces on which the singularities are defined have no global Lie group of equivalence... But when the substrate space has a canonical metric allowing a transitive equivalence group, there is still some hope to construct a "mixed" theorization allowing algebraization and spatial aggregation. For this, one has to admit that the unfolding space has a canonical flow defined on it, and that such a flow has a recurrent character, allowing the return to the organizing center (or at least, turning around, as in the Hen and Egg situation). One could hope that on such unfoldings $\tilde{U}$, there are (canonically defined) four dimensional vector valued one-forms ω_1 such that the integral $\int_\gamma \omega$ defines the spatial displacement of the point along a trajectory arc γ . In Biology, such metric controls of unfoldings quite certainly exist, and also the reconstitution of the genetic information (the return to the organizing center) in gametogenesis.

Conclusion.

Cat. Theory seems to have reached now a critical stage. After a promising start, it got a considerable sociological success due to an overoptimistic (and not always very enlightened) vulgarization. Now disappointed sectators are expressing a lot of critics, as they become aware of the fact that C.T. may not hold all its promises. As, myself, I never felt overoptimistic about the possibilities of the theory, I still cling to the belief that the conceptual - epistemological - interest of C.T. is beyond any doubt. In the contrary, its practical usefulness is still very much dubious, as the theory did never predict any new experimental result of very marked importance; we have to leave to the future to decide on this point.

Bifurcation Theory

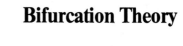

Bifurcation of a Continuum of Unstable Modes

K. Kirchgässner

1. Introductory Remarks

In this contribution we consider model equations of increasing complexity which exhibit bifurcation in a continuum of points. In spite of the wealth of solutions present there is - under certain simple geometrical conditions - a selection principle yielding only a finite number of stable solutions. The geometrical conditions mentioned guarantee essentially "supercritical" bifurcation for the continuum as an entity and thus reflect at this level the now classical result for simple eigenvalue bifurcation [2],[7].

The equations considered are intended to model certain bifurcation phenomena in fluid dynamics such as the TAYLOR- and the BENARD-problem which have been treated in detail by several authors of this workshop. Nevertheless, they apply equally well to problems in population genetics [1] and reaction diffusion equations [4],[5].

The results reported for the one dimensional case are not new, maybe the point of view we take. However, a simple generalisation to dimension two leads us into virgin territory. Some of the results depend on the order of the differential equation, some on their autonomy, some are apt to wide generalisations; we make an effort to separate these effects carefully.

2. The One-Dimensional Model Equation

Consider the ordinary differential equation

$$N(u,\lambda) \equiv u_{yy} + \lambda u - f(u) = 0 \quad , \quad y \in \mathbb{R} , \tag{2.1}$$

where λ is a real parameter and, for simplicity, f is supposed to be analytic near $u = 0$ satisfying $f(0) = f'(0) = 0$. Without further restrictions on u, (2.1) is not well posed. For this reason we confine (2.1) to bounded solutions.

The linearised equation (with respect to 0) has a 0-, 1-, 2-dimensional space of solutions for $\lambda < 0$, $\lambda = 0$, $\lambda > 0$ respectively. To solve the nonlinear equation we assume first the point of view of a bifurcationist. We consider periodic solutions with fixed period $2p > 0$ which, in addition, are even : $u(-y) = u(y)$. Then, $\lambda_n = n^2\pi^2/p^2$, $n \in \mathbb{N}$, are simple eigenvalues of the linearised part of (2.1), and $u = 0$, $\lambda = \lambda_n$ is a point of bifurcation. The nonzero solutions form locally an analytic curve. One easily verifies this formal part by showing that the power series

$$u(\varepsilon) = \varepsilon u_n^1 + \varepsilon^2 u_n^2 + \ldots$$

$$\lambda(\varepsilon) = \lambda_n + \varepsilon\lambda_n^1 + \ldots$$

is consistent with (2.1). Its convergence is not hard to prove and, that there are no other solutions follows from [2].

In view of the autonomy of (2.1), $u_c(y) \equiv u(y+c)$ is a solution as well for every $c \in \mathbb{R}$. It can be shown that all small solutions can be represented in this way. Therefore a two-dimensional manifold of non-trivial solutions emanates in every bifurcation point. Henceforth we restrict attention to the smallest eigenvalue λ_1. If $f(u) = a_p u^p$ + higher order terms, $a_p > 0$, bifurcation is supercritical ($\lambda > \lambda_1$), if $a_p < 0$ it is subcritical ($\lambda < \lambda_1$). In the first case the zero solution exchanges stability with the nonzero solution at $\lambda = \lambda_1$, whereas in the second case, the nonzero solution is unstable near $u = 0$, $\lambda = \lambda_1$ (stability is understood in a sense to be specified below). However, these claims are valid only if the perturbations have the same period as u, which is unrealistic to assume.

To obtain physically reasonable results one has to consider (2.1) in a sufficiently large space, the set of bounded solutions, say. The previous analysis shows that, for every positive λ, $(0,\lambda)$ is a point of bifurcation for a two-dimensional manifold of periodic solutions. Two questions arise: Have we found all solutions? Is there an order principle selecting certain specified patterns?

The first question can be answered even globally by phase-plane analysis. If $-\lambda u + f(u)$ has 0 as its only root (e.g. $f(u) = -u^3$) then all solutions are periodic; if however there is a nonzero root then, there are nonperiodic solutions as well. Let us denote a root of smallest modulus by $u_0(\lambda)$. In the sequel we call solutions *singular* when they are constant or nonperiodic.

To find an order principle we consider (2.1) as the time independent part of the following evolution equation

$$v_t = N(v,\lambda) \quad , \quad v_{|t=0} = v_0 \tag{2.2}$$

A solution of (2.1) is called *stable* (in the sense of LIAPUNOV) if, for every $\varepsilon > 0$ we can find a $\delta > 0$, such that $\|u-v_0\| < \delta$ implies $\|u-v(t,\cdot)\|$ less than ε for all $t > 0$. A solution is *unstable* if it is not stable. A stable solution u is said to be *asymptotically stable* if, for all sufficiently small v_0, $\lim_{t\to\infty} \|u-v(t,\cdot)\| = 0$ holds. ($\|u\| \equiv \sup \{|u(y)| / y \in \mathbb{R}\}$.

Stability is studied by using sub - and supersolutions. A smooth function $\hat{v}(t,y)$, satisfying $\hat{v}_{|t=0} \geq v_0$, is called a *supersolution* if $\hat{v}(t,\cdot) \geq v(t,\cdot)$ holds for all positive t, v solves (2.2). The notion of a *subsolution* is defined similarly by reversing the inequality signs. Note that every bounded smooth $\hat{v}$, solving $\hat{v}_t \geq N(\hat{v},\lambda)$, $\hat{v}_{|t=0} \geq v_0$, is a supersolution. (c.f. [1]). Moreover, given any two solutions u_1 and u_2 of (2.1), then $\max(u_1,u_2)$ is a subsolution and $\min(u_1,u_2)$ is a supersolution (c.f. [5]).

We apply an argument from [1]. Consider any nonconstant periodic solution u of (2.1) and set $u_c(y) \equiv u(y+c)$; then $\check{v} = \max(u,u_c) \not\equiv u$ is uniformly close to u if $|c|$ is small; take the solution v of (2.2) with $v_0 = \check{v}$ then $v(t,\cdot) \geq \max(u,u_c)$, implying $v(h,y) \geq v(0,y)$. Since $v_h(t,y) \equiv v(t+h,h)$ is a solution of (2.2) with $v_0(y) = v(h,y)$ we see that $v(t+h,y) \geq v(t,y)$ for any positive h. Hence $v(t,y)$ is increasing in t. Therefore, if $v(t,y)$ is bounded as t increases, it converges to some $v_\infty(y)$ which can be shown to be a solution of (2.1). If no bound exists v^∞ increases indefinitely. We conclude that all nonconstant periodic solutions are unstable, since the only possible v_∞ is a constant. If $-\lambda u + f(u)$ has no nontrivial root, then all solutions are unstable.

If there exist nontrivial zeroes, let us restrict attention to those of smallest modulus. Nonconstant periodic solutions are still unstable. A nonconstant singular solution which is not monotone is unstable as well; the argument is the same as in the periodic case. The only possibly stable solutions are monotone, singular solutions.

Nonconstant, monotone, singular solutions u of (2.1) (e.g. $f(u)=u^3$) are structurally unstable - they do not persist under small variations of f - (see FIFE [5] for this argument). If we apply the sub- and super-solution method we see that any v solving (2.2) and satisfying $u \leqq v_0 \leqq u_c$ stays between u and u_c for all times. Hence u is stable with respect to disturbances decaying rapidly enough at infinity. To prove stability for the constant solution, assume $\lambda > 0$, $u_0(\lambda) > 0$, and $-\lambda + f'(u_0(\lambda)) > 0$. Take $\delta, c_0 > 0$ such that $f'(1 \pm c_0)^0 \geqq \lambda + \delta$. Then

$$v(t) = (1 + c_0 e^{-\delta t}) u_0$$

$$v(t) = (1 - c_0 e^{-\delta t}) u_0$$

are super- resp. subsolutions for u_0. Hence, u_0 is even asymptotically stable. A similar argument holds if $u_0(\lambda)$ is negative. For more general conditions see [5],[6].

In the present case the order principle selecting the solution of highest symmetry consists of two effects: The diffusion which tries to flatten the solutions of (2.2) and which is independent of the non-linearity, and the existence of a solution bounding above or below all other small solutions of (2.1). The appearance of this dominating solution is strongly tied to the nonlinearity and thus to the geometry of the bifurcation picture. A precise formulation is given in the next section.

The detailed diccussion of the order principle strongly depends on the availability of super- and subsolutions and thus on the second order of (2.1). As will be seen below the qualitative description of the totality of solutions is neither tied to the autonomy nor to the order of the differential equation. The nonlinearity f may also depend on derivatives of u; for details see [6].

3. The Two-Dimensional Model Equation

The simplest generalisation of (2.1) to partial differential equations is as follows:

$$\begin{aligned} M(u,\lambda) &\equiv u_{xx} + u_{yy} + \lambda u - f(u) = 0 \\ u(0,y) &= u(1,y) = 0, \qquad (x,y) \in (0,1) \times \mathbb{R} \end{aligned} \tag{3.1}$$

The function f is assumed to be of the type discussed in the preceding section.

The discussion of the totality of "small" solutions of (3.1) is fundamentally different from (2.1) Consider the linearised part of (3.1): $\Delta u + \lambda u = 0$ with the appropriate boundary conditions. Setting $u = u_\nu(y) \sin \nu\pi x$ one sees that, for $\lambda < \pi^2$, only solutions of exponential growth in y exist. For $\pi^2 < \lambda < 4\pi^2$ all solutions are periodic whereas for $n^2\pi^2 < \lambda < (n+1)^2\pi^2$, $n > 2$, all solutions are *quasiperiodic*, i.e. they are of the form $u(x,y) = v(x,\omega_1 y,\ldots,\omega_n y)$, with $v(x,z_1,\ldots,z_n)$ being 2π-periodic in every z_j, and where $\omega_j = (\lambda - j\pi^2)^{1/2}$

To characterize all small bounded solutions it suffices surprisingly enough to prescribe, for $\lambda \in (n^2\pi^2,(n+1)^2\pi^2)$ the 2n parameters $u_\nu(0)$, $u_\nu'(0)$, $\nu=1,\ldots,n$. *There exists an $\varepsilon > 0$ such that every solution u of (3.1) with $\|u\| < \varepsilon$ is uniquely determined by the 2n parameters $u_\nu(0)$, $u_\nu'(0)$, $= 1,\ldots,n$. In particular, for $\lambda < \pi^2$, $u = 0$ is isolated among* *the bounded solutions.* (c.f. [6]).

If for every set of parameter-values such a solutions exists we have found all solutions. For $\pi^2 < \lambda < 4\pi^2$ one solves this problem as in the preceding section through bifurcation analysis. *Hence, for $\pi^2 < \lambda < 4\pi^2$ all small solutions of (3.1) are periodic in y.*

In the general case $\lambda \in (n^2\pi^2,(n+1)^2\pi^2)$, $n \geq 2$, the method of solutions still works formally. Enough solutions can be found by power series expansion with respect to a formal parameter ε such that the conjecture is substantiated that *almost all solutions of (3.1) are quasiperiodic.* However, the proof of convergence leads to small divisors and thus to a category of problems of extreme difficulty.

The existence of singular solutions can be still established under the following condition: *Suppose that, for every $\varepsilon > 0$, there is a positive δ such that for $\lambda \in (\pi^2,\pi^2+\delta)$ all solutions of (3.1) bifurcating from $(0,\lambda)$ remain in the ball $\|u\| < \varepsilon$, then there exists a singular solution (constant or nonperiodic first Fourier component u_1) in $(\pi^2,\pi^2+\delta)$. This solution is the locally uniform limit of y-periodic solutions of arbitrary large periods.*

In our case it is easy to show that a y-independent solution of one sign exists which dominates all periodic solutions near $\lambda = \pi^2$. The discussion of stability and instability with respect to solutions of

$$v_t = N(v,\lambda) \quad , \; v(0,y) = v(1,y) = 0 \qquad v|_{t=0} = v_0 \tag{3.2}$$

is literally the same as in the preceding section. In particular, all nonconstant, y- periodic or quasiperiodic solutions are unstable as well as the "non-monotone" singular solutions, (i.e. u and u_c intersect). An y- independent solution of (3.1) is asymptotically stable under the conditions stated above.

The order principle : diffusion and dominating solution, is the same as in the preceding section. The latter is tied to the geometry of the bifurcation. Our discussion of stability properties is restricted to second order elliptic equations, whereas the uniqueness result carries over to uniformly elliptic operators of arbitrary order whose coefficients may depend smoothly on x. Autonomy is used to classify all small solutions to be y-periodic for $\lambda < 4\pi^2$. However, it is expected not be essential if other methods are used. Under certain symmetry conditions the nonlinearity may depend on the derivatives of u as well.

Generalizations to higher order equations include the TAYLOR-problem of fluid mechanics. Here,for every supercritical λ two frequencies exist and thus it is conjectured that all small solutions are quasiperiodic in this case.An indication of this fact can be found in [3].The selection mechanism is not yet well understood, for experimental results see KØSCHMIEDER's contribution to this volume.

Equation (3.1) considered in $\mathbb{R}^2$ would be a model for the BENARD-problem. However, not even the uniqueness result is known. The linearized problem is highly degenerate and only some special solutions have been shown to exist for the full nonlinear problem (see SATTINGER's contribution to this volume).

4. References

1 Aronson,P.G., Weinberger,H.F. (1975), Nonlinear Diffusion in Population Genetics, Construction, and Nerve Propagation, Proc. Tulane Program in Partial Differential Equations, Lecture Notes in Math. No. 446, Springer Verlag, Berlin

2 Crandall,M.G., Rabinowitz,P.H., (1971), Bifurcation from Simple Eigenvalues, J. Functional Analysis,8,pp.321-340

3 DiPrima,R.C., Eckhaus,W. and Segel,L.A., (1971), Nonlinear Wave Number Interaction in Near-Critical Two-Dimensional Flows, J. Fluid Mech. 49, pp. 705-744

4 Fife,P.C., (1977), On Modelling Pattern Formation by Activator-Inhibitor-Systems, MRC Technical Symmary Report 1724, Univ. of Wisconsin, Math. Res. Center, Madison

5 Fife,P.C., (1977), Stationary Patterns for Reaction-Diffusion Equations, MRC Technical Summaty Report 1709, Univ. of Wisconsin, Math. Res. Center, Madison

6 Kirchgässner,K., (1977), Preference in Pattern and Cellular Bifurcation in Fluid Dynamics, to appear in: Applications of Bifurcation Theory, Proc. Advanced Seminar, Academic Press

7 Sattinger,D.H., (1971), Stability of Bifurcating Solutions by Leray-Schauder Degree, Arch.Rat.Mech.Anal., 43, pp.154-166

The Bifurcation of T-periodic Solutions into nT-periodic Solutions and TORI

D. Joseph

With 3 Figures

My lecture on bifurcation and stability of solutions which branch from forced T-periodic solutions is based on the recent work of G. IOOSS and myself [1] and on my forthcoming paper on factorization theorems [2]. In general, forced T-periodic solutions bifurcate into subharmonic solutions with a fixed period τ (τ=nT; n=1,2,3,4) independent of the amplitude or into a torus [1,3,4,5,6] containing solutions whose analytic properties are not yet fully understood. The subharmonic bifurcating solutions with n=1 are the T-periodic equivalent of a symmetry-breaking bifurcation of steady solutions with other steady solutions. The symmetry breaking flower instability of the axisymmetric climb of a viscoelastic fluid on an oscillating rod [7] which is shown in the movie "Novel Weissenberg effects" by G. S. BEAVERS and myself is one example of such a symmetry breaking T-periodic bifurcation. The solutions on the torus are very roughly the T-periodic equivalent of a Hopf bifurcation, of a steady solution into a periodic solution; like the Hopf bifurcation the solutions on the torus possess frequencies which depend on the amplitude but in the nonautonomous, T-periodic case the variation of these frequencies need not be smooth. A good example of smooth variation of frequencies on a two-dimensional torus appears to describe the observations of SWINNEY, FENSTERMACHER and GOLLUB [8,9] of the oscillatory regimes of flow which follow wavy vortices in the Taylor problem when the Reynolds number is increased.

1. Stability and Repeated Bifurcation of Solutions of Nonlinear Evolution Equations in a Single Variable [2,10]

To clarify some aspects of the mathematical nature of bifurcation theory we shall first construct a simple theory for evolution equations in R_1. In this theory we give complete and rigorous results for stability and repeated branching which actually apply to the repeated bifurcation of steady solutions in a Banach space at a simple eigenvalue. In the general problem of bifurcation of steady solutions we have in mind steady equilibrium solutions to nonlinear equations possessing different patterns of spatial symmetry. These solutions are points in a Banach space and the families with different symmetries may be projected as plane curves (bifurcation curves). In R_1 the projections and the solutions coincide and the theory simplifies enormously.

We are going to study the repeated bifurcation and stability of solutions of the evolution problem

$$V_t + F(\mu,V) = 0 \tag{1.1}$$

where $F(\mu,0) = 0$, $F(0,V) \neq 0$ when $V \neq 0$ and F together with its first two partial derivatives are continuous functions of μ, $V \in R_1$; in particular we have

$$F(\mu, V+V') = F(\mu, V) + F_V(\mu, V)V' + \frac{1}{2} F_{VV}(\mu, V)V'^2 + R(\mu, V, V')V'^3 \tag{1.2}$$

for any $V, V' \in R_1$.

Bifurcating solutions arising from autonomous evolution equations in R_1 are necessarily steady. This means that the study of bifurcation in R_1 is equivalent to finding branches of solutions of the equation

$$F(\mu, V) = 0 . \tag{1.3}$$

Suppose that $V = \varepsilon$ and $\mu = \mu(\varepsilon)$ is a solution of (1.3). Then

$$F(\mu, \varepsilon) = 0 = F_\mu(\mu, \varepsilon)d\mu + F_V(\mu, \varepsilon)d\varepsilon \tag{1.4}$$

We define a point of bifurcation to be a double point of (1.4); that is, a point through which there are two solutions of (1.4), possessing distinct tangents. At such a point

$$F_\mu = F_V = 0 \tag{1.5}$$

and

$$F_{V\mu}^2 - F_{VV}F_{\mu\mu} > 0 . \tag{1.6}$$

A disturbance w of $V = \varepsilon$ with $\mu = \mu(\varepsilon)$ satisfies the equation

$$\frac{dw}{dt} + F_V(\mu(\varepsilon), \varepsilon)w + 0(w^2) = 0 .$$

Linearizing for small disturbances $w = e^{-\gamma t}w'$, we find that $\gamma(\varepsilon) = F_V(\mu(\varepsilon), \varepsilon)$. The function $\gamma(\varepsilon)$ is nicely described by a factorization theorem for the stability of the solution $V = \varepsilon$, $\mu = \mu(\varepsilon)$:

$$\gamma(\varepsilon) = F_V(\mu(\varepsilon), \varepsilon) = -\mu_\varepsilon F_\mu(\mu(\varepsilon), \varepsilon) \equiv \mu_\varepsilon \hat{\gamma}(\varepsilon) - \mu_\varepsilon\{-\varepsilon F_{V\mu}(0,0) + 0(\varepsilon^2)\} \tag{1.7}$$

The second equality in (1.7) follows from $(1.4)_2$. The third equality is a definition of $\hat{\gamma}(\varepsilon)$ and the fourth follows from expanding $F(\mu(\varepsilon), \varepsilon)$ in powers of ε. We note that the stability of the solution $v = 0$ of (1.1) is governed by $V_t + F_V(\mu, 0)V = 0$. Then, with $V = e^{-\sigma t}V'$, we find that $\sigma = F_V(\mu, 0)$, so that $\sigma_\mu = F_{V\mu}(0,0)$. If $V = 0$ loses stability strictly as μ increases past zero then $F_{V\mu}(0,0) < 0$. It follows from (1.7) with $F_{V\mu}(0,0) < 0$ that locally, near $\varepsilon = 0$, subcritical bifurcating solutions $\mu_\varepsilon < 0$ are unstable, $\gamma(\varepsilon) < 0$, and supercritical solutions $\mu_\varepsilon > 0$ are stable, $\gamma(\varepsilon) > 0$.

A point at which $\mu_\varepsilon = 0$ is a stationary point of the bifurcation curve. A point at which μ_ε changes sign is a critical point of the bifurcation curve. If $\hat{\gamma}(\varepsilon) \neq 0$ at a critical point then $\gamma(\varepsilon)$ changes sign when μ_ε does. If $\hat{\gamma}(\varepsilon) \neq 0$ at a critical point, then $F_\mu(\mu, \varepsilon) \neq 0$ and (μ, ε) is not a point of bifurcation (see Fig. 1.1)

Now we show that points at which $\hat{\gamma}(\varepsilon_0) = 0$ and $\gamma_\varepsilon(\varepsilon_0) \neq 0$ are points of bifurcation. At such points (1.7) shows that (1.5) holds and

$$\gamma_\varepsilon(\varepsilon_0) = F_{VV}(\mu(\varepsilon_0), \varepsilon_0) + \mu_\varepsilon(\varepsilon_0)F_{V\mu}(\mu(\varepsilon_0), \varepsilon_0) = -\mu_\varepsilon(\varepsilon_0)F_{\mu V}(\mu(\varepsilon_0), \varepsilon_0) - \mu_\varepsilon^2 F_{\mu\mu}(\mu(\varepsilon_0), \varepsilon_0) . \tag{1.8}$$

It follows that at $\varepsilon = \varepsilon_0$

$$d\varepsilon^2 F_{VV} + 2\, d\varepsilon d\mu\, F_{V\mu} + d\mu^2 F_{\mu\mu} = 0 . \tag{1.9}$$

The discriminant of (1.9)

$$F_{V\mu}^2 - F_{VV} F_{\mu\mu} \geq 0 \tag{1.10}$$

is not negative. Suppose equality holds in (1.10). Then the second or third term in (1.8) must vanish which is impossible since $\gamma_\varepsilon(\varepsilon_0) \neq 0$. It follows that the inequality is strict and $(\mu(\varepsilon_0),\varepsilon_0)$ is a point of bifurcation

As one example of the foregoing, consider the equation

$$V_t + V(\mu V-9)(\mu+2V-V^2) = 0 \tag{1.11}$$

The bifurcating solutions are

(i) $V = 0 \quad \mu \in R_1$,

(ii) $\mu = V^2 - 2V$

and

(iii) $\mu = 9/V$.

The curve (ii) has two bifurcation points and one critical point at $(\mu,v) = (-1,1)$. The stability of various branches of (1.11) are indicated in Fig. 1.1. It is almost a miracle that examples of secondary bifurcation in R_1 as simple as the one just given seem not to have been discussed in the literature on bifurcation. Of course, everyone knows that $F(\mu,V) = 0$ can have multiple solutions.

Our understanding is that bifurcation in R_1 is equivalent to the continuous branching of solutions $F(\mu,V) = 0$. This conventional use of the term "bifurcation" is restrictive since it excludes isolated solutions of $F(\mu,V) = 0$ which are not ultimately connected to the solution $V = 0$. The hyperbola $\mu = 9/V$ in the third quadrant of Fig. 1.1 is just one type of isolated solution which can occur.

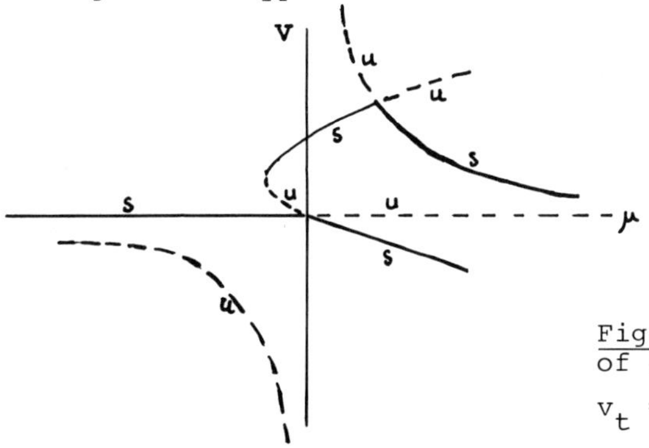

Fig. 1.1 Stability and bifurcation of equilibrium solutions of $v_t = v(9-\mu v)(\mu + 2v - v^2)$

2. Bifurcation and Stability of Solutions Branching from T-periodic Forced Solutions [1]

We turn now to the problem of stability, bifurcation and repeated bifurcation of the nonlinear, nonautonomous, evolution problem

$$\frac{dV}{dt} + F(t,\mu,V) = 0 \tag{2.1}$$

Here, $F(t,\mu,V)$ is a nonlinear, T-periodic $(F(t,\cdot,\cdot) = F(t+T,\cdot,\cdot))$, map from $R \times \mathbb{C} \times H$ into H, where H is a Hilbert space with natural scalar product $(u,v)_H = \overline{(u,v)}_H$, which carries real vectors $V \in H$ into real vecters when $\mu \in \mathbb{C}$ is real. It is further assumed that $F(t,\mu,V)$ is an analytic operator with a Fréchet expansion

$$F(t,\mu,V+w) = F(t,\mu,V) + F_V(t,\mu,V|w) + \frac{1}{2} F_{VV}(t,\mu,V|w,w) + 0(w^3) \tag{2.2}$$

for $t \in R$, V, $w \in X$ = domain $(F_V) \supset H$(compactly) and μ in a neighborhood of a real interval of $\mathbb{C}$. It is assumed that $F(t,\mu,0) = 0$ so that $V = 0$ is a solution of (2.1).

We may identify (2.1) with a partial differential equation in which V is the difference between two solutions driven by prescribed T-periodic data. One of the two solutions is T-periodic and it accounts for the appearance of t in $F(t,\mu,V)$. The solution $V = 0$ of (2.1) corresponds to a forced T-periodic solution of the original problem. When the data is steady, the same type of analysis leads to an autonomous problem

$$\frac{dV}{dt} + F(\mu,V) = 0 \ . \tag{2.3}$$

The evolution problems (2.2) and (2.3) have very different properties [2].

The stability of the solution $V = 0$ of (2.1) to small disturbances may be determined by Floquet analysis of the variational equations

$$\frac{dz}{dt} + F_V(t,\mu,0|z) = 0 \ . \tag{2.4}$$

According to Floquet theory we may determine the stability of zero by studying the Floquet exponents $-\sigma(\mu)$ of the representation $z = e^{-\sigma(\mu)t}\zeta(t)$ where $\zeta(t) = \zeta(t+T)$ and $\sigma(\mu) = \xi(\mu) + i\omega(\mu)$. These exponents are eigenvalues of the operator $J(\mu)$; that is,

$$-\sigma\zeta + J(\mu)\zeta = 0 \tag{2.5}$$

where $J = d/dt + F_V(t,\mu,0|\ )$. The operator $J(\mu)$ is a Fredhom operator (with a compact resolvent) taking T-periodic vectors in X into T-periodic vectors in H; that is $J(\mu)$: $X_T \to H_T$. The scalar product on H_T is

$$[u,v]_T = \frac{1}{T}\int_0^T (u,v)_H dt. \tag{2.6}$$

The Floquet exponents are the exponents of the Floquet multiplier $\lambda(\mu) = \exp(-\sigma(\mu)T) = \exp(-\sigma(\mu)T+2\pi ik)$. The Floquet multiplier is an eigenvalue of the monodromy operator. This operator is the analogue of the monodromy matrix for ordinary differential equations (see, for example section 7 of [7]). The monodromy matrix is the fundamental solution matrix whose values at $t = 0$ coincide with the unit matrix. This matrix can be regarded as a map $z(0) \to z(T)$. The same type of map has been defined by IOOSS [5] for evolution equations on Banach spaces. In this case, $\lambda(\mu)$ are the eigenvalues of linear compact operator $S_\mu(T)$

in X mapping $z(0)$ into $z(T)$.

When $\xi(\mu) > 0$ for all eigenvalues of $J(\mu)$, then $V = 0$ is stable. We assume (H.1) that $V=0$ loses stability strictly as μ is increased past zero: $\sigma(0) = \sigma_0 = i\omega_0$, $\xi_{,\mu}(0) < 0$. At criticality ($\mu=0$), $J(0) \equiv J_0$ and $\lambda(0) \equiv \lambda_0$ is of modulus one. We may represent the Floquet exponent $-\sigma_0 = -i\omega_0$ at criticality with

$$\omega_0 = \frac{2\pi}{T}(r+k), \quad 0 \le r < 1, \quad k = 0, \pm 1, \pm 2, \ldots$$

Without losing generality, we may set $k = 0$ [1] because repeated points on the imaginary axis of the complex σ plane correspond to unique points

$$\lambda_0 = e^{-2\pi i r/T} \tag{2.7}$$

on the unit circle of the complex λ plane. Problems with $k \neq 0$ may always be reduced to the one with $k=0$; if $\hat{\zeta}(t) = \hat{\zeta}(t+T)$ is an eigenfunction of (2.5) with $k \neq 0$ the $\hat{\zeta}(t) = \hat{\zeta}(t+T) = e^{2\pi ikt/T}\zeta(t)$ is also an eigenfunction of (2.5) with $k = 0$.

We now study bifurcation under the hypothesis (H.1) and (H.2): Assume that $-2\pi ir/T$ is an algebraically simple eigenvalue of J_0. (Then λ_0 is a simple eigenvalue of $S_0(T)$.) If $\zeta(t)$ is the eigenvalue of J_0 belonging to $\sigma_0 = 2\pi ir/T$, then $\overline{\zeta}(t)$ is another eigenvector of J_0 belonging to $\overline{\sigma}_0 = -2\pi ir/T$. Hence $z(t)=e^{-2\pi irt/T}\zeta(t)$ and $\overline{z}(t)$ both solve (2.4) when $\mu = 0$. In the analysis of bifurcation we must consider all values of λ_0 on the unit circle; that is, all values r, $0 \le r < 1$. If r is irrational then $z(t)$ and $\overline{z}(t)$ are independent and zero is a semi-simple double eigenvalue of the operator

$$\mathbb{J}_0 = \frac{d(\cdot)}{dt} + F_V(t,0,0|\cdot) \tag{2.8}$$

in a space of doubly-periodic functions $f(\frac{2\pi t}{T}, \frac{2\pi rt}{T})$. In this case, we get a bifurcating two-dimensional torus. When the amplitude ε of the torus small, the principal part of the solution on the torus is doubly periodic function with a frequency $2\pi/T$ and a frequency $\omega(\varepsilon)$ which varies with ε and is such that $\omega_0 = 2\pi r/T$ [5,11]. The variation of the frequency $\omega(\varepsilon)$ of the solutions on the torus need not be smooth. The analytical properties of solutions on bifurcating tori in Navier-Stokes and other problems are not well understood.

The complement of the set of irrational numbers on $0 \le r < 1$ is the set of rational fractions m/n, $m<n$. Of course these fractions are dense on $(0,1)$. It would be unfortunate if the bifurcation results depended in any important way on the difference between the irrational values of r and the rational fractions. Fortunately, this difference does not exist as a general feature; only the values $r = 0, 1/2, 1/3, 2/3, 1/4, 3/4$ are special; they lead to subharmonic periodic solutions with the property that their period is a fixed multiple of T, independent of ε.

We can organize the motivate the study of subharmonic bifurcation at a rational fraction in the following way. A subharmonic solution $z(t)$ is an nT-periodic function; hence

$$z(t+nT)=e^{\frac{-2\pi ir}{T}(t+nT)}\zeta(t+nT) = e^{-2\pi irn}z(t) = z(t).$$

if and only if

$$e^{-2\pi i r n} = 1 = \lambda_0^n = \bar{\lambda}_0^n . \qquad (2.9)$$

It follows that the subharmonic solution z(t) can exist with $0 \le r < 1$ if and only if λ_0 is the nth root of unity and $r = m/n$, $m = 0,1,2,\ldots;$ $n = 1,2,\ldots,$ $m < n$.

To construct subharmonic bifurcating solutions we introduce the domain space X_{nT} = domain $\mathbb{J}_0$ and the range space H_{nT} with scalar product defined by (2.6) with nT replacing T. There is an adjoint $\mathbb{J}_0^*$ relative to $[\cdot,\cdot]_{nT}$ and, if

$$z^* = {}^{-2\pi imt/n}\ \zeta^*(t), \qquad (2.10)$$

where

$$2\pi i\frac{m}{n}\ \zeta^* + J_0^*\ \zeta^* = 0,\ \zeta^*(t) = \zeta^*(t+T),$$

is complex, then z^* and $\bar{z}^*$ span the null space of $\mathbb{J}_0^*$. It is easy to verify that $[z,\bar{z}^*]_{nT} = 0$ and we may take $[z,z^*]_{nT} = 1$.

We may take z and z* as real-valued when $\lambda_0 = e^{-2\pi im/n}$ is real; that is when $n = 0$, $\lambda_0 = 1$ and $m/n = 1/2$, $\lambda_0 = -1$. When $m = 0$, $z = \zeta(t+T)$ is real and T-periodic; that is n=1. When $n/m = 1/2$, $z = e^{-\pi it/T}\zeta = e^{\pi it/T}\bar{\zeta}$ is real and 2T-periodic. In these two cases, and only these two, $\mathbb{J}_0$ has a one-dimensional null space.

Now I shall indicate how the subharmonic solutions can be constructed by analytic perturbation theory. The detailed demonstrations are given in [1]. Assume that there are subharmonic bifurcating solutions $V = U(t,\varepsilon) = U(t+nT,\varepsilon)$, $\mu = \mu(\varepsilon)$ which are analytic in some neighborhood $I(\varepsilon)$ of the origin. Then introducing the notation $(\cdot)_n = d^n(\cdot)/d\varepsilon^n$ and for short, $F_V(\mu,U|U_1) = F_V(U_1)$, etc, which suppresses the dependence of the operators on t, $U(t,\varepsilon)$ and $\mu(\varepsilon)$, we find that

$$\frac{dU}{dt} + F(t,\mu,U) = 0\ , \qquad (2.11)$$

$$\frac{dU_1}{dt} + F_V(U_1) + \mu_1\ F_\mu = 0\ , \qquad (2.12)$$

$$\frac{dU_2}{dt} + F_V(U_2) + F_{VV}(U_1,U_1) + 2\mu_1\ F_{V\mu}(U_1) + \mu_1^2\ F_{\mu\mu} + \mu_2 F_\mu = 0\ , \qquad (2.13)$$

$$\frac{dU_3}{dt} + F_V(U_3) + F_{VVV}(U_1,U_1,U_1) + 3\ F_{VV}(U_1,U_2) \qquad (2.14)$$

$$+ 3\mu_1\ F_{V\mu}(U_2) + 3\mu_1\ F_{VV\mu}(U_1,U_1) + 3\mu_2\ F_{\mu V}(U_1)$$

$$+ 3\mu_1^2\ F_{V\mu\mu}(U_1) + \mu_2\mu_1\ F_{\mu\mu} + \mu_3\ F_\mu = 0\ .$$

Existence and stability properties of the subharmonic bifurcating solutions near $\varepsilon = 0$ are determined by these equations and the equation for U_4.

When $\varepsilon = 0$, $F_\mu(0,0) = F_{\mu\mu}(0,0) = 0$ and the first two terms of

(2.12), (2.13) and (2.14) may be replaced by $\mathcal{J}_0 U_n$, n = 1,2,3. Since, $J_0U_1 = 0$ and zero is a simple eigenvalue of $\mathcal{J}_0$ (when n=1 or n=2) or a semi-simple double eigenvalue of $\mathcal{J}_0$ we have

$$U_1 = az + \overline{az} \ . \tag{2.15}$$

When n=1 or 2, $z=\bar{z}$ and $a = \bar{a}$ is determined by the normalization associated with the definition of ε. In all other cases this normalization gives one relation between a and $\bar{a}$ and the ultimate values of these quantities is determined by the conditions for solvability. These conditions arise from the Fredholm alternative for $\mathcal{J}_0$:

Let (H.1) and (H.2) hold. Then there is $u \in X_{nT}$ solving

$$\mathcal{J}_0 \ u = f \in H_{nT} \tag{2.16}$$

if and only if

$$[f,z^*]_{nT} = [f,\bar{z}^*]_{nT} \ , \tag{2.17}$$

If f is real-valued one of the conditions (2.17) implies the other and

$$u = bz + \overline{bz} + \omega \tag{2.18}$$

where $\omega = \mathcal{J}_0^{-1} f$ is unique.

It follows from this statement of the Fredholm alternative that

$$U_\ell = a_\ell z + \bar{a}_\ell \bar{z} + w_\ell , \ \ell > 1$$

and, using (2.15),

$$\mathcal{J}_0 w_2 + F_{VV}(az + \overline{az}, az + \overline{az}) + 2\mu_1 \ F_{V\mu}(az + \overline{az}) = 0 \ . \tag{2.19}$$

(2.19) is solvable if and only if

$$a^2[F_{VV}(z,z),z^*]_{nT} + \bar{a}^2[F_{VV}(\bar{z},\bar{z}),z^*]_{nT} + 2|a|^2[F_{VV}(z,\bar{z}),z^*]_{nT}$$
$$+ 2\mu_1 a \ \sigma_\mu = 0 \tag{2.20}$$

where, by a standard perturbation result using (2.5) and (2.17)

$$\sigma,_\mu = [F_{V\mu}(z),z^*]_{nT} \ . \tag{2.21}$$

We are assuming that the real part $\xi,_\mu$ of $\sigma,_\mu$ is negative at criticality; that is, V = 0 loses stability strictly as μ crosses zero.

To evaluate (2.2), we make use of the following computational lemma. Let $g(t) = f(\varepsilon) \exp(2\pi imt/n) = g(t+mT)$. Then

$$[f,z^*]_{nT} = [g,\zeta^*]_{nT} = 0 \ . \tag{2.22}$$

(2.22) may be proved by direct computation. Application of (2.22) to (2.20) leads to the conclusion that

$$\mu_1 = 0 \text{ for all n except n = 1 and n = 3 } . \tag{2.23}$$

We may conclude that if solutions with n = 2,4,5,6,7,... bifurcate and $\mu_2 \neq 0$ they bifurcate on one side of criticality. In general,

$$\mu_1 \neq 0 \text{ when } n = 1 \text{ or } n = 3. \tag{2.24}$$

The T-periodic and 3T-periodic solutions bifurcate on both sides of criticality.

Supposing now that (2.23) holds we find that (2.14) is solvable if and only if

$$3\mu_2\,\sigma_\mu\,a + 3[F_{VV}(U_1,U_2),z^*]_{nT} + 3[F_{VVV}(U_1,U_1,U_1),z^*]_{nT} = 0. \tag{2.25}$$

When $n \geq 5$ (2.25) we find, using (2.22), that (2.25) is in the form

$$\mu_2\sigma_\mu + |a|^2 f = 0$$

which has no solution except for the special case in which f/σ_μ is real. Hence, in general, there is no nT periodic bifurcation for $n \geq 5$.

Equation (2.25) must be studied when n=2 and n=4. Equation (2.20) must be studied when n=1 and n=3. I have already noted when n=1 or n=2, z is real and $a=\bar{a}$ is determined by normalization. In this case, the analysis of bifurcation and the stability of bifurcation follows along by now classical lines. For n=3, (2.20) is a cubic equation and for n=4, (2.25) is a quartic equation for the real or imaginary part of a. (We may eliminate, say, the imaginary part by enforcing a normalizing condition connected to the definition of the amplitude ε.) Having once determined the number of real roots of the cubic equation (2.20) or the quartic equation (2.25) for allowed values of a we must then verify that the higher order perturbation equations are solvable. This will be true at all orders if it is true at one order beyond (2.20) or (2.25).

The results of the study of bifurcation and stability of solutions branching from T-periodic ones is given in Fig. 2.1, see page 9.

3. Factorization Theorems for the Stability of nT-bifurcating Solutions

We now relax the assumption that the amplitude ε of the $\tau = nT$ periodic bifurcating solutions is small and undertake the study of the stability of these solutions under the hypothesis H.3: $U(\tau,\varepsilon) = U(t+nT,\varepsilon)$, $\mu(\varepsilon)$ is a $\tau = nT$-periodic bifurcating solution which is analytic on some possibly large interval $I(\varepsilon)$. I do not require that $I(\varepsilon)$ be a neighborhood of the origin so that the factorization theorems will apply to isolated solution branches and to branches of solutions which arise from repeated bifurcations.

The stability of $U(t,\varepsilon)$ is governed, in the linearized approximation, by

$$\frac{dw}{dt} + F_V(t,\mu(\varepsilon),U(t,\varepsilon)|w) = 0 \tag{3.1}$$

where $F_V(t,\cdot,\cdot|\cdot)$ is T-periodic and $U(t,\varepsilon)$ is τ periodic. The spectral problem associated with (3.1) may be obtained from the Floquet representation $w = e^{-\gamma t}\Gamma(t)$ where $\Gamma(t)$ is $\tau = nT$ periodic and

$$-\gamma(\varepsilon)\Gamma + \mathbb{J}(\varepsilon)\Gamma = 0 \tag{3.2}$$

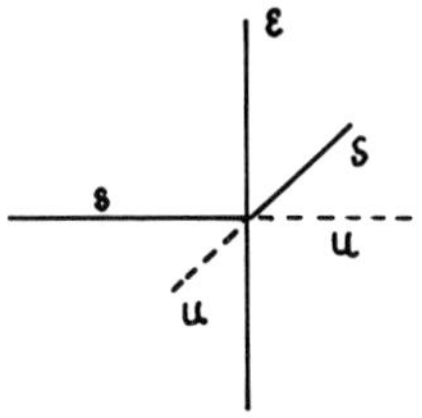

n=1. A single one-paramerter family of T-periodic solutions bifurcate on both sides of criticality at a simple eigenvalue

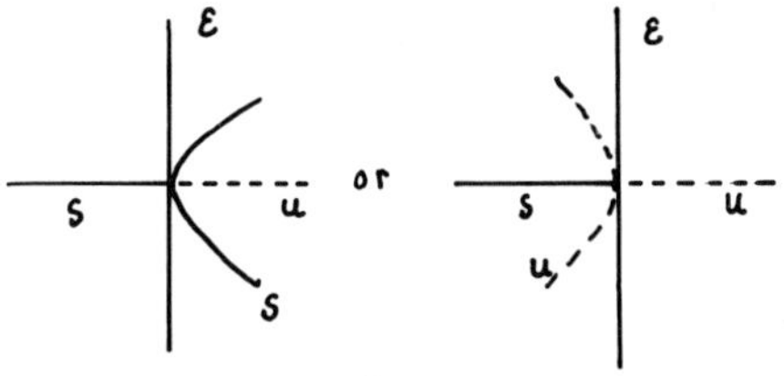

n=2. A single one-parameter family of 2T-periodic solutions bifurcate on one side of criticality at a simple eigenvalue

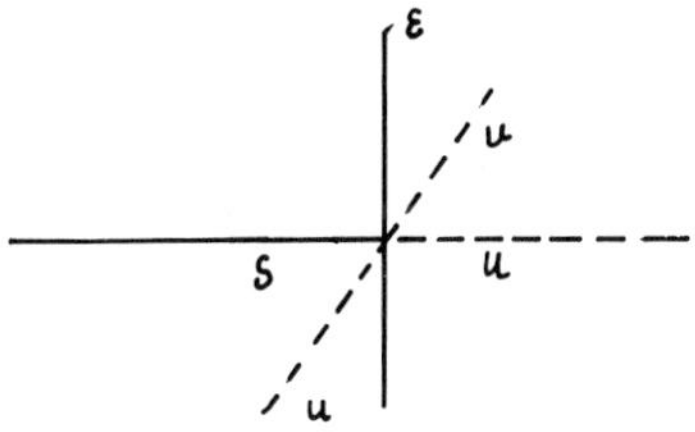

n=3. A single one-parameter family of 3T-periodic solutions bifurcates on both sides of criticality at a semi-simple double eigenvalue

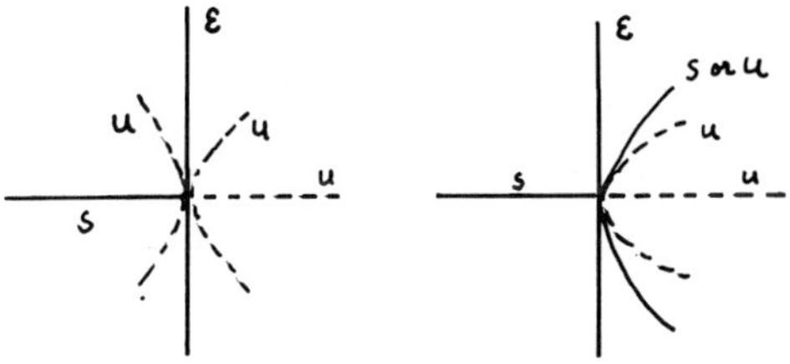

n=4. There are three alternatives for the bifurcation of 4T-periodic solution at a semi-simple double eigenvalue: (a) Two solutions bifurcate, each on a different side of criticality and both are unstable, (b) Two solutions bifurcate on the same side criticality and one is unstable, the stability of the other being determined by details varying from problem to problem, (c) no 4T-periodic solutions bifurcate

Fig. 2.1 Bifurcation of forced T-periodic solutions. The period of nT-periodic solutions with n=1,2,3,4 is independent of the amplitude. For all values of $r \neq 0$, 1/2, 1/3, 2/3, 1/4, 3/4 a torus bifurcates on one side of criticality. The supercritical torus is stable and the subcritical torus is unstable. The frequencies of the solutions on the torus vary continuously with amplitude but the variation need not be differentiable.

where $\mathbb{J}(\varepsilon)(\cdot) = (\cdot)_{,t} + F_V(t,\mu(\varepsilon),U(t,\varepsilon)\ (\cdot))$ maps X_{nT} into H_{nT} and satisfies the hypothesis

H.4: $\mathbb{J}(\varepsilon)$ is a Fredholm operator with a compact resolvent from X_{nT} into itself.

Since $\mathbb{J}(\varepsilon)$ has a compact resolvent its spectrum is of eigenvalues of at most finite multiplicity and there is an adjoint $\mathbb{J}^*(\varepsilon)$ into H_{nT} such that

$$-\overline{\gamma}(\varepsilon)\Gamma^* + \mathbb{J}^*\ \Gamma^* = 0 \tag{3.3}$$

where $\overline{\gamma}$ is the conjugate of γ. The number of independent eigenvectors Γ_i belonging to $\gamma(\varepsilon)$ is the dimension of the null space of $-\gamma(\varepsilon) + \mathbb{J}(\varepsilon)$

which is the same as the dimension of the null space of the adjoint operator $-\bar{\gamma}(\varepsilon) + \mathbb{J}^*$.

The last hypotheses which we need to prove the factorization is

H.5: $\gamma(\varepsilon)$ is an algebraically simple eigenvalue of $\mathbb{J}(\varepsilon)$ for all $\varepsilon \in I(\varepsilon)$ except possibly on an exceptional set of isolated points across which $\Gamma(t,\varepsilon)$ and $\Gamma^*(t,\varepsilon)$ are continuous. H.4 implies that $\gamma(\varepsilon)$ is continuous across points in the exceptional set. We may normalize

$$[\Gamma(\varepsilon),\Gamma^*(\varepsilon)]_{nT} = 1 \tag{3.4}$$

at all points where $\gamma(\varepsilon)$ is simple, and by continuity also on points in the exceptional set.

Suppose H.3, H.4, H.5 hold and assume that

$$[U_\varepsilon(\varepsilon),\Gamma^*]_{nT} \neq 0 \ . \tag{3.5}$$

Then there is a unique continuous function $\hat{\gamma}(\varepsilon)$ defined on all $I(\varepsilon)$ such that

$$\gamma(\varepsilon) = \mu_\phi(\varepsilon)\hat{\gamma}(\varepsilon) \tag{3.6}$$

where

$$\hat{\gamma}(\varepsilon) = -[F_\mu(\mu(\varepsilon),U(\varepsilon),\Gamma^*]_{nT}/[U_\varepsilon,\Gamma^*]_{nT} \ . \tag{3.7}$$

Moreover,

$$\Gamma = b(\varepsilon)(U_\varepsilon + \mu_\varepsilon q) \tag{3.8}$$

where $b(\varepsilon)$ is a normalizing factor for Γ and

$$q(t,\varepsilon) = q(t + nT,\varepsilon) \tag{3.9}$$

is uniquely determined by

$$\hat{\gamma}U_\varepsilon + F_\mu(\mu(\varepsilon),U(\varepsilon)) + \{\gamma-\mathbb{J}\}\, q = 0 \tag{3.10}$$

and

$$[q,\Gamma^*]_{nT} = 0 \ . \tag{3.11}$$

Proof: (2.12) may be written as

$$\mathbb{J}U_\varepsilon + \mu_\varepsilon F_\mu(\mu(\varepsilon),U(\varepsilon)) = 0 . \tag{3.12}$$

Since Γ^* satisfies (3.3) we have

$$-\mu_\varepsilon[F_\mu,\Gamma^*]_{nT} = [\mathbb{J}U_\varepsilon,\Gamma^*]_{nT} = [U_\varepsilon,\mathbb{J}^*\Gamma^*]_{nT} = \gamma(\varepsilon)[U_\varepsilon,\Gamma^*]_{nT} \ . \tag{3.13}$$

Equation (3.12) holds at all points where $\gamma(\varepsilon)$ is an algebraically simple eigenvalue and also, by continuity, across points in the exceptional set where $\gamma(\varepsilon)$ is not algebraically simple. Solving (3.13) for $\gamma(\varepsilon)$ we find (3.7). Now combining (3.2) and (3.8) we get

$$- \mu_\varepsilon \hat{\gamma}(\varepsilon)(U_\varepsilon + \mu_\varepsilon q) + \mathbb{J}U_\varepsilon + \mu_\varepsilon \mathbb{J}q = 0$$

Elimination of JU_ε with (3.12) leads to

$$\mu_\varepsilon\{\hat{\gamma}U_\varepsilon + F_\mu(\mu,U) + [\gamma - \mathbb{J}]_q\} = 0 \tag{3.14}$$

The coefficient of μ_ε in (3.14) vanishes when $\mu_\varepsilon \neq 0$ and, by continuity, even when $\mu_\varepsilon = 0$ at a point. This proves (3.10). Since $-\gamma+\mathbb{J}$ is Fredholm and (3.13) holds, (3.14) is uniquely solvable with $[q,\Gamma^*] = 0$ wherever $\gamma(\varepsilon)$ is algebraically simple. So we get a unique q when $\gamma(\varepsilon)$ is algebraically simple and, by continuity, at isolated points in the exceptional set. This proves the factorization theorem for nT-periodic bifurcating solutions.

Remark 1: If $\mathrm{re}\hat{\gamma}(\varepsilon_0) \neq 0$ at a point ε_0 where $\mu(\varepsilon)$ changes sign (such points are called critical points or turning points), then $\mathrm{re}\ \gamma(\varepsilon)$ changes sign as ε crosses ε_0. This remark suggests that in most problems the bifurcating solutions will gain or lose stability across a critical point

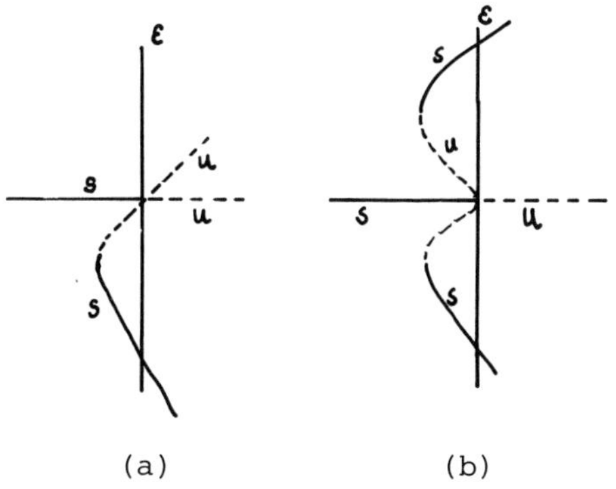

Fig. 3.1 Recovery of stability at critical points of bifurcation curve. (a) is the conjectured form of the global extension of 3T-periodic bifurcating solutions (b) is the conjectured form of any unstable subcritical one-sided bifurcating solution; say, n=4.

Remark 2: In most problems $\gamma(\varepsilon)$ is an algebraically simple eigenvalue of $\mathbb{J}(\varepsilon)$ for nearly all values of ε. The stability of the bifurcating solution $U(t,\varepsilon)$ is controlled by the eigenvalue $\gamma(\varepsilon)$ with the smallest $\mathrm{re}\ \gamma(\varepsilon)$. The factorization may be used to calculate the stability of the bifurcating solution at $\varepsilon=0$ when $\gamma_0 = 0$ is a simple eigenvalue of $\mathbb{J}_0$; that is, when n=1 or n=2. In this case we find that when $\varepsilon \to 0$; $\mu \to 0$,

$$(\Gamma_0, \overset{*}{\Gamma}_0) \to (z(t), z^*(t)) = (z(t+nT), z^*(t+nT)),$$

$$U(t,\varepsilon) \to \varepsilon z + O(\varepsilon^2),$$

and

$$F_\mu(t,\mu,U) \to \varepsilon F_{\mu V}(t,0,0\ z).$$

Since $[z,z^*]_{nT} = 1$, $[U_\varepsilon,\Gamma^*]_{nT} \to 1$ and using (3.7) and (2.21), we find that when ε is small

$$\hat{\gamma}(\varepsilon) \simeq - \varepsilon[F_{\mu V}(t,0,0\ z),z^*]_{nT} = - \varepsilon\sigma'_\mu . \tag{3.15}$$

Equations (3.6) and (3.15) implies the local ($\varepsilon \to 0$) statement of stability for T-periodic and 2T-periodic bifurcating solutions: subcritical solutions are unstable and supercritical solutions are stable.

When n=3 or n=4 the analysis of the stability of the bifurcating solutions requires that one construct a perturbation analysis of a semi - simple double eigenvalue to separate the branches of $\gamma(\varepsilon)$. Without such an analysis it would not be possible to specify the linear combinations of $(z,\bar{z})$ and $(z^*,\bar{z}^*)$ which give the limiting $\varepsilon \to 0$ value of $\Gamma(\varepsilon)$ and $\Gamma^*(\varepsilon)$.

Factorization theorems can be used to characterize points of secondary and repeated bifurcation [2] at a simple eigenvalue. Factorization theorems for autonomous problems may also be proved [2,7]. It is interesting that the factorization theorem for periodic bifurcating solutions of the Hopf type show that the eigenvalue $\gamma=0$ is always an algebraically double eigenvalue of the appropriate operator. This algebraically double eigenvalue is geometrically simple in the general case. In the special case the derivative of the frequency $\omega(\varepsilon)$ of the Hopf solution with respect to ε vanishes when $\gamma(\varepsilon)$ does. ($\varepsilon=0$ is a special case). If $\omega_{,\varepsilon}(\varepsilon) = \gamma(\varepsilon) = 0$ then $\gamma(\varepsilon)$ is a semi-simple double eigenvalue [2].

Acknowledgement: This work was supported by the U.S. Army Research Office and by a grant (GK 19047) from the U.S. National Science Foundation.

References

1. G. Iooss and D. D. Joseph, Arch Rational Mech. Anal. (to appear)

2. D. D. Joseph, Arch. Rational Mech. Anal. (to appear)

3. R. J. Sacker, Ph.D thesis. New York Univ. IMM-NYU report 333 (1964)

4. D. Ruelle and F. Takens, Comm. Maths. Phys. 20, 167 (1971)

5. G. Iooss, Arch. Rational Mech. Anal. 58, 35 (1975)

6. J. E. Marsden and M. McCracken, The Hopf Bifurcation and its Applications (Springer, Appl. Math Sciences 19, New-York-Heidelberg-Berlin, 1976)

7. D. D. Joseph, The Stability of Fluid Motions II. Chap XIII. (Springer, Tracts in natural philosophy. New York-Heidelberg-Berlin, 1976)

8. H. L. Swinney, P. R. Fenstermacher and J. P. Gollub, Transition to Turbulence in Circular Couette Flow. Paper given at the Symposium on Turbulent Shear Flow, April 1977, University Park, PA.

9. J. P. Gollub and H. L. Swinney, Phys. Rev. Letters, 35, 927 (1975)

10. S. Rosenblat, Arch. Rational Mech. Anal. (to appear)

11. D. D. Joseph. Springer Lecture Notes in Mathematics, 322, 130 (1973)

Cooperative Effects in Fluid Problems

D. H. Sattinger

With 3 Figures

When a layer of fluid is heated from below convection sets in when the temperature drop across the layer exceeds a certain critical value. Then, under certain circumstances, the convective motions which take place are organized in cellular patterns - rolls or hexagonal solutions being the most common. [1],[3],[6]

This phenomenon has long interested researchers in fluid mechanics, but it has recently begun to attract attention from physicists working in solid state physics and statistical mechanics because it is an excellent example of "cooperative effects" in a physical system - that is, a situation in which a physical system spontaneously passes from a disorganized or homogeneous state, to one exhibiting internal symmetry. [2]

Before the onset of convection, the fluid motion is invariant under the entire group of rigid motions; but after the onset of convection the motion is invariant only under a subgroup - a crystallographic subgroup of the group of rigid motions. An analogous situation is the phenomenon of freezing in statistical mechanics [4].

The mechanisms governing the onset of convection are well understood and there is a good quantitative agreement between theory and experiment. The mechanisms governing pattern selection are, in my opinion, still not completely clear, despite the excellent work of a number of researchers (esp. BUSSE, SEGEL, and STUART).

The mathematical model for convection usually studied is the elliptic system of partial differential equations known as the Boussinesq equations:

$$\Delta u_k + \delta_{k3}\theta - \frac{\partial p}{\partial x_k} = \frac{1}{P_r} u_j \frac{\partial u_k}{\partial x_j}$$

$$\Delta\theta + \mathcal{R} u_3 = u_j \frac{\partial \theta}{\partial x_j}$$

$$\frac{\partial u_j}{\partial x_j} = 0 .$$

Here u_1, u_2, u_3 are the Cartesian components of the fluid velocity, θ is the temperature deviation, and p is the pressure.

$\mathcal{R}$ is the Rayleigh number, $\mathcal{R} = \frac{\alpha g h^3}{\nu k}(T_0 - T_1)$

P_r = Prandtl number = ν/k .

I will write this more concisely as

$$G(\lambda, u) = 0$$

since the specific structure of the equations will not play a role in the arguments to follow. These equations are covariant with respect to the group of rigid motions

$$T_g \, G(\lambda, u) = G(\lambda, \, T_g u)$$

$$g = \{0, a\}$$

$$T_g u = T_g \begin{bmatrix} u_1(\underline{x}) \\ u_2(\underline{x}) \\ u_3(\underline{x}) \\ u_4(\underline{x}) \\ u_5(\underline{x}) \end{bmatrix} = \begin{bmatrix} O & \begin{array}{c} 0 \\ \end{array} \\ \hline \begin{array}{cc} 0 & \end{array} & \begin{array}{cc} 1 & \\ & 1 \end{array} \end{bmatrix} \begin{bmatrix} u_1(g^{-1}\underline{x}) \\ u_2(g^{-1}\underline{x}) \\ u_3(g^{-1}\underline{x}) \\ u_4(g^{-1}\underline{x}) \\ u_5(g^{-1}\underline{x}) \end{bmatrix}$$

The mathematical analysis of the onset of convection consists of an analysis of the bifurcation of nontrivial solutions of

$$G(\lambda, u) = 0$$

in the vicinity of the critical point $(\lambda_c, 0)$. The problem is reduced, via an alternative method, to a finite dimensional problem

$$F_i(\lambda, \, z_1, \ldots, z_n) = 0 \qquad i = 1, \ldots, n$$

where $n = \dim \ker G_u(\lambda_c, 0)$. These are called the bifurcation equations.

The direct determination of the functions F_i via the alternative method is at best an extremely complicated numerical procedure, but these complexities can be circumvented by group theoretic methods. This approach has the additional advantage that it is completely general - that is the structure of the bifurcation equations depends only on the covariance of the problem and not on the particular structure of the functional $G(\lambda, u)$.

The problem of pattern selection, then, is to determine which bifurcating cellular solutions are stable - rolls, squares, hexagons, etc. These cellular solutions are doubly periodic functions in the plane:

$$u(\underline{x} + \underline{\omega}) = u(\underline{x}) \qquad \text{for all} \quad \underline{\omega} \in \Lambda$$

where Λ is a lattice.

We consider 3 lattices in the plane

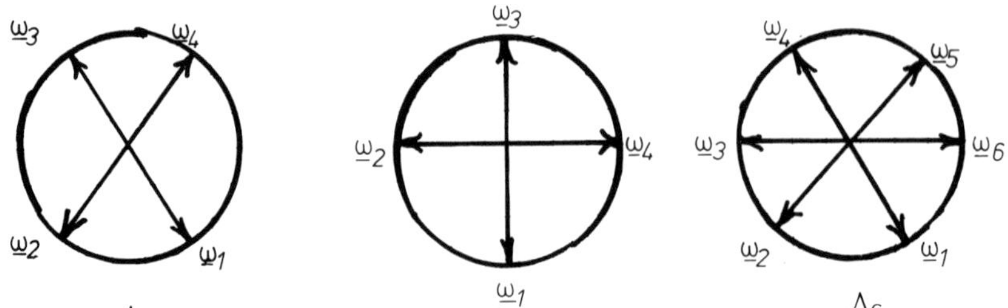

The bifurcation equations can be reduced to the following forms

Λ_2, Λ_4:

$$x_1(\lambda + a\, x_1^2 + b\, x_2^2) = 0$$
$$x_2(\lambda + a\, x_2^2 + b\, x_1^2) = 0$$

Λ_6:

$$x_1(\lambda + a\, x_1^2 + b(x_2^2 + x_3^2)) = 0$$
$$x_2(\lambda + a\, x_2^2 + b(x_3^2 + x_1^2)) = 0$$
$$x_3(\lambda + a\, x_3^2 + b(x_1^2 + x_2^2)) = 0$$

The parameters a, b depend on the lattice and on the external physical parameters of the problem (<u>viz.</u> Rayleigh and Prandtl numbers, coefficient of surface tension, etc.).

A straight-forward analysis of the bifurcation equations yields the following necessary conditions for stability

Λ_2, Λ_4

stable rolls	$b < a < 0$
stable squares/ rectangles	$a+b < 0$, $a-b < 0$

Λ_6

stable rolls	$b < a < 0$
stable hexagons	$a < b$, $a+2b < 0$

These necessary conditions are obtained by testing the solution against all disturbances in the same lattice class.

Can squares and hexagons be simultaneously stable? In order to compare simultaneously the stability of solutions in different lattices we must determine the dependence of a and b on the lattice.

<u>Theorem</u>: <u>Let the lattice</u> Λ <u>be generated by basic vectors</u> ω_1 <u>and</u> ω_2 <u>making an angle</u> θ <u>with one another,</u> $0 < \theta < \pi/2$. <u>There exists a function</u> $q(\theta)$ <u>such that</u>

1) $q(\theta) = A_0 + A_2 \cos 2\theta + A_4 \cos 4\theta + \dots$

2) $a = 3q(0)$, $b = 6q(\theta)$.

<u>The coefficients</u> $A_0, A_2, \dots$ <u>depend only on the basic physical parameters of the problem</u>.

It is at this stage that the internal structure of the equations G becomes important. The internal structure determines the weight of $q(\theta)$ - that is, the highest non-vanishing term. If G transforms as a scalar then $q(\theta) = \text{const}$.

<u>Example</u>

1) $\Delta^3 u - R\, \Delta_2 u = \alpha u^2 + \beta u^3$

$$\Delta_2 = \partial_x^2 + \partial_y^2$$
$$\Delta = \Delta_2 + \partial_z^2$$
$$q(\theta) = A_0$$

2) $\Delta^3 u - R\,\Delta_2 u = \alpha(\nabla u)^2$

$$q(\theta) = A\cos 2\theta + B$$

In the case of convection problems, BUSSE has considered an approximation to the full Boussinesq equation for which $q(\theta)$ has weight two - that is

$$q(\theta) = A\cos 2\theta + B\ .$$

When q has weight two the necessary conditions for stability can be translated into conditions on A and B. The results may be displayed in the following diagram

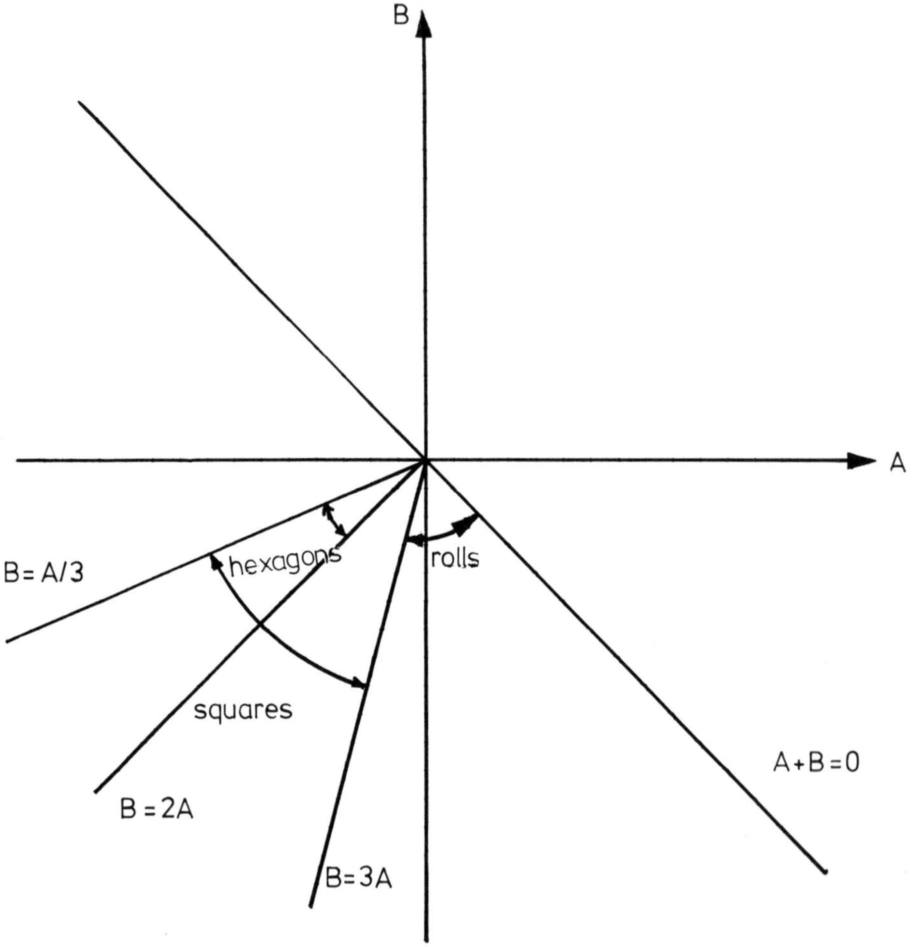

The diagram is perhaps a bit complicated so I will explain it. In the sector which lies between the lines $B = 3A$ and $A+B = 0$, rolls are the only possible doubly periodic solutions which can be stable. That is, if for a given set of external physical conditions the pair (A,B) lies in this sector, then rolls are the only solutions which can be stable when tested against all disturbances in the three lattice classes. In the sector $2A < B < 1/3A < 0$ hexagons may be stable while rolls are unstable. Squares, however, are also stable in this region, as are rectangles in the rectangular lattices with angle $\theta > \pi/3$. In general, θ-rectangles may be stable in the sector

$$A(1 - 2\cos 2\theta) < B < -(1 + 2\cos 2\theta/3)A \; .$$

Some conclusions may be drawn immediately about the infinite layer model. Firstly, rolls are the only cellular solutions which are uniquely selected on the basis of a formal (linearized) stability analysis relative to disturbances within the same lattice class. None of the other patterns are uniquely selected on this basis. Of course, the possibility is still open that one of these patterns is uniquely selected if one tests their stability against still larger classes of disturbances. This would entail a spectral analysis of a linear operator in the plane with doubly periodic coefficients - a two dimensional analogue of HILL's equation. The following is immediate.

Corollary 2.4 If the lattice function is constant then $A = 0$ in Theorem 2.3 and rolls are the only possible stable supercritical class of cellular solutions: hexagons, rectangles, and squares, though they exist, are unstable.

Corollary 2.4 applies in particular to the case of scalar equations (that is, when B_1 is just the field of complex numbers). Then rolls are the only stable class in the vicinity of the bifurcation point - all other patterns are unstable. These remarks apply, for example, to the bifurcation models for phase transitions investigated by RAVECHE and STUART [4]. They do not apply to convection problems since the Boussinesq equations do not transform as a scalar equation.

The above diagram refers to supercritical convection. Let me discuss the phenomenon of subcritical convection. The quadratic terms in the bifurcation equations always vanish in the square and rhombic case, but in the hexagonal case the quadratic term may be non-zero. Therefore, as every good bifurcation-theorist knows, there is the possibility of subcritical hexagonal solutions. It can be shown by group-theoretic arguments alone that this "transcritical" branch of hexagonal solutions is always unstable on both sides of the branch point. However, on the subcritical branch there is only one unstable mode, and this branch may "turn back" and regain stability. In this way we obtain the following diagram.

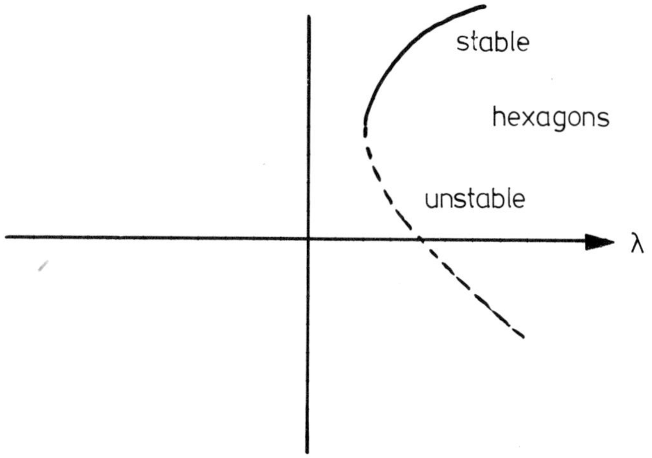

I emphasize that no general claim is being made that the subcritical hexagonal branch always regains stability as in the diagram.

Nevertheless, one possible mechanism for the selection of hexagonal solutions is that, in a given situation, they are the only stable subcritical solutions.

In the pure Bénard problem the convection is always supercritical; but transcritical convection can occur in problems of generalized convection: for example, when the material properties such as viscosity, specific heat, or thermal diffusivity vary with temperature; when the density temperature relation is nonlinear; or when surface tension drives the convection. (See the recent book by JOSEPH on the stability of fluid motions.)

So far I have discussed pattern selection mechanisms in an infinite plane layer. One may ask the same questions for other geometries. For example the onset of convection in a spherical mass is important in geophysical applications. Here, too, we can approach the problem from a general group theoretic view and our analysis therefore applies to any physical problem in which one is concerned with bifurcation in in the presence of $O(3)$.

We again denote the bifurcation equations by $F_i(\lambda, z_1, \ldots, z_n)$. Assume $\ker G_u(\lambda_c, 0) = \eta_0$ transforms irreducibly under the representation T_g of $O(3)$. Then $\dim \eta_0 = 2\ell + 1$, since the irreducible representations of $O(3)$ are of dimension $2\ell + 1$. We must construct functions F_i which transform according to an irreducible representation of $O(3)$. If we assume η_0 is irreducible then the linear part of the mapping F_i is a scalar multiple of the identity. To get the other terms we proceed as follows. We begin with the infinitesimal generators of the rotation group J_1, J_2, J_3 . These satisfy the commutation relations

$$[J_i, J_j] = \epsilon_{ijk} J_k$$

where ϵ_{ijk} is the completely antisymmetric tensor. Now put

$$J^{\pm} = \pm J_2 + iJ_1 \ , \quad J^3 = -iJ_3$$

These operators satisfy the commutation relations

$$[J^+, J^-] = 2J^3 \qquad [J^3, J^{\pm}] = \pm J^{\pm} \ .$$

The $J^{\pm}$ are the familiar ladder operators of quantum mechanics. The spare η is real, but the complexified space $\eta + i\eta$ has a basis f_m such that

$$J_3 f_m = m f_m \qquad J_{\pm} f_m = \beta_{\pm m} f_{m \pm 1}$$

$$\beta_m^{\ell} = \sqrt{(\ell - m)(\ell + m + 1)} \ .$$

We now represent η as the vector space of linear polynomials in the variables $z_{-\ell}, \ldots, z_{\ell}$. Thus $J_3 z_m = m z_m$, $J_{\pm} z_m = \beta_{\pm m} z_{m \pm 1}$. Denote by $K[z_{-\ell}, \ldots, z_{\ell}]$ the ring of polynomials in $z_{-\ell}, \ldots, z_{\ell}$. K is isomorphic to the algebra of symmetric tensors over η . We extend J_3 and $J_{\pm}$ to be derivations over K :

$$J(\alpha f + \beta g) = \alpha J f + \beta J g$$

$$J(fg) = f J g + (J f) g \ .$$

It is natural to extend the infinitesimal generators J as derivations on the ring, since they are Lie derivatives.

Now here is how we construct the polynomials F_m . We require the polynomials F_m to transform as the z_m , viz.

$$J_3 F_m = m F_m \qquad J_{\pm} F_m = \beta_{\pm m} F_{m \pm 1} \ .$$

Consider first quadratic terms:

$$\begin{aligned} J_3 z_j z_k &= (J_3 z_j) z_k + z_j (J_3 z_k) \\ &= (j + k) z_j z_k \ . \end{aligned}$$

If F_m is to be a sum of quadratic terms and is to satisfy $J_3F_m = mF_m$ we must have

$$F_m(z_{-\ell},\dots,z_\ell) \quad \sum_{i+j=m} a_{ijm} z_i z_j \quad .$$

Furthermore we want $J_+F_\ell = \beta_\ell F_{\ell+1} = 0$. The two equations $J_3F_\ell = \ell F_\ell$ and $J_+F_\ell = 0$ have a unique solution if ℓ is even. For odd ℓ there are no quadratic terms and one must go to cubic terms. For even ℓ we get a linear system of equations for the coefficients of F_ℓ , and F_ℓ can be determined up to a scalar multiple. Once F_ℓ is obtained we obtain the F_m for $m < \ell$ by successive applications of the ladder operator J_- .

For $\ell = 1$ there is only one cubic covariant mapping F , but for $\ell = 3$ there are two independent mappings. This fact presents the possibility for a mechanism for pattern selection, as in the planar case. I hope to discuss the details in a future paper.

References

1. Busse, F. "The Stability of finite amplitude cellular convection and its relation to an extremum principle", Jour. Fluid Mech. 30 (1967), 625-650.

2. Haken, H. "Cooperative phenomena in systems far from thermal equilibrium and in nonphysical systems", Reviews of Modern Physics, 47 (1975), 67-121.

3. Joseph, D. Stability of Fluid Motions I,II, Springer-Verlag, Berlin, 1976.

4. Raveché, H.J. and Stuart, C.A. "Towards a molecular theory of freezing", Jour. Chem. Phys. 63 (1975).

5. Sattinger, D.H. "Group representation theory, bifurcation theory, and pattern formation", Jour. Funct. Anal.

6. Sattinger, D.H. "Selection rules for pattern formation", Arch. Rat. Mech. Anal.

7. Sattinger, D.H. "Group representation theory and branch points of nonlinear functional equations", SIAM Jour. Math. Anal.

Instabilities in Fluid Dynamics

Transition to Turbulence in a Fluid Flow

H. L. Swinney, P. R. Fenstermacher and J. P. Gollup

With 9 Figures

1. Introduction

1.1 The Phenomena

We have used the technique of laser Doppler velocimetry to study the flow of a fluid contained between concentric cylinders with the inner cylinder rotating. It is convenient to describe the flow as a function of the (dimensionless) Reynolds number R, which can be defined as $R = \omega_{cyl} r_i (r_o - r_i)/\nu$, where ω_{cyl} is the angular frequency of the inner cylinder, r_i and r_o are the radii of the inner and outer cylinders respectively, and ν is the kinematic viscosity. For small R the flow is purely azimuthal, but as TAYLOR [1] showed in 1923, above a critical Reynolds number R_c the simple azimuthal flow is no longer stable and there is a transition to a flow with a horizontal vortex pattern superimposed on the azimuthal flow. A photograph of the flow with the Taylor vortices is shown in Fig.1(a); the flow pattern is rendered visible by a suspension of small flat flakes [2] which align with the flow. The Taylor vortex flow has been extensively studied in recent years by DONNELLY [3], KOSCHMIEDER [4], SNYDER [5], and others. At a higher, well defined R there is a transition from the time-independent Taylor vortex flow to a time-dependent flow with transverse waves superimposed on the horizontal vortices, as shown in Fig.1(b) and (c). Although the wavy flow has been observed in many experiments, the most thorough study has been the photographic investigations of COLES [6], who found that as R was increased beyond the onset of the wavy regime, the flow became noisier and noisier and ultimately turbulent (see Fig.1(d)).

The transitions to the Taylor vortex and wavy regimes are fairly well understood, both experimentally and theoretically. The present study is concerned with the transition from the wavy regime to turbulent flow, which is not well understood, either experimentally or theoretically.

1.2 Present Study

We have measured the time dependence of the radial component of the local fluid velocity using laser Doppler velocimetry [7]. The velocity is recorded digitally and then Fourier transformed to obtain velocity power spectra. The velocity spectra at dif-

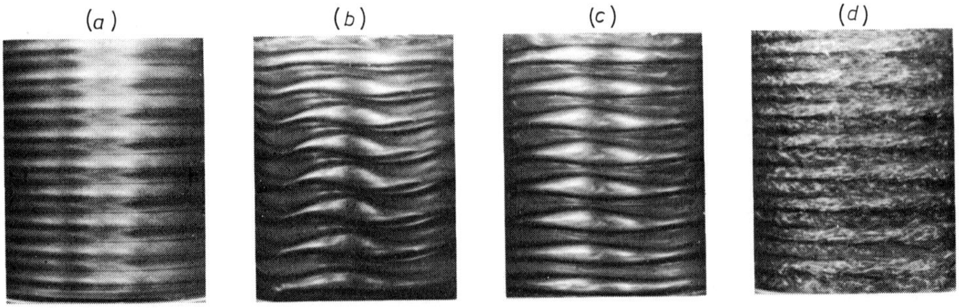

Fig.1 Photographs of flows in a fluid contained between concentric cylinders with the inner cylinder rotating [2]. The cell specifications are in Table 1. (a) Taylor vortex flow, $R/R_c = 1.1$. (b) Wavy flow, $R/R_c = 5.5$. (c) Wavy flow, $R/R_c = 14.5$. (d) Aperiodic flow, $R/R_c = 43$

ferent Reynolds numbers reveal the existence of several dynamical regimes which are not apparent in photographic studies or torque measurements and hence were not observed in previous studies. A periodic dynamical regime is observed with spectra characterized by one or more sharp frequency components at commensurate frequencies, while another regime, termed quasiperiodic, is characterized by spectral components at incommensurate frequencies. At a higher, well defined Reynolds number there is a transition to an aperiodic or chaotic regime where the spectrum contains no sharp frequency components and the corresponding velocity autocorrelation function decays. The observed chaotic regime is of course not fully developed turbulence, but we take its aperiodicity to be our operational definition of turbulence.

Before describing the experiment we review briefly models that have been proposed to describe the transition to turbulence.

2. Models for the Transition to Turbulence

In this paper we consider the transition to turbulence in systems which pass through a regime characterized by sharp temporal and spatial frequencies rather than a continuous spectrum; those shear dominated transitions which occur catastrophically with no intermediate periodic regime will not be discussed. Examples of the class of systems of interest here are, in addition to a fluid between rotating cylinders, Rayleigh-Bénard convection and its many variants [8] and a symmetrically heated rotating fluid [9].

In 1944 LANDAU conjectured that the transition to turbulence may occur as an infinite sequence of discrete transitions to new periodic states, with each transition adding a new frequency to the motion[10]. The Landau conjecture is not supported by experiment since turbulent flows are not characterized by discrete spectra.

A contrasting picture of the onset of turbulence was first suggested 14 years ago by LORENZ [11], whose highly truncated model of a symmetrically heated rotating fluid yielded a few transitions to different periodic regimes followed by a transition to aperiodic behavior. MCLAUGHLIN and MARTIN [12] found a similar sequence of events in their numerical modeling of convection, which was designed to explain the sharp onset of noise in the sensitive heat flux measurements of AHLERS [13] on liquid helium.

RUELLE and TAKENS [14] have argued that a sharp transition to aperiodic flow after a few periodic regimes is a consequence of the most general aspects of the nonlinear hydrodynamic equations, and other systems of highly nonlinear equations should show similar behavior. Ruelle and Takens predicted, using abstract topological reasoning, that after a fluid passes through at most three or four transitions to periodic regimes, then there will be a transition to a qualitatively different type of regime, where the behavior is truly random and the velocity autocorrelation function decays rather than oscillating.

These general hypotheses regarding the transition to turbulence by a sequence of instabilities do not supplant detailed analyses of particular systems. For a fluid between concentric rotating cylinders DAVEY, DIPRIMA, and STUART [15] and EAGLES [16] have calculated the properties of the flow in the neighborhood of the onset of the wavy regime, but the analysis has not yet extended beyond the onset of the wavy regime.

3. Experimental Procedure

3.1 Fluid Flow Cell

The cell specifications are given in Table 1. The inner cylinder is made of stainless steel and the outer cylinder is glass. The height of the inner cylinder is 30 cm, although in the present work the cell was filled with fluid only to a height of 6.25 cm. The cell is mounted on an XYZ stage so it can be moved with respect to the laser beam probe. The cell is contained inside a box which has a proportional temperature controller that maintains the temperature at 27.50 ± 0.05° C.

The inner cylinder is powered through a belt drive by a synchronous motor which is driven by a stable oscillator. The rotation frequency of the inner cylinder (which

Table 1 Concentric cylinder fluid flow cell

Inner cylinder radius, r_i	2.224 cm
Outer cylinder radius, r_0	2.536 cm
Ratio of radii, r_i/r_0	0.877
Fluid height, h	6.25 cm
Height/gap, $h/(r_0 - r_i)$	20.0
Fluid kinematic viscosity (water at 27.5° C)	8.45×10^{-3} cm^2 s^{-1}
Number of axial vortices	17
Number of waves around annulus in wavy state	4
Diameter of spherical LDV seed particles (approx. 10^{-4} wt.%)	0.48 μm
Sample volume for velocity measurement	~(0.013 cm)^3
Critical Reynolds No., R_c (calculated from Table 3A in [3])	119

ranged from 0.2 to 10 Hz in these experiments) is measured directly with an electronic timer and has an r.m.s. fluctuation of less than 0.3%.

The fluid is water, seeded with 0.48 μm diameter polystyrene spheres for the laser Doppler velocimetry studies and with Kalliroscope suspension [2] for the photographic studies.

3.2 Laser Doppler Velocimetry (LDV) Optical System

The optical system (which in LDV terminology is called a reference beam backscatter system) is shown schematically in Fig.2(a). The collimated incident laser beam (power= 0.1W; wavelength = 0.488 μm) is focused into the fluid with a 5.5 cm focal length lens. The light backscattered at a 160° angle is collected by the same lens, passes through a pinhole aperture and beam splitter, and is focused onto a pinhole in front of the photomultiplier detector. The linear dimension of the scattering volume in the radial direction is less than one-tenth of the gap between the cylinders. The scattering volume is located midway between the inner and outer cylinders for most of the measurements. The incident wavevector $\vec{k}_0$ and the scattered wavevector $\vec{k}_s$, both in a vertical plane which passes through the cylinder axis, are at equal angles with respect to the horizontal plane, so the momentum transfer-vector $\vec{q} = \vec{k}_0 - \vec{k}_s$ is in the horizontal plane; hence the scattered light is Doppler shifted by an amount

$$\omega_D = \vec{V}\cdot\vec{q} = V_r q = [(4\pi n/\lambda_0)\sin\tfrac{1}{2}\theta]V_r \qquad (1)$$

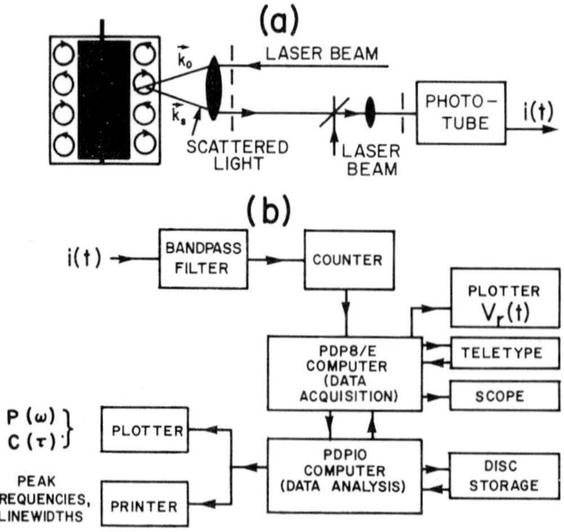

Fig.2 Schematic of laser Doppler velocimetry system. (a) Optical System. (b) Digital data acquisition and analysis electronics

where n is the refractive index (1.333), λ_0 is the laser wavelength, θ is the scattering angle (160°) and $V_r(r,t)$ is the radial component of the fluid velocity. The Doppler shifted scattered optical field is mixed with unshifted laser light at the photocathode, and the photocurrent i(t), being proportional to the square of the incident optical field, oscillates at the Doppler shift frequency. Thus measurements of this frequency in successive time intervals yield the time dependence of the radial component of the velocity of the fluid at a well defined point in the cell.

3.3 Data Acquisition and Analysis Electronics

The data acquisition system is diagrammed in Fig.2(b). The photocurrent is bandpass-filtered and the oscillations (typically 100KHz in the present experiment) are counted for a time Δt which is long compared to a period of oscillation, but short compared to the time scale of changes in the velocity. At the end of each time interval Δt the number of oscillations for that interval is deposited in the memory of a minicomputer (DEC PDP8/E). After the accumulation of 8192 points at successive time intervals, the resultant Doppler shift record, directly proportional to $V_r(t)$, is plotted and then transferred to a larger computer (DEC PDP10). Then the velocity power spectrum and the velocity autocorrelation function are computed and plotted, and another program calculates the positions and linewidths of the spectral lines. The immediate feedback of the spectral information serves as a valuable guide in the course of an experiment.

3.4 Spatial State

COLES [6] found that the number of Taylor vortices and tangential waves is not a unique function of the Reynolds number. The particular spatial state reached at a given R in general depends on the Reynolds number history of the system. We have restricted our study primarily to a single spatial state, one with 17 axial vortices and, in the region where the waves exist, 4 azimuthal waves. Although several different states other than the 17 vortex/4 wave state were found to be accessible, depending on the Reynolds number history of the system, the 17/4 state, once achieved, was stable indefinitely while the Reynolds number was varied throughout the range studied. Switching between states was avoided by using a short fluid height and accelerating slowly when changing Reynolds number. Thus it was possible to study the Reynolds number dependence of the dynamics of a single spatial state, without being concerned with the state switching problem. Although the present measurements were made primarily on the 17 vortex state, a few measurements were also made on the 15 vortex/4 wave state, and the same dynamical regimes were observed for these two spatial states.

4. Experimental Results

4.1 Time-Independent Flow

In the laminar state below R_c, V_r is zero, while in the Taylor vortex state (above R_c) V_r is periodic in the axial coordinate z but still time-independent. GOLLUB and FREILICH [17] have studied the Reynolds number dependence of the amplitude of V_r in the Taylor vortex regime and have found that near R_c the $(R-R_c)^{\frac{1}{2}}$ dependence predicted by Landau is confirmed and the observed next order term in the perturbation expansion is consistent with a $(R-R_c)^{3/2}$ dependence as predicted by DAVEY [18].

4.2 Periodic Flow

The second instability, at $R/R_c=1.2$, marks the onset of time dependence in V_r, which in the wavy regime oscillates as the azimuthal waves pass the fixed point probed by the laser beam. Figure 3 shows a portion of a record of $V_r(t)$ and the corresponding power spectrum (on a logarithmic scale) in the wavy regime at $R/R_c=5.7$. [All power spectra are normalized so that $\int_0^{\omega_{max}} P(\omega)d\omega=\langle(\Delta V_r)^2\rangle$, and the frequencies ω are expressed in units of the inner cylinder frequency; thus $\omega\equiv\omega(\text{rad/s})/\omega_{cyl}(\text{rad/s})$.] The power spectrum contains a single fundamental which we call ω_1; this is the frequency of the azimuthal waves passing the point of observation. Although many harmonics are visible in the spectra obtained in the wavy regimes, the higher harmonics are several orders of magnitude weaker in intensity than the fundamental and are visible only because of the high signal to noise ratio of the spectra. The relative amplitude of the peaks (particularly

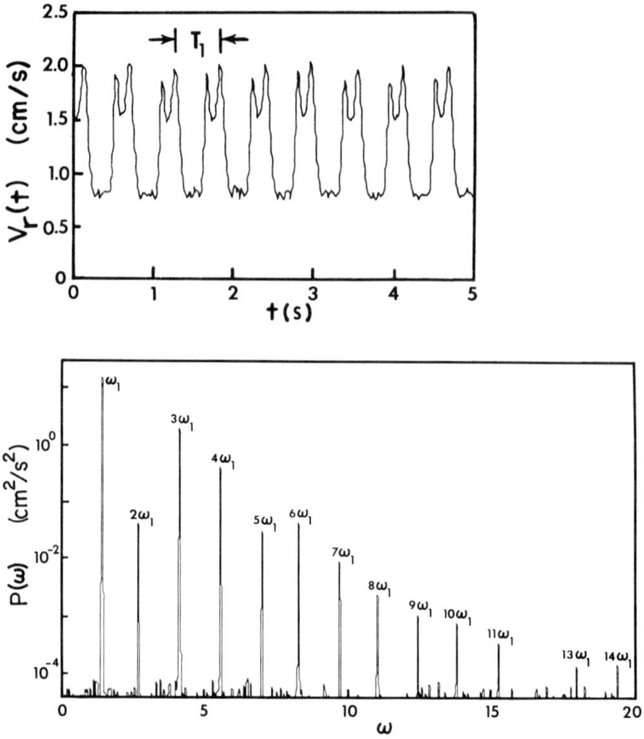

Fig.3 The radial component of the velocity, $V_r(t)$, and the corresponding power spectrum, $P(\omega)$, at $R/R_c=5.7$. The velocity plot shows only part of the 151s long record that was Fourier-transformed to obtain the spectrum. T_1 is the period of the azimuthal waves that pass the point of observation and ω_1 is the corresponding frequency (relative to ω_{cyl}; hence ω is dimensionless)

ω_1 and $2\omega_1$) is a strong function of z, as one would expect from the structure of vortex flow.

4.3 Quasiperiodic Flow

As the Reynolds number was increased a low frequency component, ω_2, appeared in some spectra but was absent in other spectra obtained at the same Reynolds number. This component is the only spectral feature that was not reproducible from run to run, and its physical significance is not understood. It may be a vacillation preceding a change in spatial state such as that observed by COLES [6].

At a higher, well defined Reynolds number, $R/R_c=10.0\pm0.2$, a previously undetected spectral component appears at a frequency $\omega_3\simeq(2/3)\omega_1$, as illustrated in Fig.4 for $R/R_c=13.3$. The component at ω_3 reproducibly appears when R is increased through $R/R_c=10.0$, and disappears when R is decreased through $R/R_c=10.0$. Its amplitude decreases continuously to zero as $R/R_c=10.0$ is approached from above.

At larger Reynolds numbers the amplitude of ω_3 begins to decrease, and at $R/R_c=19.8\pm0.1$ ω_3 disappears from the spectrum, as shown in Fig.5. In the regime with ω_3 present the ratio ω_3/ω_1 increases monotonically as shown in Fig.6; hence ω_3 and ω_1 appear to be incommensurate. Thus, the regime with ω_3 present is an example of quasiperiodic flow.

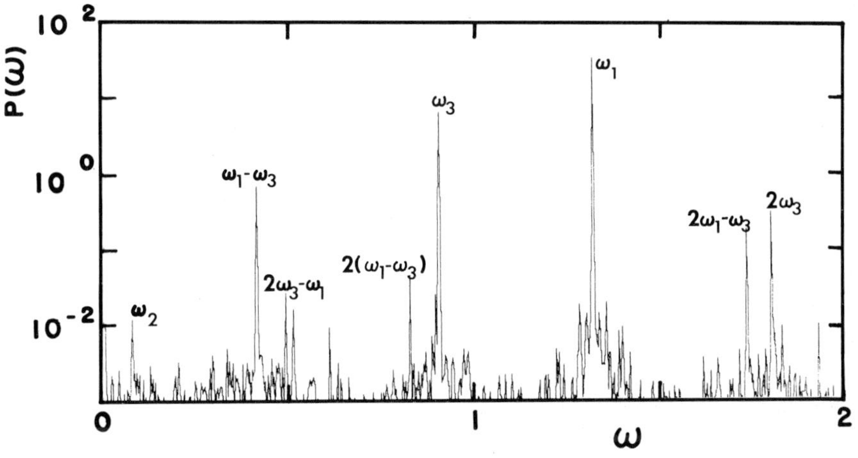

Fig.4 The power spectrum of the radial component of the velocity at $R/R_c=13.3$

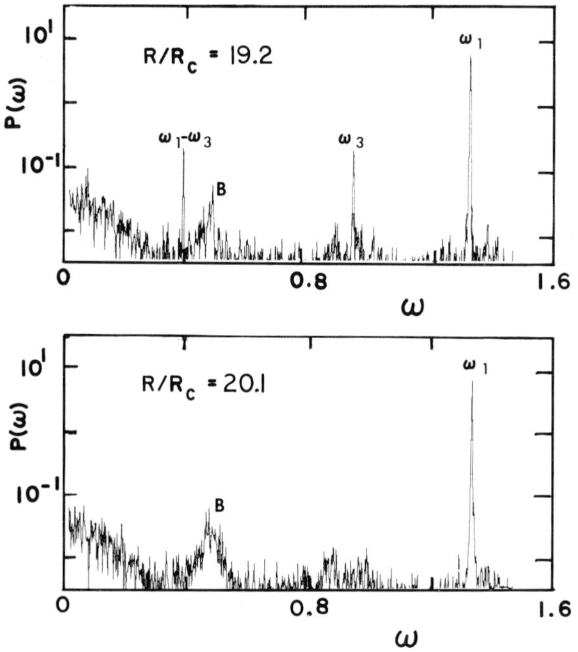

Fig.5 These spectra illustrate the disappearance of ω_3 at $R/R_c=19.8\pm0.1$. Note the broad component B at $\omega\simeq0.45$ in both spectra

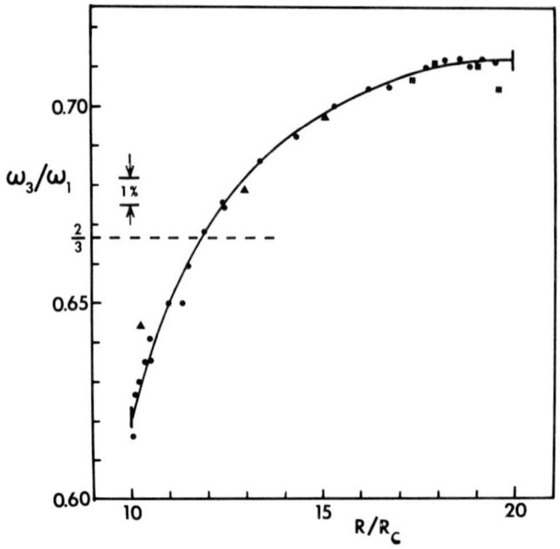

Fig.6 The dependence of ω_3/ω_1 on Reynolds number. The different symbols indicate different sets of measurements, obtained over a period of more than one year

4.4 Chaotic Flow

As the Reynolds number is increased through R/R_c=22.4±0.2 a dramatic change in the spectrum occurs, as Fig.7 illustrates: the amplitude of the sharp frequency component ω_1 decreases to zero, leaving only broad components located approximately at multiples of $\omega_1/3$. Thus R/R_c=22.4 marks a transition from a nearly periodic flow to a qualitatively different type of behavior, one characterized by a continuous spectrum or, equivalently, a decaying velocity autocorrelation function. The transition from a nearly periodic to chaotic flow is quite marked in the autocorrelation function as shown in Fig.8.

Spectra have been recorded up to R/R_c=45 and no further transitions have been observed.

5. Summary and Conclusions

The dependence of the observed fundamental frequency components on the Reynolds number is shown in Fig.9. The velocity was measured at several different heights and radii within the fluid, and the observed frequencies and transition Reynolds numbers were found to be independent of the laser probe position. The measured widths of ω_1 and ω_3 and their harmonics were instrumentally limited even in the highest resolution spectra, where $\Delta\omega$=0.0007 (half-width at half-maximum). Our values of ω_1 agree well with those obtained previously by COLES [6], while the components ω_3 and B have not been previously observed. Another component, ω_2, was present in a few spectra and is perhaps a slowly decaying transient.

Photographs of the flow in the vicinity of R/R_c=10.0, where ω_3 appears, and R/R_c= 19.8, where ω_3 disappears, show no radical change in the flow pattern, while the disappearance of ω_1 at R/R_c=22.4 can be seen as the loss of azimuthal waves. The transitions marked by the appearance and disappearance of ω_3 and the disappearance of ω_1 exhibit no hysteresis within the experimental resolution in R. Since B is broad and its amplitude increases slowly with increasing R, it is difficult to determine precisely the R at which this component appears.

In conclusion, we have observed three distinct dynamical regimes (periodic, quasiperiodic, and aperiodic) which are characteristic of nonlinear mathematical models with only a few degrees of freedom. The observed behavior is qualitatively consistent

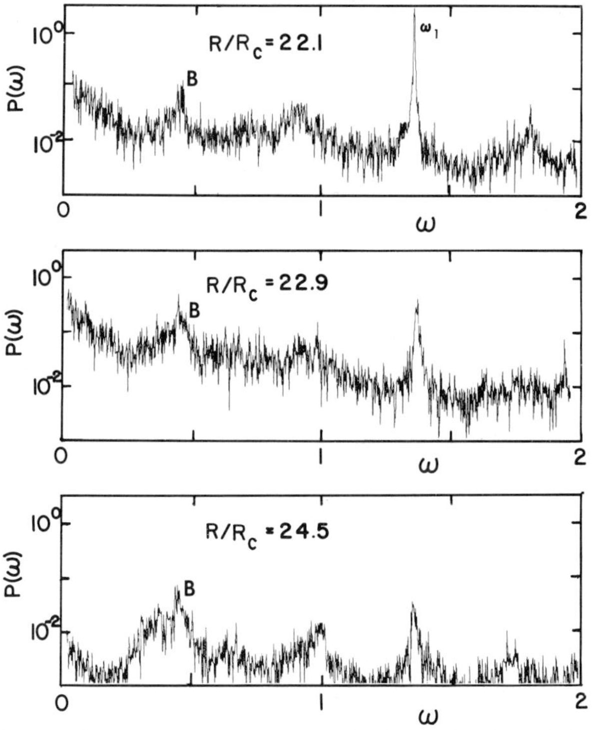

Fig.7 These spectra at successively higher Reynolds numbers illustrate the disappearance of the sharp frequency component at ω_1, which occurs at $R/R_c=22.4\pm0.2$. For $R/R_c<22.4$, two components are centered at ω_1, an intense narrow component and a broad weak component. As R/R_c is increased to 22.4, the amplitude of the narrow spectral line rapidly goes to zero, and for $R/R_c>22.4$ only the broad component remains

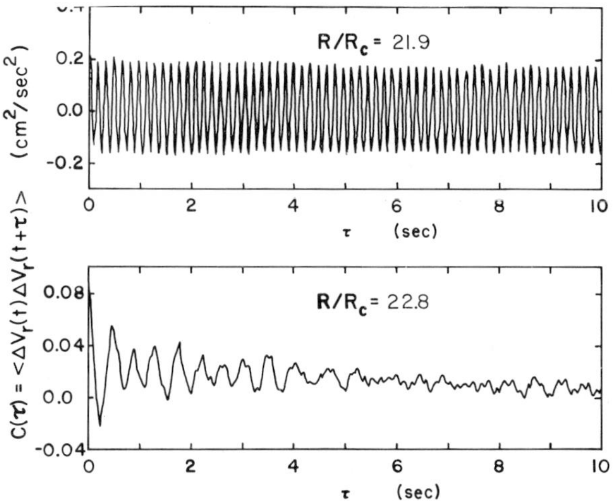

Fig.8 The velocity autocorrelation function before and after the disappearance (at $R/R_c=22.4$) of the sharp frequency component at ω_1. Note that the ordinate scales are different for the two graphs

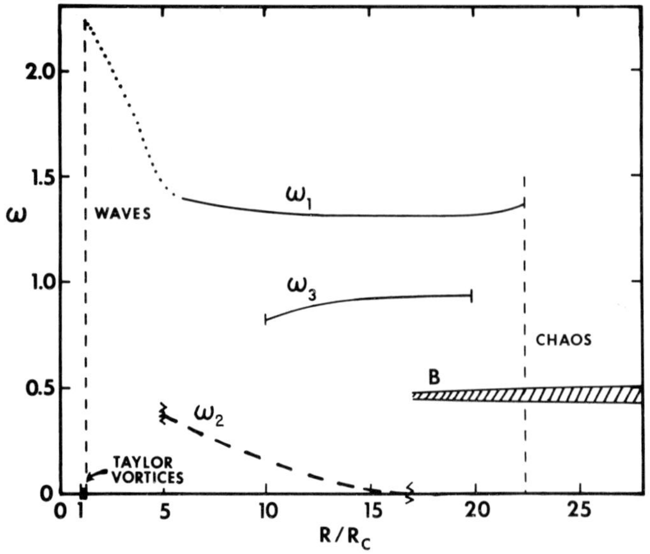

Fig.9 The fundamental frequency components observed in the power spectra of the radial component of the velocity. The time-independent Taylor vortex regime extends over a relatively small range in Reynolds number, from $R/R_c=1$ to 1.2, a region we have not studied in the present experiment. We have also not studied systematically the region $R/R_c<5.6$ indicated by the dotted portion of the ω_1 curve. The values of ω_1 and ω_3 at any given Reynolds number reproduced from run to run with an accuracy of 1% or better (see the data for ω_3/ω_1 in Fig.6), but, as discussed in the text, ω_2 was absent in most runs and when present its value was not reproducible. No sharp frequency components were observed for $R/R_c>22.4$

with the ideas of Lorenz and Ruelle and Takens, who suggested that a few nonlinearly coupled modes are sufficient to produce aperiodic motion. However, there is a small amount of broadband noise (component B) in the periodic regime that precedes the aperiodic regime. Whether this noise is important in the dynamics is not yet clear.

These experiments are continuing and a more detailed report will be presented elsewhere. Also, a similar laser Doppler velocimetry study is underway on the transition to turbulent convection in a fluid between parallel plates heated from below (Bénard system), and preliminary results indicate that this system also exhibits distinct dynamical regimes [19].

Acknowledgments

It is a pleasure to acknowledge helpful discussions with R. C. DiPrima, R. J. Donnelly, D. Joseph, E. L. Koschmeider, J. B. McLaughlin, and D. Ruelle.

This research was supported by the National Science Foundation.

References

1. G.I. Taylor, Phil. Trans. Roy. Soc. (London) A223, 289 (1923)
2. For flow visualization we used Kalliroscope AQ1000 rheoscopic liquid, Kalliroscope Corp., Cambridge, Mass. USA
3. R.J. Donnelly, K.W. Schwarz, and P.H. Roberts, Proc. Roy. Soc. (London) A283, 531 (1965)
4. E.L. Koschmieder, Adv. Chem. Phys. 32, 109 (1975)
5. H.A. Snyder, Int.J. Non-Linear Mech. 5, 659 (1970)

6. D. Coles, J. Fluid Mech. 21, 385 (1965)
7. Preliminary results were reported by J.P. Gollub and H.L. Swinney in Phys. Rev. Lett. 35, 927 (1975)
8. See, for example, the review by R. H. Rogers in Rept. Prog. Phys. 39, 1 (1976)
9. R. Hide and P.J. Mason, Adv. in Phys. 24, 1 (1975)
10. L. Landau, C. R. Acad. Sci. (USSR) 44, 311 (1974). See also L.D. Landau and E. Lifshitz, Fluid Mechanics (Addison-Wesley, Reading, Mass. 1959) p.103.
11. E.N. Lorenz, J. Atm. Sci. 20, 448 (1963)
12. J.B. McLaughlin and P.C. Martin, Phys. Rev. A12, 186 (1975)
13. G. Ahlers in Fluctuations, Instabilities and Phase Transitions ed. by T. Riste (Plenum, New York 1975) p. 181
14. D. Ruelle and F. Takens, Comm. Math. Phys. 20, 167 (1971)
15. A. Davey, R. C. DiPrima, and J. T. Stuart, J. Fluid Mech. 31, 17 (1968)
16. P.M. Eagles, J. Fluid Mech. 49, 529 (1971)
17. J.P. Gollub and M. Freilich, Phys. Fluids 19, 618 (1975)
18. A. Davey, J. Fluid Mech 14, 336 (1962)
19. J.P. Gollub, S.L. Hulbert, G.M. Dolny, and H.L. Swinney, in Photon Correlation Spectroscopy and Velocimetry, ed. by E.R. Pike and H.Z. Cummins (Plenum, New York 1977)

Instabilities in Fluid Dynamics

E. L. Koschmieder

With 7 Figures

1. Introduction

The investigation of instabilities in fluid dynamics is concerned with the following question: What happens to infinitesimal disturbances of a specific fluid flow, do the disturbances either grow or do they decay? In the case that the disturbances grow we call the fluid unstable. If the disturbances decay we call the fluid stable. For the occurrence of instability it is not necessary that all possible disturbances grow, just one growing disturbance means instability. Neither is it necessary that all disturbances are infinitesimal. However, the study of infinitesimal disturbances permits the use of the linear approximation of the Navier-Stokes equation and then leads in many cases to unambiguous analytical predictions about the conditions under which a fluid flow is unstable for a certain kind of disturbance. There are numerous types of instabilities in fluid dynamics, ranging from the very familiar instability of a water jet to the famous and cumbersome problem of turbulence. We will restrict the discussion here to two classical instabilities, Bénard convection and Taylor vortex flow. These two instabilities relate most closely to Synergetics.

2. Bénard Convection

Probably the best known feature of Bénard convection are the famous hexagonal cells which BÉNARD [1] discovered in 1900 when he heated a shallow fluid layer uniformly from below. A photograph of such cells is shown in Fig. 1. Until a publication of BLOCK [2] in 1956 it was believed that the hexagonal cells were due to the unstable stratification of the fluid, caused by the heating from below. This means that cold heavy fluid is on top of warm light fluid. The stability of such an arrangement was first studied in a fundamental paper by RAYLEIGH [3] in 1916. Rayleigh established that the onset of convection does not occur with any arbitrarily small negative vertical temperature gradient, but only if a critical value of a characteristic dimensionless parameter, now called the Rayleigh number

$$R = \frac{\alpha g \Delta T d^3}{\nu \kappa} \tag{1}$$

is exceeded.

The other basic result of Rayleigh's theory concerns the pattern of the convective motion. At the critical Rayleigh number the problem is degenerated, any pattern that covers an infinite plane regularly is a possible solution of the equations. Such patterns are straight parallel rolls, triangular cells, square cells, rectangular cells of all side ratios, hexagonal cells, circular concentric rolls, etc.

Rayleigh's theory has been refined in numerous subsequent studies which cannot all be discussed here. However, two fundamental questions still remain in limbo.

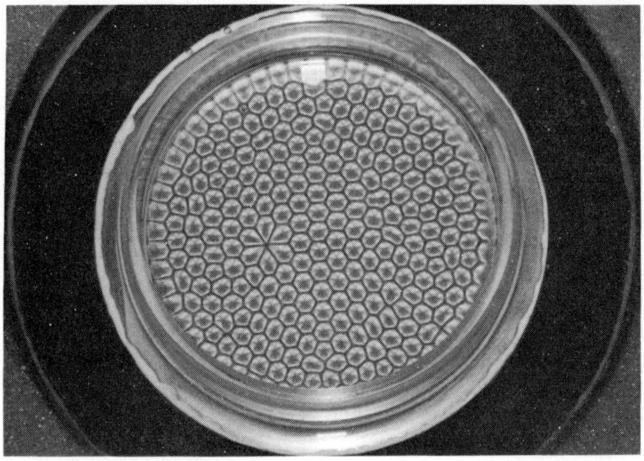

Fig. 1 Hexagonal convection cells on a uniformly heated copper plate and under an air surface. Fluid silicone oil. Visualization with aluminum powder. After [11]

Namely first, which of the infinite number of theoretically possible patterns will actually be established at the critical Rayleigh number; and second, what is the solution of the equations when the Rayleigh number exceeds R_c, when consequently the nonlinear terms can no longer be neglected.

First a discussion of the pattern selection. As mentioned above it was believed for a long time that the hexagonal patterns occured as the result of a natural selection mechanism, simply because the hexagonal pattern was the one which was observed in the experiments. However BLOCK [2] showed, that the hexagons observed by Bénard were most likely caused by surface tension effects, more precisely by the variation of surface tension with temperature. Subsequent theoretical investigations by PEARSON [4] and NIELD [5] and others have made it a certainty that the hexagons observed under a socalled free surface are caused by surface tension effects. A modern experimental investigation of the formation of such hexagons has been made by KOSCHMIEDER [6]. So if the hexagons are not the consequence of heating from below, which pattern is then selected under conditions which correspond to the assumptions made in the linear theory of convection caused by heating from below? For an answer to this question one has looked at the nonlinear terms of the Navier-Stokes equation. Rolls seem to be a likely choice as was shown by SCHLÜTER, LORTZ and BUSSE [7], but hexagons and square cells cannot be excluded definitely. Sophisticated mathematical investigations of this problem, e.g. SATTINGER [8] and KIRCHGÄSSNER [9], apparently still retain the option between different patterns. This might be due to the fact that the question of the pattern selection is ambiguous per se if it is studied in context with a fluid layer of infinite horizontal extent. An experimental verification can, of course, only be made in a finite container with lateral boundaries.

The onset of convection in a bounded circular container is shown in Fig. 2a, and Fig. 2b, following KOSCHMIEDER [10], see also the review article KOSCHMIEDER [11]. The experimental setup used for the experiment shown in Fig. 2 corresponds exactly to the assumptions made in theory, the fluid is heated uniformly from below and cooled uniformly from above. Surface tension effects are eliminated by a rigid transparent lid which is in contact with the fluid. The Boussinesq approximation is satisfied. One convective roll can be seen along the wall already at $R/R_c = 0.5$, two

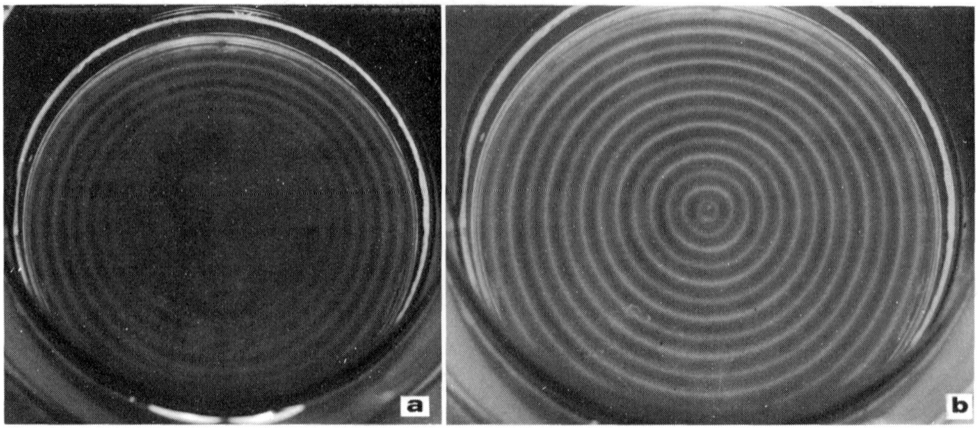

Fig. 2a Onset of convection in a circular container at $R = 0.95\ R_c$. Steady state. Fluid under a glass lid. After [11]

Fig. 2b Onset of convection in a circular container. Just critical

adjacent rolls at $R/R_c = 0.8$, six rolls (Fig. 2a) at $R/R_c = 0.95$, ten rolls at $R/R_c = 0.98$ and the entire layer is covered with circular concentric rolls at $R = R_c$ (Fig. 2b). Heating proceeds very slowly, each of the (subcritical) rings coming from the walls can be maintained indefinitely, if the applied vertical temperature is held constant. Corresponding observations in a rectangular container have been made by BERGÉ [12] and OERTEL [13].

It thus appears that the form of the convective motion in a bounded container is determined by the shape of the lateral wall. Theoretical investigations of convection in circular containers by CHARLSON and SANI [14] and JONES, MOORE and WEISS [15], and in rectangular containers DAVIS [16] support this statement. However this conclusion is heatedly debated by others, who interpret the frequently occuring irregular patterns in bounded containers as proof of the fact that the boundaries do not determine the form of the convective motion. It must be noted though that in any apparatus the first thing that one observes are irregular patterns. Symmetric pattern appear only when everything feasible has been done to make top and bottom temperature of the fluid as uniform as possible. Concerning the irregular patterns it must be noted also that these suffer from the elementary defect that they are not reproducible.

In order to settle the pattern selection problem once and for all a definite theoretical analysis of the steady pattern of convection in a shallow fluid layer in a bounded container is needed, with the fluid satisfying the Boussinesq approximation. After this question has been settled more intricate problems, such as non-Boussinesq effects, can be of great interest.

Let us now turn to the second question that was raised above, namely what is the solution of the Bénard convection problem under supercritical conditions ($R > R_c$)? The predictions of linear theory are best expressed by Fig. 3. According to linear theory there is for any $R > R_c$ a band of possible unstable wavenumbers a. In other words the solution is not unique. Which of the possible solutions is then actually utilized by the fluid when the temperature, i.e. the Rayleigh number, is increased quasi-steadily above R_c? The results of such an experiment are shown in Fig. 4a-c. As can be seen, the number of convective rings decreases continuously, there are only 12 rings on Fig. 4a, while there were 13 rings at R_c (Fig. 2b). The twelfth ring has shrunk in Fig. 4b, it can be maintained in that form indefinitely if ΔT is maintained. Increasing ΔT further makes the twelfth ring to disappear likewise, see Fig 4.c. What we observe, in the quasi-steady case, is an increase of the socalled wavelength, which is the ratio of the width of two rolls divided by the fluid depth. Wavelength λ and wavenumber a are related through the formula $a = 2\pi/\lambda$. For a couple of other experimental results see [11]. There is a total of 13 (often only qualitative) observations that the wavelength increases with R, for detail see [11]. There is, to my knowledge, not a single observation of a steady stable wavelength with $\lambda < \lambda_c$ at $R > R_c$. The results of experiments reported by BUSSE and WHITEHEAD [17] concerning wavelengths shorter than λ_c are misleading, since the fluid in these experiments was far from equilibrium.

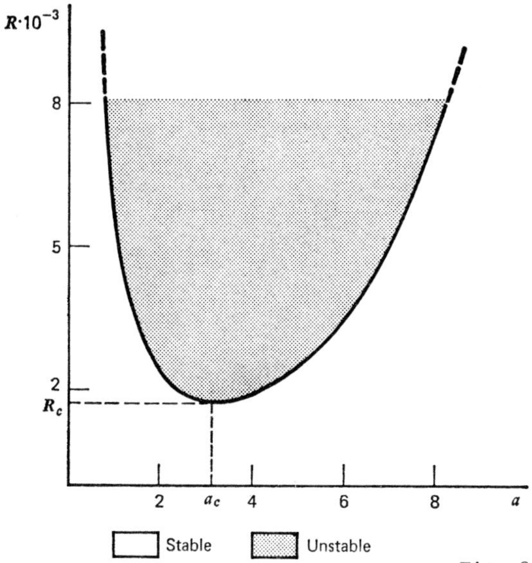

Fig. 3 The marginal curve of linear theory

Theoretical results obtained so far flatly contradict the experimental observations. There are five investigations which predict that the wavelength should decrease with increased $R > R_c$, for detail see [11]. There is also a paper by BUSSE [18] in which a new range for stable twodimensional flow is given (a modification of the marginal curve Fig. 3). However, as mentioned, no stable steady wavelengths with $\lambda < \lambda_c$ have been observed. And if there is a stable range of wavelengths $\lambda > \lambda_c$ for a given $R > R_c$, not just one wavelength, then this range is much smaller than the range predicted by Busse. To summarize, there is, at present, no theory which would be in qualitative agreement with the experimental results concerning the wavelength of supercritical convective motions.

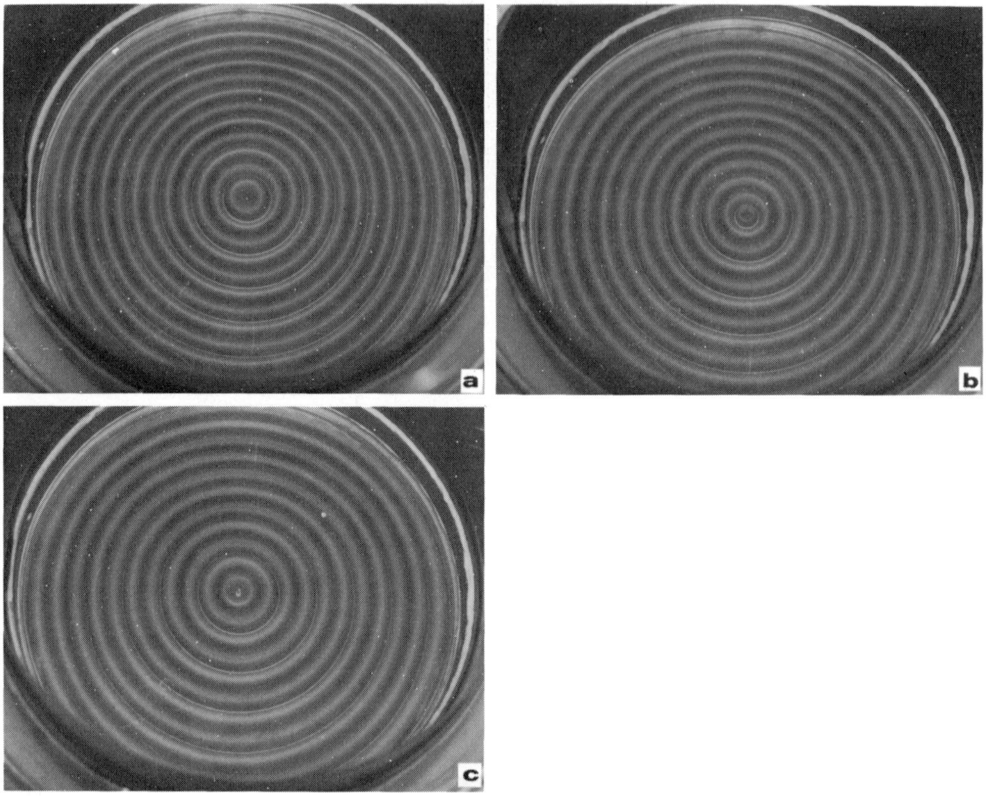

Fig. 4a Supercritical convective motion at $R = R_c$, twelve rings. The fine bright rings are settled aluminum powder

Fig. 4b Supercritical convective motion at $R = 3.06\ R_c$. Twelfth ring shrunk

Fig. 4c Supercritical convective flow at $R = 3.47\ R_c$. Eleven rings

3. Taylor Vortex-Flow

Let us now turn our attention to the other instability which is of particular interest in context with Synergetics, namely the Taylor vortex flow. This instability was first described in a classical paper by TAYLOR [19] in 1923, who developed the theory of the instability and who verified his theory with unambiguous experiments. Taylor vortices form in a fluid in the gap between two vertical, circular concentric cylinders when the inner cylinder is rotated with sufficient speed. The outer cylinder can be rotated as well, either in the same direction as the inner cylinder or in opposite direction and vortices will still form at appropriate rotation rates. We will, however, discuss here only the simple case with a resting outer cylinder. The instability is caused by an unstable radial distribution of angular momentum. The onset of the instability occurs when a critical value of a nondimensional parameter called the Taylor number

$$T = \frac{4\Omega^2 d^4}{\nu^2} \qquad (2)$$

is exceeded. A picture of Taylor vortices is shown in Fig. 5. The Taylor vortices are toroidal rings, the fluid moving outwards at the location of the fine dark lines in Fig. 5 and inwards at the location of the heavy dark lines. Two adjacent vortices have radial circulations of opposite direction. Superposed on the radial circulation is a strong overall azimuthal flow, originating from the rotation of the inner cylinder. So fluid parcels actually move on a helical path within the toroid of a Taylor vortex.

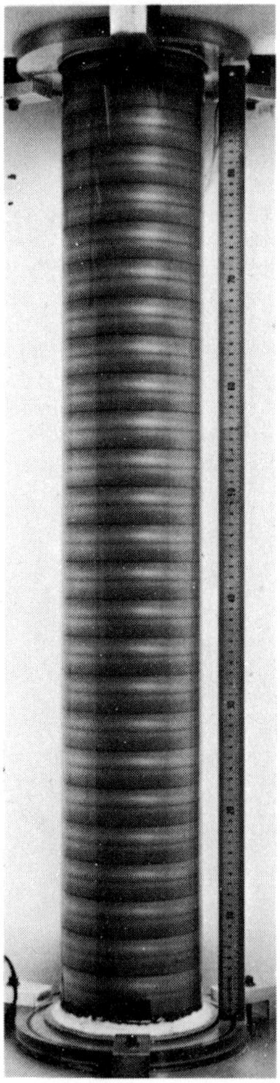

Fig. 5 Supercritical Taylor vortex flow. Outer cylinder at rest. Visualization with aluminum powder

Since Taylor's original work numerous theoretical and experimental studies of Taylor vortex flow have been published which cannot all be discussed here in detail. The results of linear theory are in good agreement with experimental observations. There is no degeneracy of the flow pattern as in Bénard convection since the geometry of the apparatus is clearly defined, although the fluid column is always assumed to be of infinite length, in order to avoid consideration of end effects. Experiments, however, are made with fluid columns of finite length. The instability in finite columns does not occur spontaneously in the entire fluid column the moment the critical Taylor number is reached, as follows from linear theory for an infinite column, but rather develops at both column ends (regardless of the specific type of end boundary) at subcritical Taylor numbers and progresses inwards until the entire column is filled with vortices at the critical T. This behavior corresponds exactly to the onset of instability in Bénard convection.

Let us now compare supercritical Taylor vortex flow with supercritical Bénard convection. Linear theory yields an interval of unstable wavenumbers for a given supercritical T, the interval being the same as in the marginal curve of Bénard convection

(Fig. 3), as was shown by CHANDRASEKHAR [20]; (a reference where much information concerning the linear theory of the Taylor instability can be found). The interval in possible unstable wavenumbers means that supercritical Taylor vortex flow should not be unique. The size, i.e. the wavelength, of axisymmetric supercritical Taylor vortices (extrapolated to an infinite fluid column length), has been measured by BURKHALTER and KOSCHMIEDER [21]. They found that the wavelength of the vortices does not vary with the Taylor number, provided T is increased quasi-steadily. So, while the wavelength of Bénard convection increases with increased Rayleigh number, the wavelength of (axisymmetric) Taylor vortices remains constant if the Taylor number is increased. As we observe, both instabilities exhibit a definitely different behavior in the supercritical range.

Are any of the other wavelengths which are unstable according to the marginal curve ever found at some supercritical T? In other words, is supercritical Taylor vortex flow really nonunique? Indeed it is, as was established by COLES [22] for doubly periodic (or wavy) Taylor vortices, by SNYDER [23] for axisymmetric vortices and studied in detail by BURKHALTER and KOSCHMIEDER [24]. Fig. 6 shows a photograph of Taylor vortex flow in the same apparatus, with the same fluid at exactly the same supercritical rotation rate ($T = 9.1\ T_c$). Obviously the wavelengths of the vortices in the three photographs are different. The left column shows Taylor vortices (with $\lambda < \lambda_c$) after a sudden start of the inner cylinder, the center picture shows the quasi-steady experiment exhibiting the critical wavelength, and the flow on the right (with $\lambda > \lambda_c$) is from a socalled filling experiment, in which the fluid is filled in from below while the inner cylinder rotates at a constant (supercritical) rotation rate. All three flows in Fig. 6 are perfectly steady and stable, actually stable against finite azimuthal disturbances as was shown by KOSCHMIEDER [25]. A summary of such experiments showing the nonuniqueness of Taylor vortex flow is shown in Fig. 7, where the experimental results are compared with the marginal curve of linear theory and an (extrapolated) curve of the range of axisymmetric flow obtained in a nonlinear study by KOGELMAN and DIPRIMA [26]. The striking nonuniqueness of Taylor vortex flow is in contrast to the behavior of Bénard convection, where there

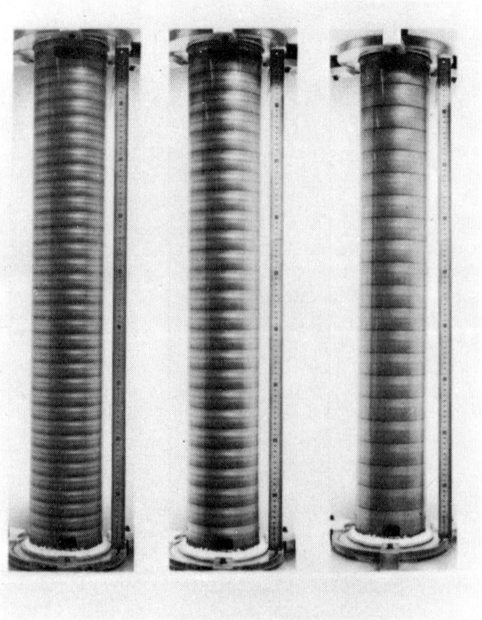

Fig. 6 Nonuniqueness of Taylor vortex flow at $T = 9.1\ T_c$. After [23]

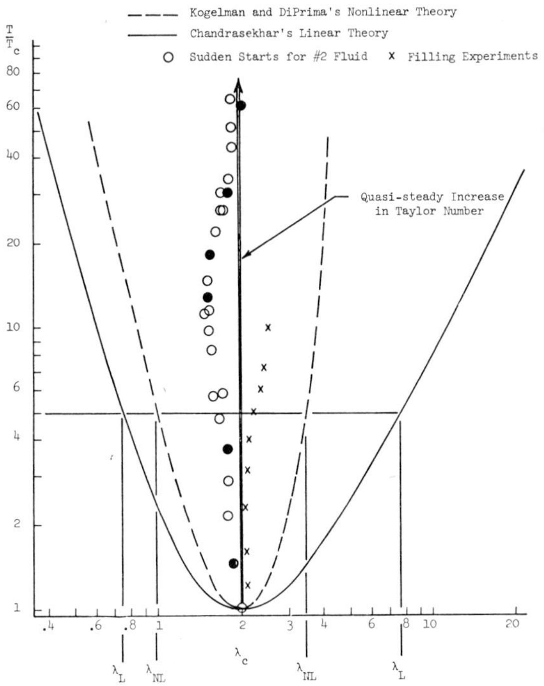

Fig. 7 Theoretically predicted and actually observed range of wavelengths of supercritical Taylor vortex flow. After [23]

is no evidence for a large range of stable wavelengths as a given supercritical R.

If we look at the experimental facts from the point of view of theory we are faced with two problems. Namely, first providing an explanation for the constancy of the wavelength of axisymmetric vortices with increased T; and second explaining the dependence of the wavelength on initial conditions and giving a narrower range of λ of the possible axisymmetric vortices. There are many other most interesting problems associated with Taylor vortex flow, see for example SWINNEY [27].

4. Conclusions

Synergetics tries to find and develop a common ground between various apparently very diverse instabilities occuring in the fields of physics, fluid dynamics, chemistry, biology etc. Fluid dynamics is but a very small part of this venture. We have considered above only two out of the large number of instabilities in fluids. It is nevertheless encouraging to find such a remarkably close similarity between Bénard convection and Taylor vortex flow in the linear regime, in spite of the fundamentally different apparatus used to create both instabilities. It is, on the other hand, disappointing to note basic differences in the reaction of both instabilities to supercritical conditions. It appears as if still very much work has to be done before a universal description of instabilities is feasible.

References

1. H. Bénard, Rev. Gen. Sci. Pure Appl. 11, 1261 (1900)
2. M. J. Block, Nature 178, 650 (1956)
3. Lord Rayleigh, Phil. Mag. 32, 529 (1916)
4. J. R. Pearson, J. Fluid Mech. 4, 489 (1958)
5. D. A. Nield, J. Fluid Mech. 19, 341 (1964)
6. E. L. Koschmieder, J. Fluid Mech. 30, 9 (1967)
7. A. Schlüter, D. Lortz, and F. Busse, J. Fluid Mech. 23, 129 (1965)
8. D. H. Sattinger, this book
9. K. Kirchgässner, this book
10. E. L. Koschmieder, Beitr. Phys. Atmos. 39, 1 (1966)
11. E. L. Koschmieder, Adv. Chem. Phys. 26, 177 (1974)
12. P. Bergé, to be published
13. H. Oertel, private communication
14. G. S. Charlson and R. L. Sani, Int. J. Heat Mass Transfer 13, 1479 (1970)
15. C. H. Jones, D. R. Moore, and N. O. Weiss, J. Fluid Mech. 73, 353 (1976)
16. S. H. Davis, J. Fluid Mech. 30, 465 (1967)
17. F. H. Busse and J. A. Whitehead, J. Fluid Mech. 47, 305 (1971)
18. F. H. Busse, J. Math. Phys. 46, 140 (1967)
19. G. I. Taylor, Phil. Trans. Roy. Soc. (London) A 223, 289 (1923)
20. S. Chandrasekhar, Hydrodynamic and Hydromagnetic Stability Oxford, 1961
21. J. E. Burkhalter and E. L. Koschmieder, J. Fluid Mech. 58, 547 (1973)
22. D. Coles, J. Fluid Mech. 21, 385 (1965)
23. H. A. Snyder, J. Fluid Mech. 35, 273 (1969)
24. J. E. Burkhalter and E. L. Koschmieder, Phys. Fluids 17, 1929 (1974)
25. E. L. Koschmieder, Phys. Fluids 18, 499 (1975)
26. S. Kogelman and R. C. Diprima, Phys. Fluids 13, 1 (1970)
27. H. L. Swinney, this book

Instabilities in Astrophysics

Instabilities in Stellar Evolution

R. Kippenhahn

With 10 Figures

In the following I first give a brief review of stellar evolution, and I then pick out situations in which the whole star or a part of it becomes unstable. As it occurs at present the process of star formation in our galaxy appears to be a process by which the system (our galaxy), out of irregular motion forms a very regular pattern and therefore seems to belong to these phenomena which are covered by synergetics.

1. An Average Star's Life Story

Stars spend their lifes in galaxies. There they are formed, there they circle around the common centre of gravity some hundred times before in their orbit they end as compact or even as collapsed objects which fill the stellar graveyard. Many galaxies show the well known spiral structure, which is due to the bright blue stars which ionize and illuminate the interstellar gas along the spiral arms. Since bright blue stars have not enough nuclear fuel to live for much more than 1 million years we know that they must be young objects. Consequently the spiral arms are the places where just recently stars have been formed. Along the spiral arms the process of star formation recently has been triggered: Here, interstellar matter has been brought to a critical density at which gravity overcomes the pressure and matter falls together. Numerical calculations for the spherical-symmetric approximation give a rough idea of the process.

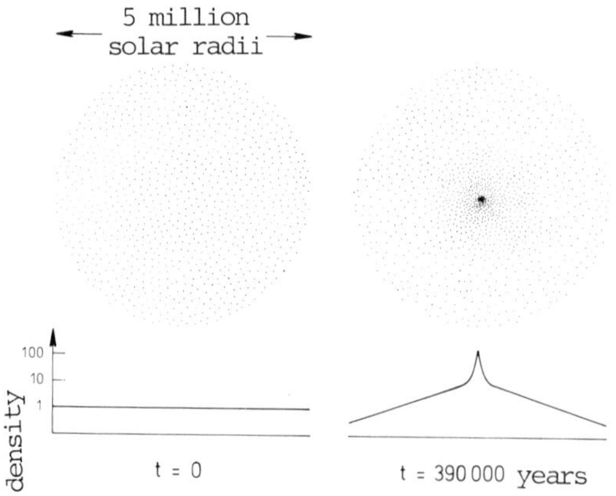

Fig. 1

In Fig.1 on the left-hand side the initial uniform density of a cloud consisting of one solar mass, is displayed at the start of such a numerical calculation. The right-hand side gives the situation 390 000 years later; the density is now higher in the centre. All the energy gained from contraction is radiated away, so that the contraction is practically isothermal. Later the central region becomes optically thick, and then the increasing internal pressure of the central region stops the infall. A core is formed which is practically in hydrostatic equilibrium and on top of which the remaining mass is raining. Fig.2 shows this stage, the central core is plotted on an enlarged scale on the right.

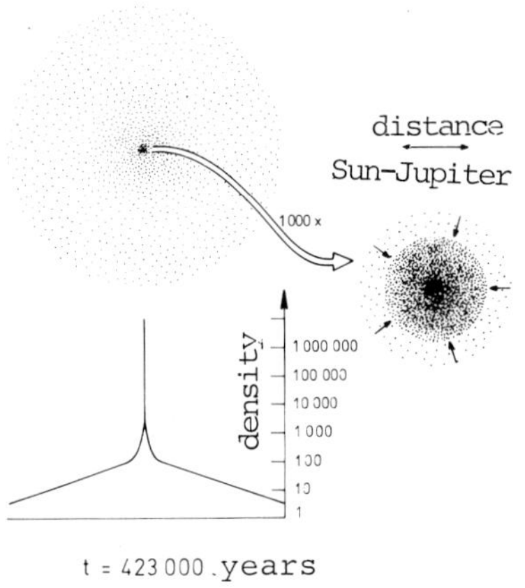

Fig.2

At the stage of Fig.2 hydrogen is still in the molecular state throughout the region. But when the core reaches a temperature of about 2000° dissociation takes place, and as we shall show later an instability occurs and the core collapses after some time (Fig.3). A second core is

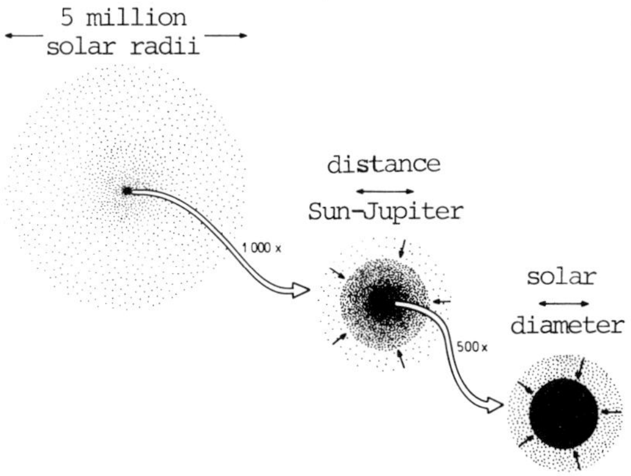

Fig.3

formed which has already the dimensions of the present sun, and the outside remaining mass in the later stages will fall on top of this core which then becomes a normal star. (Figs.1,2,3 are based on computations by LARSON [5].) Most of the stages of star formation cannot be directly observed. While the core has already formed the envelope is still obscuring the central object. Only infrared radiation reaches the observer or radiation in the radio range. We observe such objects for instance in the Orion nebula, which tells us that star formation is still going on there.

The numerical treatment of the problem becomes much more complicated if the rotation of the original cloud is taken into account. Then certainly binary stars and stars with planetary systems occur.

After the star is formed several time scales determine the further life. The shortest is the free-fall time scale τ_{ff} , or the pressure time scale, which is the time necessary for a pressure wave to go through the star. This is the time by which a distorted star tries to achieve hydrostatic equilibrium. For the sun it is of the order of some hours. Then there is a time for thermal adjustment, the Kelvin-Helmholtz time scale τ_{KH} ; it is some hundred million years for the sun, and finally there is the time scale τ_{n} of nuclear evolution, some billion years for the sun (but only some million years for stars of ten solar masses and more, which form the bright blue stars). For all stars we have the following hierarchy of time scales:

$$\tau_{ff} \ll \tau_{KH} \ll \tau_{n} \quad .$$

Most of its time the star spends in the phase of central hydrogen-burning. The changes due to consumption of nuclear fuel are so slow (since is so big) that the stars are always hydrostratically and thermally adjusted. Then after the exhaustion of hydrogen higher nuclear processes take over. The transition of one kind of nuclear burning to the other takes place on the time of thermal adjustment τ_{KH}. A rough idea of the evolution of the star after the onset of hydrogen burning is illustrated in Figs.4 and 5 for a star of 7 solar masses at different times measured from the ignition of hydrogen.

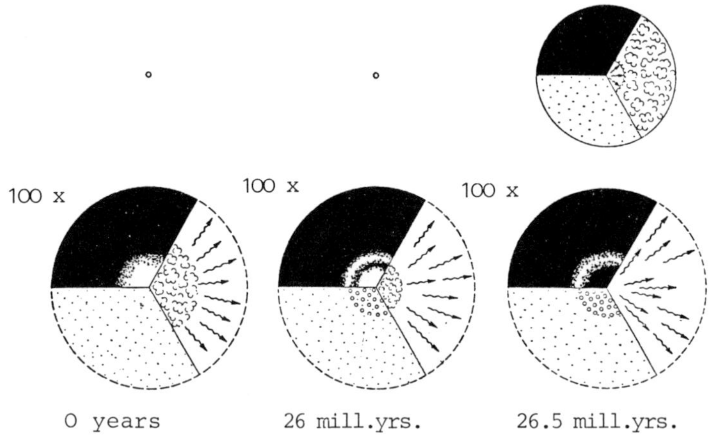

Fig.4 Different stages of the evolution of a star of seven solar masses as calculated by KIPPENHAHN, WEIGERT [4]. In the upper part the star is always plotted with the same scale, below the central region on an enlarged scale. The enlargement factors given refer to the uppermost picture. The ages given for subsequent stages give the times after the onset of hydrogen burning (left). Three sectors describe the internal structure. Right sector: energy transport which is either convection (cloudy regions) or radiation (arrows). Lower left section: chemical composition, dots for the original hydrogen rich material, open circles for helium, filled circles for carbon. Upper left section: areas of nuclear energy generation (white).

One can see how the star goes from hydrogen to helium burning and how a more complicated chemical structure is built up. One can further see that the radius of the star increases after the exhaustion of hydrogen. The star becomes a red giant.

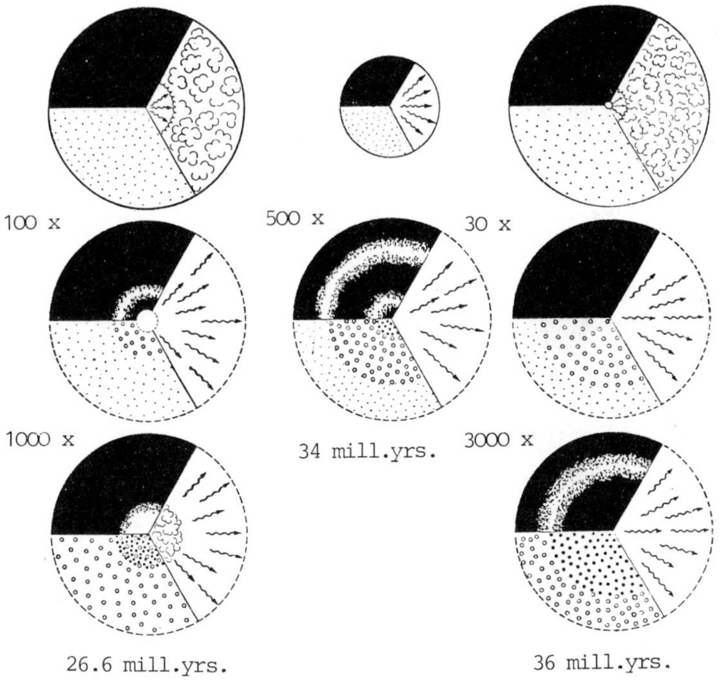

Fig.5 Continuation of Fig.4

Sometimes evolved stars come into a phase during which their equilibrium is unstable. They pulsate around their equilibrium state, their period being of the order of some days which for these stars is of the order of τ_{ff} . One can observe these pulsating stars, they are the well-known cepheids.

Further evolution becomes more and more difficult to be followed on the computer, and after some still not well understood transitions like the supernova explosion the star or what is left of it settles down as a highly condensed object. Although we are not able to follow the evolution continuously through these late stages we can classify the final products according to their mass. Stars with masses of less than 0.1 solar masses will never reach a temperature of nuclear burning and they will cool off before the onset of hydrogen burning and become black dwarfs, which might more resemble Jupiter than a star. Stars between 0.1 and 1.44 solar masses can become white dwarfs as we observe them on the sky. The bright star Sirius has a faint white dwarf companion. White dwarfs have consumed part of their nuclear fuel and they now cool off slowly. Stars which have a slightly higher mass, probably up to 2 solar masses, can form neutron stars (bodies in which the mass of the order of one sun is compressed to a sphere of some kilometers diamter). There is evidence that neutron stars are remnants of supernova explosions, a striking example is the pulsar in the Crab nebula. If bodies with even higher masses try to settle down as condensed objects due to general relativity they cannot find hydrostatic equilibrium and fall together for ever, forming a black hole.

As already mentioned we now discuss instabilities connected with that scenario of stellar evolution.

2. Spirals in Galaxies

Detailed theories have been developed to predict the formation of spiral structure in rotating discs consisting of stars. (For reviews see LIN, [6], WIELEN, [8].) A more direct access to the problem comes from computer simulation for N gravitating mass points with a given total angular momentum and a given random velocity which corresponds to a "temperature" of the cloud of stars. Spirals start to form if the "temperature" is sufficiently small, as can be seen from Fig.6 which is taken from HOHL [3]. Spirals do not always consist of the same stars, the stars cross the spirals, the spirals are density waves the relative

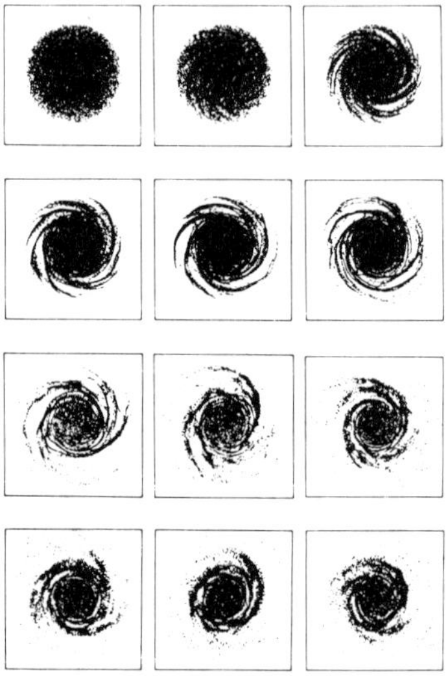

Fig.6 Computer simulation of the evolution of a rotating disc of stars (HOHL, [3]. Under favourable conditions spirals are formed

amplitudes of which are rather small. The drastic event is what happens to the interstellar gas which moves through the potential well, given by the gravity field of the stars. There the gas compressed and star formation is triggered. This can be directly seen in Fig.7 where superimposed to the optical picture of the spiral galaxy M 51 there are the lines of maximum radio intensity at a wave length of 1415 MHz. These radio maxima indicate the regions of maximum density of the interstellar matter. It is a good approximation to assume that the spiral pattern does not rotate while stars and interstellar matter are moving counter-clock-wise through these stationary patterns. As one then can see in average the gas first reaches its maximum density and then some time later the bright blue stars have formed which now illuminate the optical spiral arm. The time between maximum density and star formation seems to be of the order of 6 million years.

It is not yet clear how star formation is really triggered. An old stability criterion by Jeans says that for a given density a sufficiently big mass can become unstable and collapse. But it might well be that the thermal properties of the interstellar gas make it possible that matter in the spiral wave first undergoes a phase transition from a hot low density phase to a cool high density stage, and then the Jeans instability takes over.

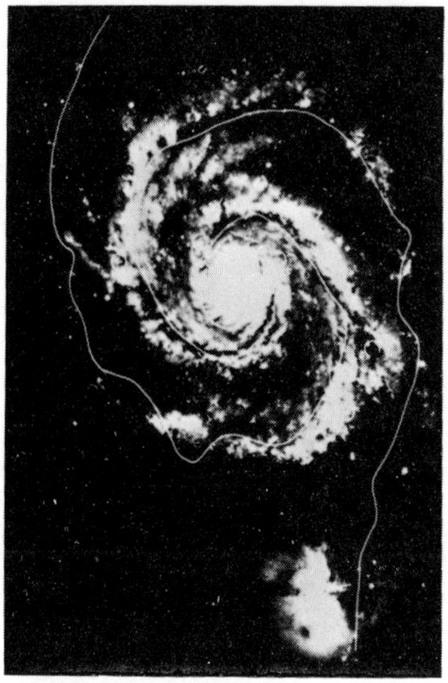

Fig.7 The spiral galaxy M 51. Superimposed white solid lines which indicate the maximum density of the interstellar gas as derived from radio observations (after Mathewson et al.[7])

3. Spherical-Symmetric Instabilities of Stars in Hydrostatic Equilibrium

Stars contract and expand, and a very simplified way of describing this is to assume that these radial motions go uniformly throughout the star in the following sense: If a mass element changes its distance r from the center by dr, then throughout the star we have $dr/r = x = \text{const.} = dR/R$ where R is the radius of the star. This is called homology. It is never exactly fulfilled in stars, but the main features of stars can already be understood on the basis of this simplifying assumption. We now consider matter at a certain depth at distance r from the center (Fig.8).

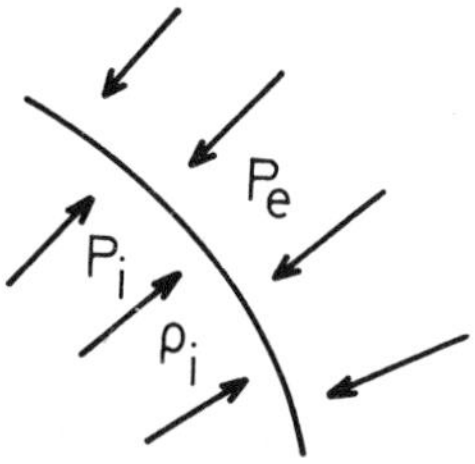

Fig.8 Equilibrium between external pressure at a sphere around the star's center

At a sphere of radius r there is equilibrium between the outer pressure P_e and the internal pressure P_i. The former comes from the weight of matter above the latter from the pressure of the matter below. During homologous contraction P_e changes like R^{-4}. This can be derived directly from hydrostatic equilibrium, but it can also easily be understood directly. During contraction the weight of the outer layers changes like R^{-2} since the matter comes closer to the center of gravity. This weight has to be distributed over the sphere which due to the change of its radius shrinks during contraction. This gives another factor of R^{-2}. The internal pressure (if the compression is sufficiently fast, compared to τ_{KH}) together with the internal density ρ_i obeys the adiabatic condition $P_i \sim \rho_i^{\gamma}$, where γ is the ratio of the specific heats, and since during homologous contraction $\varrho_c \sim R^{-3}$ one can easily see that after compression the higher internal pressure pushes back only if $\gamma > 4/3$. If this condition is violated then after compression the external pressure exceeds the internal pressure and the star is falling together. For a normal monoatomic gas with $\gamma = 5/3$ there is no danger; but ionization or dissociation can lower γ below its critical value, and this type of instability caused by the dissociation of hydrogen in the protostar is responsible for the formation of a second core as indicated in Fig.3.

We now apply the homology formalism to a small sphere around the star's center. We then can write down the variation of pressure and density in that central sphere during compression or contraction by

$$\frac{dP}{P} = -4\frac{dr}{r} = -4x \quad , \quad \frac{d\varrho}{\varrho} = -3x \quad , \quad \frac{d\varrho}{\varrho} = \alpha\frac{dP}{P} - \delta\frac{dT}{T} \qquad (1)$$

where we have added a third equation which comes from a general equation of state $\varrho \sim P^{\alpha}/T^{\delta}$. For the perfect gas we have $\alpha = \delta = 1$ but our general equation of state also includes for instance non-relativistic Fermi-Dirac degeneracy of the electron gas ($\alpha = 3/5, \delta = 0$). Elimination of dP/P and x from eq. (1) gives

$$\frac{dT}{T} = \frac{4\alpha - 3}{3\delta}\frac{d\varrho}{\varrho} \qquad (2)$$

and this equation tells us how the temperature of the central region of a star changes during contraction. The coefficient in equation (2) gives the slope in a logarithmic density temperature plane as indicated in Fig.9. Stars after their formation start on the lower left-hand side,

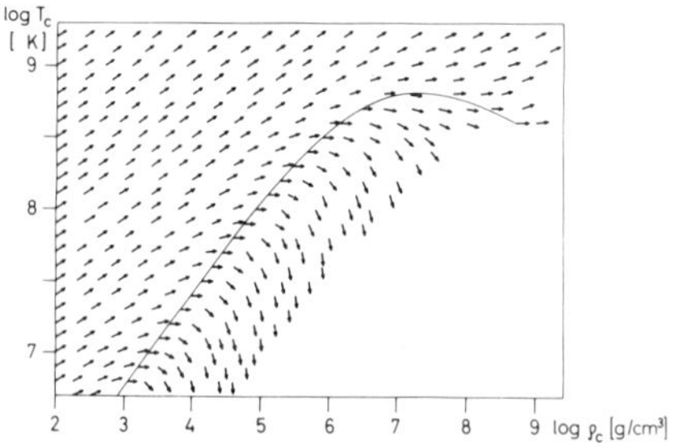

Fig.9 The log ϱ - log T plane for the center of a star. The arrows indicate the evolution during contraction derived from eq.(2). On the solid line the arrows become horizontal due to degeneracy

and since there the perfect gas equation holds; the coefficient is 1/3. Therefore while the star is slowly contracting it is heating up until at a temperature of 10 million degrees its center reaches the temperature of hydrogen burning. Then the stellar center stays there until the hydrogen fuel is consumed and then it contracts again until it reaches the temperature of helium burning which is ten times higher. As one can see for not too high temperatures but high densities the little arrows become horizontal and even point downwards. This is due to degeneracy and stars which come into this area do not heat up any more in their central region, at least as long as our simplifying assumption of homology remains true. One can show that the more massive a star, the higher up it starts on the left-hand side of the diagram. Consequently sufficiently massive stars will never get caught by degeneracy. They can heat up and reach higher and higher stages of nuclear burning.

We can use this formalism to derive another type of instability. For this purpose we make use of the first law of thermodynamics which up to now had not yet been used in our derivation. We write it in the form

$$dQ = du + P\,d(1/\varrho) \equiv c_p\,T\left[\frac{dT}{T} - \left(\frac{d\ln T}{d\ln P}\right)_{ad}\frac{dP}{P}\right]. \tag{3}$$

If we then eliminate the pressure variation via eq.(1), we can write it in a form

$$dQ = c_*\,dT\ ,\qquad c_* = \left[1 - \left(\frac{d\ln T}{d\ln P}\right)_{ad}\frac{4\delta}{4\alpha-3}\right]c_p\ . \tag{4}$$

Here c_* is a specific heat, but while we are normally used to specific heats under the boundary condition of constant pressure or constant volume, here the boundary condition is that the matter under consideration is in the center of a star and mechanically coupled with the outer part of that star. For the perfect monoatomic gas ($\alpha = \delta = 1$, $(d\ln T/d\ln P)_{ad} = 0.4$) the effective specific heat c_* is negative. This is good because it stabilizes nuclear burning: Thermonuclear reaction rates increase with temperature. Therefore if in the center of a star the nuclear energy output is accidentally slightly increased, then the additional heat input causes cooling and reduces the nuclear energy production. But this is not the case for degenerate material since if $\delta \to 0$ the effective specific heat becomes positive, and indeed one can then have unstable nuclear burning. When in its future the sun will have exhausted its hydrogen and will later then ignite helium, the helium burning will start with a thermal run-away due to the positive specific effective heat.

4. Non-Spherical Symmetric Perturbation in Stars in Hydrostatic Equilibrium

As one can see from Figs.4 and 5, convection can occur in stars at different stages of evolution. In contrary to the normal treatment of convection by Boussinesq approximation, the compressibility of the stellar material has to be taken into account. A heuristic derivation can be made by taking a blob of material and lifting it upwards and let it adjust its pressure with the neighborhood. Then the question is whether the material is hotter or cooler than the surrounding. In the first case the layer is unstable because then the matter is driven even further away from its origin by buoyancy forces; in the other case it is stable. The condition for this has been derived by K. Schwarzschild according to which the layer is stable if the temperature gradient does not exceed its adiabatic value

$$\frac{d\ln T}{d\ln P} < \left(\frac{d\ln T}{d\ln P}\right)_{ad} \tag{5}$$

In this case a blob lifted into a certain height will oscillate periodically around a mean position. If the molecular weight μ varies with depth - and this is very often the case in stars where heavier atoms are built up in the central regions - (5) has to be replaced by

$$\frac{d\ln T}{d\ln P} < \left(\frac{d\ln T}{d\ln P}\right)_{ad} + \frac{d\ln\mu}{d\ln P} \tag{6}$$

which has been derived by Ledoux and which takes into account that a layer also can be stable even if it violates the Schwarzschild criterion, if the gradient increases in the direciton of gravity.

But there is an interesting over-stability in the case of a layer which violates the Schwarzschild criterion, but fulfills the Ledoux criterion:

$$\left(\frac{d\ln T}{d\ln P}\right)_{ad} < \frac{d\ln T}{d\ln P} < \left(\frac{d\ln T}{d\ln P}\right)_{ad} + \frac{d\ln \mu}{d\ln P} \quad . \tag{7}$$

In this case the blob which is oscillating is hotter than its surrounding if it is above its mean position. But then, if we take into account dissipation, it will cool a bit. Consequently, when it oscillates back and goes through the mean position, it will be cooler than the neighborhood and therefore shoots through the mean position slightly faster than it would without dissipation. Consequently, in the next half-cycle it will heat up and shoot through the mean position again slightly faster. Therefore dissipation makes the amplitude of the oscillation increase, and over-stability occurs. It is still not yet known whether this mechanism is important for stars. There are certainly extended regions in stars which fulfil the condition (7).

5. Pulsations

Another type of instability occurs in most cases if the surface temperature of a giant star reaches a temperature of about 5.900° (BAKER, KIPPENHAHN, [1]. This is the stage at which the star becomes a cepheid. In order to understand the mechanism which drives the instability we have a very simple model. In Fig.10 the gas in an isolated container covered by a piston mimics the average mass element in the star held

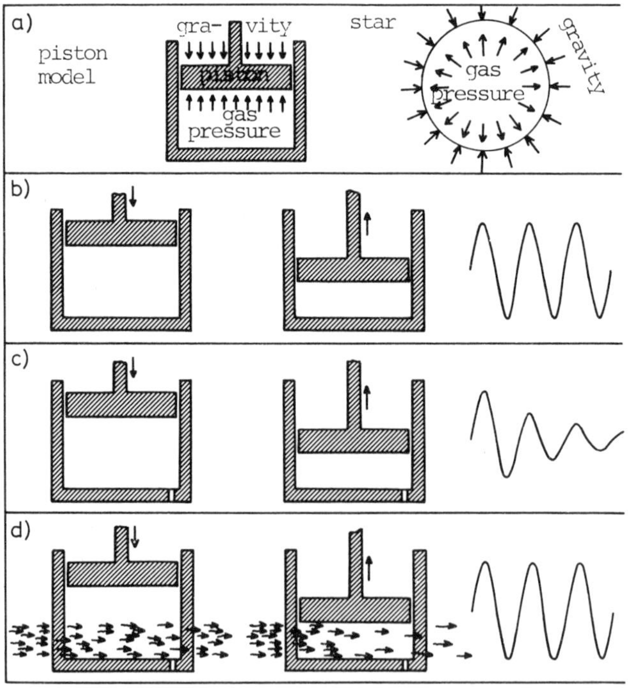

Fig.10 A simple model for Cepheid pulsation.
a) equilibrium, b) adiabatic pulsation, c) damping due to heat leakage, d) excitation due to an absorption coefficient which is highest at maximum compression

together by the weight of the layers above. If the gas below the piston is thermally completely insulated, then during oscillation of the piston the gas undergoes an adiabatic change, and no energy is lost. The model oscillates with constant amplitude. If we put the model into a surrounding heat bath and make a little hole in the insulation, and if the mean temperature inside is equal to the temperature outside, then during the oscillation of the piston small amounts of heat are periodically flowing inwards and outwards during each period, and as a second order effect damping occurs. This is the reason why stars normally do not pulsate. If they were somehow to be forced into oscillation and left alone, then thermal damping (which is much more important in stars than friction) would within some hundred or thousand years damp the oscillations. Cepheids therefore need a driving mechanism and presently there is general agreement that this is due to the absorption coefficient in the outer layers of a star. The absorption coefficient is a rather complicated function, depending on temperature and density. In order to mimic the stellar material with our piston model we assume that in addition the gas is exposed to a radiation field which somehow goes through the piston model and a certain part of that radiation is absorbed by the gas and converted to heat. If then during the oscillation of the piston the absorption coefficient of the gas is biggest during maximum compression and smallest during maximum expansion, then the gas absorbes most when it is compressed and this energy is then used for expansion. This mechansim can drive the piston model against the damping due to heat leakage and this mechanism also drives the star. In the outer layers of a star the ionisation of hydrogen and of helium is such that a fairly large part of the outer layers have indeed the property that the absorption coefficient increases during adiabatic compression, and this drives the cepheids. Up to now one has not only analyzed the stability with linearized theories, but also followed the non-linear oscillation numerically.

I have chosen here just a few simple examples of stability problems which arise in the theory of stellar structure. Many more types of instabilities occur when rotation is taken into account. Not only that stars or protostellar clouds can break into binaries - a problem which stimulated bifurcation theory - but also for slowly rotating stars the stability problem of given angular velocity distributions is not yet completely solved. Actually, neither the time scale by which a nonstable state goes into another state nor the final state itself is known for these types of problems. For a review see FRICKE, KIPPENHAHN 2 .

References

1. N.Baker, R.Kippenhahn: Astrophys.J. 142, 868 (1965)
2. K.Fricke, R.Kippenhahn: Ann.Rev.Astron.Astrophys. 10, 45 (1972)
3. F.Hohl: IAU Symp. No. 38, 368 (1970)
4. R.Kippenhahn, A.Weigert: Sterne u. Weltraum 3, 173 (1964)
5. R.B.Larson: M.N.R.A.S. 145, 271 (1969)
6. C.C.Lin: Lectures in Appl.Math. Vol.9, ed. J.Ehlers, Am.Math.Soc., Providence, Rhode Island (1967)
7. D.S.Mathewson, P.C.van der Kruit, W.N.Brouw: Astron.Astrophys. 17, 468 (1972)
8. R.Wielen: Mitt.Astron.Ges. 30, 31 (1971)

Solitons

Solitons
I. Basic Concepts

R. K. Bullough and R. K. Dodd

With 1 Figure

1. Introduction

Solitons are mathematical objects which arise as solutions of certain non-linear wave equations. This class is a rapidly growing one and the reasons for current interest are threefold. First, methods have been found for finding these solutions analytically and therefore exactly, and ideas on their "completeness" have emerged. Second solitons have mathematical properties which make them particle-like and in those contexts of non-linear physics where they arise they are the natural elementary excitations. Third those contexts of non-linear physics are proving to embrace the whole of modern physics itself.

In this lecture we want to do two things: to introduce the soliton solutions of a number of physically important non-linear wave equations in a very simple way; and to illustrate by some relatively simple examples the range of application of solitons to modern non-linear physics. Abstract soliton theory requires considerable mathematical technique. In a second lecture (II) we shall develop some of that technique to show the remarkable mathematical structures which apparently underlie the whole soliton concept.

2. Solitons as Solitary Waves

A working definition of a soliton is that of a solitary wave with a particular collision property. Therefore we need the idea of a solitary wave. This is typically a bell-shaped disturbance which not only does not break up but has an unchanging shape: the disturbance $u(x,t)$ depends on the argument $x-ct$ where c is a constant. We consider plane waves - there is one space variable x as well as the time t.

The distortionless disturbance $u(x-ct)$ (sometimes called a wave of permanent profile) moves up the x-axis with speed c. It is a solution of the linear dispersionless wave equation

$$u_t + c\, u_x = 0 \,. \tag{1}$$

In contrast the linear equation

$$u_t - c\, u_{xxx} = 0 \tag{2}$$

has harmonic wave solutions $u^k = \cos(\omega t - kx)$ with dispersion $\omega = ck^3$ and speed ck^2. The large k modes run away from the small k modes and an initial bell shaped disturbance disperses (breaks up). The harmonic waves are the only distortionless solutions of (2).

The simplest non-linear wave equation is perhaps

$$u_t + 6u\, u_x = 0 \tag{3}$$

(the number 6 is conventional: notice that by scaling x, t and u itself the 6 can be scaled to any value). This is the equation of a 'simple wave' with an implicit solution

$$u = u(x-6ut) \quad , \tag{4}$$

that is the solution depends upon u itself. This solution shocks. Points where u is large and positive have large effective speed 6u along x and u develops a rising front where the large values of u pile up.

It is possible to balance the non-linearity in (3) against the dispersion in (2) to get a solitary wave u(x-ct). We take the Korteweg-de Vries equation (KdV) [1]

$$u_t + 6uu_x + u_{xxx} = 0 \tag{5}$$

which has the solitary wave solution

$$u(x,t) = 2\xi^2 \operatorname{sech}^2 \left[\xi(x-4\xi^2 t)\right] \tag{6}$$

in which $c = 4\xi^2$ and ξ is a constant. This is a single bell shaped pulse which propagates along x without changing shape. It was the first solitary wave recorded [2]. It is a one-parameter solution and the speed c is proportional to its amplitude $2\xi^2$. This is a common but by no means necessary characteristic of a soliton. In fact (6) happens to be both a solitary wave and a soliton.

Evidently at $x = -\infty$ we can 'superpose' N distinct solutions (6) characterised by parameters $\xi_1, \ldots, \xi_N$ by spacing these out an infinite distance apart without overlap (there is no superposition principle for solutions of non-linear wave equations although we are about to find one for solitons). Conveniently we order the solutions 1, 2,...,N from the left with $\xi_1 > \xi_2 > \ldots > \xi_N$. Then (since solutions can be placed a finite distance apart with only exponential damage) the larger pulses overtake the smaller ones and the set undergoes a complicated collision.

For an arbitrary non-linear wave equation it would be difficult to say more. Non-linear scattering takes place between pulses and in perturbation theory, for example, terms would grow dramatically in number. The behaviour of the KdV is very simple however (though perturbation theory is not the ideal vehicle to show this - although see [3]): at $x = +\infty$ there emerges the N pulses (6) in the reverse order $\xi_N, \xi_{N-1}, \ldots, \xi_1$ [4]. The only change is that the argument $\xi_i\,(x - 4\xi_i^2 t - x_i) \to \xi_i(x - 4\xi_i^2 t - x_i - \delta_i)$ (x_i is the position of the i-th pulse at the starting time t = o). The small "phase shift" δ_i is made up of strictly pairwise shifts: $\delta_i = \sum_{j\neq i} \delta_{ij}$ and there are no 3-body shifts. Total phase shift is conserved: $\sum_i \delta_i = \sum_i \sum_{j\neq i} \delta_{ij} = 0$. This is one of an infinite number of conserved quantities associated with the simpler solitons (see II).

We can interpret this result of collisions in two ways: either solitons pass through each other with small pairwise interactions inducing the shifts δ_{ij}; or they bump like particles transferring energy, momentum and therefore amplitude, without losing their essential form. Either point of view is tenable: computer solutions of a soft collision between solitons show that a trough always exists between the two disturbances and amplitude leaks mysteriously between one side of the trough and the other. In a hard collision the pulses overlap and the trough disappears however (for some rather specialised illustrations of the troughs in soft collisions see [5, 6]).

Not all solitary waves are solitons. For example the non-linear diffusion equation, the Burgers equation [7],

$$u_t + 6uu_x - b\,u_{xx} = 0 \quad , \quad b > 0 \tag{7}$$

has a permanent profile solution

$$u(x,t) = u_1 + (u_2 - u_1)[1 + \exp\{3b^{-1}(u_2 - u_1)(x - ct)\}]^{-1} \tag{8}$$

with $c = \frac{1}{2}(u_1 + u_2)$. It is a constant step-like shock, with asymptotes u_1 and u_2 as $x \to \mp\infty$, moving along x. Two such shocks with parameters (u_3,u_2) and (u_2,u_1) and $u_3 > u_2 > u_1$ collide: they fuse and emerge as the single confluent shock (u_3,u_1) as $t \to \infty$ [3].

Steplike solutions like (8) are called "kinks". Their derivatives are bell-shaped, and may be solitary waves. These solitary waves may be solitons and in this case the kinks also have the soliton collision property. In this case we may also loosely refer to the kinks as solitons.

The supreme example of the kink of this kind is the kink solution of the sine-Gordon equation (s-G)

$$u_{xx} - u_{tt} = \sin u \ . \tag{9}$$

The kink is

$$u(x,t) = 4\tan^{-1} \exp\{(x - vt)(1 - v^2)^{-\frac{1}{2}}\} \ ; \ 1 > v \geq 0 \ . \tag{10}$$

The derivative of the kink is

$$- u_t = 2v[1 - v^2]^{-\frac{1}{2}} \operatorname{sech}\{(x - vt)(1 - v^2)^{-\frac{1}{2}}\} \tag{11}$$

and the speed is v. The kink is a "2π-kink": $u \to 0$ as $x \to -\infty$, $u \to 2\pi$ as $x \to +\infty$, and u_x, u_{xx}, etc. $\to 0$ as $|x| \to \infty$. Two such kinks satisfy $u \to 4\pi$ as $x \to +\infty$. They collide and pass through each other if $v_1 > v_2$ and v_1 is the parameter of the kink on the left at $t = 0$.

Notice that 0, 2π, 4π, ... are zeros of sin u. If we Lorentz transform (9) to the rest frame that equation is the ordinary differential equation $u_{xx} = \sin u$ with a first integral $\frac{1}{2}u_x^2 + \cos u = \text{constant}$. If the constant is one, the zeros of u_x are zeros of sin u. These are points in the 2-dimensional phase space (u,u_x) characterised by "zero 'velocity' u_x means zero 'force' u_{xx}". They are therefore equilibrium points.

It is easy to see they are unstable equilibrium points since, e.g. $\frac{1}{2}u_x^2 + \cos u \approx \frac{1}{2}u_x^2 + 1 - \frac{1}{2}u^2$ for small u and trajectories are rectangular hyperbolas near $u = 0$. The points π, 3π, etc are stable: trajectories near π are ellipses about this value. The trajectory $\frac{1}{2}u_x^2 + \cos u + 1$ describes a motion between $u = 0$ and $u = 2\pi$ determined from $u_x = 2 \sin \frac{1}{2} u$ or

$$\int^{\frac{1}{2}u} \frac{dv}{\sin v} = x + \text{constant} \tag{12}$$

so $u = 4\tan^{-1} e^{x+\delta}$. This is therefore the trajectory of a kink.

It is worthwhile understanding the unstable equilibrium points. We take the related "ϕ-four" equation

$$\phi_{xx} - \phi_{tt} = - \phi + \phi^3 \tag{13}$$

which in the rest frame is $\phi_{xx} = - \phi + \phi^3$ with kink trajectories $\frac{1}{2}\phi_x^2 + \frac{1}{2}\phi^2 - \frac{1}{4}\phi^4 = \frac{1}{4}$ between $\phi = -1$ and $\phi = +1$. The kinks are

$$\phi = \pm \tanh\left\{ \frac{1}{\sqrt{2}} \frac{x - vt}{(1 - v^2)^{\frac{1}{2}}} \right\} \tag{14}$$

after Lorentz transformation to the moving frame. The two signs allow us to define a kink (+) and an antikink (-): ϕ goes from -1 to +1 in the kink and from +1 to -1 in the antikink as x goes from $-\infty$ to $+\infty$. The points $\phi_x = 0$, $\phi = \pm 1$ are unstable equilibrium points at the maxima ± 1 of the 'potential' $\frac{1}{2}\phi^2 - \frac{1}{4}\phi^4$. However the ϕ-four

Hamiltonian which yields (13) is

$$H = \int_{-\infty}^{\infty} [\tfrac{1}{2}\phi_t^2 + \tfrac{1}{2}\phi_x^2 + \tfrac{1}{4} - \tfrac{1}{2}\phi^2 + \tfrac{1}{4}\phi^4]dx \ . \tag{15}$$

The true potential energy is of opposite sign and the points $\phi = \pm 1$ are stable equilibrium points energetically. Thus the kink (14) takes the system from a stable equilibrium point to a stable equilibrium point of the same energy. These points are distinct degenerate vacuum states in particle physics, for example [8].

Although the kink (10) of the s-G has the same property there is an important difference between the two cases. The ϕ-four has only the two singular points $\phi = \pm 1$ and a kink must be followed by an antikink. The s-G has singular points at 0, $\pm 2\pi$, $\pm 4\pi$, etc. and kinks or antikinks can follow kinks or antikinks. This is obviously necessary for soliton solutions. The s-G does have soliton solutions whilst the ϕ-four does not. In so far as the ϕ-four has been much used in theoretical physics (Landau-Ginsberg expansions, the laser phase transition analogy, bifurcation theory [9], displacive phase transitions in 1-D crystals [6] and 'central peak phenomena' [10], and particle physics [11] have all been concerned with ϕ-four type potentials) it may seem surprising that the ϕ-four does not have soliton solutions. Approximate soliton-like kinks like (14) seem however [6] to be important physical entities.

Another useful property of (14) is that it has a mass (= 2/3). This is a consequence of the Lorentz invariant form of (13) rather than any soliton or solitary wave property. The solitary wave character keeps this mass finite however and the kink as a particle model has a finite self-energy. The kink (10) has mass 8. Notice how 'close' to the ϕ^4 is the s-G (the negative sign corresponds to displacing the origin by π):

$$\sqrt{6}\,(u_{xx} - u_{tt}) = -\sin\sqrt{6}\,u \approx -\sqrt{6}\,u\,[1 - u^2] \ .$$

3. Multisoliton Solutions

We give here the analytical forms of the so called multisoliton solutions of the s-G equation (9). These take the forms [12]:

2π-kink

$$u(x,t) = 4\tan^{-1}\exp\Theta_1 \ ; \tag{16}$$

4π-kink

$$u(x,t) = 4\tan^{-1}\frac{\sinh\frac{1}{2}(\Theta_1 + \Theta_2)}{(a_{12})^{\frac{1}{2}}\cosh\frac{1}{2}(\Theta_1 + \Theta_2)}$$

$$a_{12} = (a_1 - a_2)^2(a_1 + a_2)^{-2} \ ; \tag{17}$$

0π-kink (or "breather")

$$u(x,t) = 4\tan^{-1} r\sin\Theta_I\,\mathrm{sech}\,\Theta_R$$

$$r = a_R\,a_I^{-1}$$

$$\Theta_R = \tfrac{1}{2}a_R[(1 + \frac{1}{a_R^2 + a_I^2})x + (1 - \frac{1}{a_R^2 + a_I^2})t + x_R]$$

$$\Theta_I = \tfrac{1}{2}a_I[(1 - \frac{1}{a_R^2 + a_I^2})x + (1 + \frac{1}{a_R^2 + a_I^2})t + x_I] \tag{18}$$

$2N\pi$-kink

$$\cos u(x,t) = 1 - 2(\frac{\partial^2}{\partial x^2} - \frac{\partial^2}{\partial t^2})\ln f(x,t)$$

$$f(x,t) = \det||M|| \ . \tag{19}$$

The $N\times N$ matrix $||M||$ has elements

$$M_{ij} = 2(a_i + a_j)^{-1}\cosh\{\tfrac{1}{2}(\Theta_i + \Theta_j)\} \ ,$$

$$\Theta_i = \pm\,\gamma_i(x - v_i t) + x_i$$
$$a_i^{\,2} = (1 - v_i)(1 + v_i)^{-1}$$
$$\gamma_i^{\,2} = (1 - v_i^{\,2})^{-1} \;, \tag{20}$$

in (19). In (16) Θ_1 is given by Θ_i as in (20). In (17) Θ_1, Θ_2 are similar: a_1 and a_2 are the two (real) parameters of the 2-parameter solution (17) related to velocities v_1, v_2 as in (20). Asymptotically (17) is [12]

$$u(x,t) \sim 4\,\tan^{-1}\,\exp[\Theta_1 + \eta_1^{\pm}] + 4\,\tan^{-1}[\Theta_2 + \eta_2^{\pm}]$$

as $x \to \pm\,\infty$:

$$\eta_1^{\pm} = -\,\eta_2^{\pm} = \pm\,\tfrac{1}{2}\,\ln a_{12}\;. \tag{21}$$

The 0π or "breather" solution (18) is of a type not discussed before. It is a 2-parameter solution which does not break up. Instead it has an internal oscillation which progresses and is modulated by a moving external envelope. It is obtained from (19) by choosing a_1 and a_2 complex with $a_1 = a_2^* = a_R + i\,a_I$ and $x_1 = x_2 = x_R + i\,x_I$. Its energy is $16\gamma_R\{r^2(1 + r^2)^{-1}\}^{\frac{1}{2}} = 16\gamma_R \sin\mu$ (for $a_1 = a\,e^{i\mu}$): $\gamma_R = (1 - v_R^{\,2})^{-\frac{1}{2}}$ and $v_R = (1 - a_R^{\,2} - a_I^{\,2})(1 + a_R^{\,2} + a_I^{\,2})^{-1}$. The 0π acts like a soliton: it can collide with 2π-kinks, 2π-antikinks and other breathers and passes through them with only a phase shift. It can be thought of as a bound kink-antikink pair in which ϕ moves from 0 to 2π- and back to 0 as x moves from $-\infty$ to $+\infty$. It therefore has a total kink angle of zero.

The N-kink solution (19) contains all the other solutions as special cases. It was derived in the form equivalent to (19) [13,14] by a difficult analogy drawn between the KdV and s-G equations. HIROTA [15] had given a multisoliton solution of the KdV. We now know how to obtain this solution by the so-called inverse scattering method [16,17]. I develop the theory of the ISM in II. The ISM solves the initial value problem $u(x,0) = f_1(x)$, $u_t(x,0) = f_2(x)$; $u \to 0 \pmod{2\pi}$, u_x, u_{xx}, etc. $\to 0$ as $|x| \to \infty$; for the s-G. It shows that the solution consists of two parts: the multisoliton part like (19) and the 'background'. The background diffuses away so that asymptotically only the multisoliton part is significant. In this sense all initial conditions ultimately produce only solitons: these solitons are 2π-kinks, 2π-antikinks or breathers. All kinks or antikinks must travel at different speeds (they either attract or repel each other); but any number of breathers can travel at the same speed as any one kink or antikink. The kinks have a 'topological charge' 2π and mutually repel: antikinks have charge -2π and kink-antikink pairs attract; breathers have no charge. There are no other solutions of the s-G apart from the background and the solution is "complete" in this sense.

4. Applications of Solitons to Physics

A less than definitive list of applications of solitons in physics already includes

i) Gravity waves in deep and shallow water.
ii) The theory of plasmas and the interaction of radiation with plasmas.
iii) Superconductivity: the theory of Josephson junctions.
iv) Fermi liquid theory: spin waves in the A- and B-phases of liquid ^{3}He below 2.6 mK.
v) Ferromagnetics: Bloch wall motion.
vi) Resonant and non-resonant non-linear optics and laser physics.
vii) Non-linear crystal physics: theory of dislocations, anharmonic crystals; recurrence phenomena in thermal transport and non-ergodic behaviour (Fermi-Pasta-Ulam problem [18]); displacive and other phase transitions and central peak phenomena; linear conductors (like TTF-TCNQ).
viii) Theory of fundamental particles.
ix) Astrophysics: solitons in the solar corona have been suggested (the Great Red Spot in Jupiter has been called a soliton also!).

Examples which embrace almost all of this list are treated in [6]. A large number of illustrative diagrams is given there.

The key equations are:

The KdV $$u_t + 6uu_x + u_{xxx} = 0 \tag{22}$$

The modified KdV $$u_t + 6u^2u_x + u_{xxx} = 0 \tag{23}$$

The non-linear Schrödinger eqn. (NLS) $$i\, u_t + 2u|u|^2 + u_{xx} = 0 \tag{24}$$

The s-G $$u_{xt} - \sin u = 0 \tag{25}$$

Other important equations are

The Hirota eqn. $$i\, u_t + 3i\alpha|u|^2\, u_x + \rho\, u_{xx} + i\, \sigma u_{xxx} + \delta|u|^2u = 0 \qquad (\alpha\rho = \sigma\delta) \tag{26}$$

The reduced Maxwell-Bloch (RMB) eqns.
$$u_x = -\mu s$$
$$v_x = Ew + \mu u$$
$$w_x = -Ev$$
$$E_t = v \qquad (\mu = \text{const.}) \tag{27}$$

The Boussinesq eqn. $$u_{tt} - (12uu_x + u_{xxx})_x = 0 \tag{28}$$

The 3-wave interaction (decay type)
$$u_{1,x} + c_1\, u_{1,t} = iq\, u_2\, u_3^*$$
$$u_{2,x} + c_2\, u_{2,t} = iq\, u_1\, u_3$$
$$u_{3,x} + c_3\, u_{3,t} = iq\, u_1^*\, u_2 \tag{29}$$

The Toda lattice $$u_{n,tt} = e^{-(u_n - u_{n+1})} - e^{-(u_{n-1} - u_n)} \tag{30}$$

The Kadomtsev-Petviashvili eqn.
$$\tfrac{3}{4}\beta^2\, u_{yy} + \{\alpha u_t + \lambda u_x + \tfrac{1}{4}(u_{xxx} + 6uu_x)\}_x = 0\,. \tag{31}$$

All these equations have multisoliton solutions. The majority are in the form of systems of non-linear evolution equations (NEEs) $u_t = K[u]$ where K is a functional of u. These are characterised by requiring only the initial data u(x,0) to determine the motion. This is why the s-G is placed in characteristics form in (25) (by transforming to new variables $x \pm t$). In II we work from the NEE form.[1]

We quote a few physical applications: the KdV to deep and shallow water waves, lattice recurrences, plasma ion acoustic waves; the modified KdV to Alfven waves in a cold collisionless plasma; the NLS to 1-dimensional self-focussing, self-phase modulation, Langmuir turbulence in plasmas [19] (Zakharov's caverns [20]), laser-plasma interactions (optical filament formation); the s-G to spin waves, Josephson junctions, non-linear optics (self-induced transparency [21,22], lattice theory and

[1] Eqn. (28) takes the form $u_t = p$, $u_x = q$, $q_x = r$, $r_x = y$, $p_t = q_x + y_x + 12(ur + q^2)$ Other examples are similar.

particle physics [11]; the RMB to self-induced transparency [22]; the Boussinesq to hydrodynamics and plasmas; the 3-wave to stimulated Raman back scattering in plasmas; the Toda lattice as a soluble lattice with hard core and harmonic limits; the Kadomtsev-Petviashvili equation as a two dimensional non-stationary problem in a weakly dispersive medium [23,24]. We refer to [6] for more comprehensive references on the other topics.

5. Some Particular Applications of Solitons in Physics

We shall develop four particular examples in more detail. These are the application of the multisoliton solutions of the NLS to optical filament formation in laser irradiated neutral dielectrics and the application of the s-G to spin waves, optical pulses and Josephson junctions. The three applications of the s-G can be taken in one jump!

Consider the neutral dielectric: in a (scalar) field $E(\underset{\sim}{x},t)$ the dipole $P(\underset{\sim}{x},t)$ is

$$P(\underset{\sim}{x},t) = \alpha\, E(\underset{\sim}{x},t) + \alpha_{NL}\, \{E(\underset{\sim}{x},t)\}^3 + \ldots\ldots \qquad (32)$$

and α and α_{NL} are the constant linear and first non-linear susceptibilities. Maxwell's equation is linear and is

$$\nabla^2 E(\underset{\sim}{x},t) - c^{-2}\partial^2 E(\underset{\sim}{x},t)/\partial t^2 = 4\pi n c^{-2}\, \partial^2 P(\underset{\sim}{x},t)/\partial t^2 \qquad (33)$$

where n is the atomic number density. We look for complex envelope solutions $\varepsilon(x,y,z,t)$ modulating carrier waves $e^{i(\omega t-kz)}$:

$$E(\underset{\sim}{x},t) = \varepsilon(x,y,z,t)e^{i(\omega t-kz)} + c.c\,. \qquad (34)$$

We impose the linear dispersion relation $\omega^2 = c^2k^2 - 4\pi n\alpha\omega^2$. We equate coefficients of all terms in $e^{i(\omega t-kz)}$ (in principle one should go on and obtain a coupled sequence of equations for the complex envelopes of the harmonics). We find

$$\varepsilon_{xx} + \varepsilon_{yy} + \omega^2 c^{-2}\, 12\pi n\alpha_{NL}|\varepsilon|^2\varepsilon + 2\, i\, c^{-2}[\omega\varepsilon_t(1 + 4\pi n\alpha) - c^2k\varepsilon_z] = 0 \qquad (35)$$

We look for steady state solutions (ε does not depend on t). By suitable scaling

$$\varepsilon_{xx} + \varepsilon_{yy} + 2\,|\varepsilon|^2\,\varepsilon - i\,\varepsilon_z = 0 \qquad (36)$$

which is the 2-dimensional NLS equation in (x,y) and a 'time' (z).

In one space dimension (x) the NLS (36) has the 1-soliton solution

$$\varepsilon(x,z) = \frac{2\eta\,\exp\{4i(\xi^2 - \eta^2)z + 2\,i\,\xi x + i\delta\}}{\cosh[2\eta(x - x_0) - 8\eta\xi z]} \qquad (37)$$

It contains a carrier which corrects the linearised carrier in (35). Solitons (37) are typical of the so-called envelope solitons. This particular one can be seen in experiments on deep water [25]. As an electric field it carries intensity $4\eta^2\,\mathrm{sech}^2[2\eta(x - x_0) - 8\eta\xi z]$. This is the intensity across a wave-guide like channel induced in the medium by the intense field making an angle $-\tan^{-1}4\xi$ to the z-axis. An arbitrary laser profile $\varepsilon(x,0)$ breaks up into a number of such soliton channels as z increases.

The NLS is special amongst the key equations (22) - (25) in that it describes a single complex field. A consequence is that the speed (given by ξ in (37)) does not determine the amplitude (given by 2η). However in II we shall see $(\xi + i\eta)$ is an important eigenvalue of a related linear scattering problem: in the case when this eigenvalue is purely imaginary, as may be the case for the s-G and modified KdV for example, both speed and amplitude are determined by the number ξ.

To order E^3 (32) and (33) are the ϕ-four equation (13) for the real field $\phi \equiv E$. The ϕ-four can always be approximated by the NLS. Notice that the driving term of (32) and (33) is $\alpha E + \alpha_{NL} E^3 \to + \phi + \phi^3$. The linear term does not matter since it is removed by the linear dispersion relation. The sign of the ϕ^3 matters since $-\phi^3$ changes (24) to $i\, u_t + u_{xx} - 2u|u|^2 = 0$. This equation has a solitary wave solution but no multisoliton solutions.

The envelope solitons (37) look rather like breathers. The ϕ-four has no breather but the associated envelope solitons of the NLS can play this role. The NLS also has its own breather solution and breathers can modulate carriers. The NLS is in one sense a stopping point: if one looks for envelope solutions $U\, e^{i(\omega t - kx)}$ of the NLS one finds U satisfies a NLS equation! The NLS turns up as the archetypical equation describing weakly non-linear strongly dispersive systems: the KdV describes weakly non-linear weakly dispersive systems.

The second application is more complicated. We consider first a collection of spin $\frac{1}{2}$ systems ($n\ cc^{-1}$) in a constant magnetic field B_0 along $-z$. We use the gyromagnetic ratio $\gamma\hbar = e\hbar m_e^{-1} c^{-1}$ (g-factor = 2). The Larmor frequency is $\omega_L = \gamma B_0$. The $m_S = \pm\frac{1}{2}$ states have energies $\pm\frac{1}{2}\hbar\omega_L$. An inhomogeneous transverse R.F. field $B_x(\underset{\sim}{x},t)$ flips the spins and these flips propagate as spin waves. From the Hamiltonian density

$$H(\underset{\sim}{x}) = \tfrac{1}{2}\hbar\omega_S \sigma_z(\underset{\sim}{x}) - \gamma\hbar\, \sigma_x\, B_x \tag{38}$$

and commutation relations $\underset{\sim}{\sigma} \times \underset{\sim}{\sigma} = 2i\hbar\underset{\sim}{\sigma}\, \delta(\underset{\sim}{x} - \underset{\sim}{x}')$ one easily finds

$$\dot{\underset{\sim}{\sigma}} = \underset{\sim}{\omega} \times \underset{\sim}{\sigma},\ \underset{\sim}{\omega} = (-\gamma B_x, 0, \omega_L)\ . \tag{39}$$

We define the Bloch vector density $\underset{\sim}{r}(\underset{\sim}{x},t)$ by the expectation value

$$\underset{\sim}{r}(\underset{\sim}{x},t) = \langle\underset{\sim}{\sigma}(\underset{\sim}{x},t)\rangle\ .$$

Then

$$\dot{\underset{\sim}{r}} = \underset{\sim}{\omega} \times \underset{\sim}{r}\ . \tag{40}$$

The transverse magnetic dipole is $\frac{1}{2}\gamma\hbar\, r_1(\underset{\sim}{x},t) = \frac{1}{2}\gamma\langle\sigma_x(\underset{\sim}{x},t)\rangle$ and Maxwell's equations are

$$\nabla^2 B_x - c^{-2} B_{x,tt} = 4\pi n c^{-2}\tfrac{1}{2}\gamma r_{1,tt}\ . \tag{41}$$

With the choice of ω in (39) equations (40) and (41) form a system of coupled non-linear partial differential equations governing the non-linear propagation of spin waves. We call this the Bloch-Maxwell (BM) system. It is linearised by noting that $r_{1,t} = -\omega_L r_2$, $r_{2,t} = \omega_L r_1 + \gamma B_x r_3$, and for constant inversion r_3 $r_{1,tt} = -\omega_L^2 - \omega_L\gamma B_x r_3$ - the usual pseudo-Bose system.

A second equivalent problem arises in the propagation of 10^{-9}sec. optical pulses through media with a resonant non-degenerate transition. We take n 2-level atoms cc^{-1} each with the resonant frequency ω_S. Pulses are envelopes modulating carriers of frequency $\omega \approx \omega_S$. This justifies the 2-level atom approximation. This atom is a 2-state system with a spin representation: spin up (down) is occupation of the upper (lower) state. The equations take precisely the BM form with [6,22]

$$\underset{\sim}{\omega} = (-2p\hbar^{-1}E(\underset{\sim}{x},t), 0, \omega_S)\ . \tag{42}$$

The electric field replaces the magnetic field in the Maxwell equation

$$\nabla^2 E - c^{-2} E_{tt} = 4\pi c^{-2} p\ r_{1,tt} \tag{43}$$

(the electric dipole proves to be $pr_1 = p\langle\sigma_x\rangle$ with p the dipole matrix element).

A third equivalent problem arises in the Josephson junction of large area (such a junction is typically $\lesssim$ 1 mm. long). The junction is a 2-state system (the two sides of the junction). If a voltage V is applied between these two sides the two states differ in energy by 2eV (2e is the charge of the Cooper pair). The sides will couple by some tunneling parameter K (say). The junction equation will be (40) [6] with

$$\underset{\sim}{\omega} = (2K\hbar^{-1},\ 0,\ 2eV\hbar^{-1}) \ . \qquad (44)$$

Consider the plane Josephson sandwich with superconductors top and bottom separated by a thin uniform layer of oxide of effective thickness (including penetration depth) d. If V is applied across the oxide, a field $E = d^{-1}V$ exists there in the direction of the normal (called z) to the plane of the junction (the x-y plane). Consider plane E-waves propagating along x and carrying a magnetic field B along y.

The quantity $r_3 \equiv \langle\sigma_z\rangle$ is the difference in occupation number densities either side of the junction. The current density component j_z along z is thus $j_z = \beta\, r_{3,t}$ (for some constant β). This drives the (transverse) wave equation in the usual way:

$$E_{xx} - \bar{c}^{-2} E_{tt} = 4\pi\beta\bar{c}^{-2}k^{-1}r_{3,tt} \ . \qquad (45)$$

We are in a material medium so $\bar{c} = ck^{-\frac{1}{2}}$. This arises through the displacement current $C\,V_{,t}$: $C = k/4\pi d$ is the capacitance of the junction per unit length.

There are two differences from the two BM systems considered previously: r_3 rather than r_1 drives the wave equation, and V not K will depend on t in $\underset{\sim}{\omega}$ (we do not argue here that K should be more complicated: our Bloch equations (40) follow FEYNMAN [26]. See [6].). We reach the BM problem with the simple switch of components $1\leftrightarrow 3$, $2\leftrightarrow 2$!

The BM system

$$E_{xx} - \bar{c}^{-2} E_{tt} = 4\pi\beta\bar{c}^{-2}k\ r_{3,tt} \qquad (46a)$$

$$\underset{\sim}{r}_{,t} = \underset{\sim}{\omega} \times \underset{\sim}{r} \quad (\underset{\sim}{\omega} = (2K\hbar^{-1},\ 0,\ 2ed\ E\hbar^{-1})) \qquad (46b)$$

does not have multisoliton solutions [6]. The reduced Maxwell-Bloch system which replaces (46a) by the one-way going (single characteristic)wave equation $E_x + \bar{c}^{-1}E_t = 2\pi\bar{c}^{-1}\beta k^{-1}r_{3,t}$ can be scaled to (27) [22,27] and does have multisoliton solutions. The elimination of the backward going characteristic is admissible in the short optical pulse problem because typically one is concerned with the very low densities of metal vapours ($n \sim 10^{11}\ cc^{-1}$). For the large area junction the numbers scarcely permit this approximation (although the multisoliton RMB behaviour will come through approximately).

In the Josephson junction problem one usually assumes instead that $r_3 \approx 0$ everywhere (little charge imbalance on the two sides). Then the Bloch equation (46b) becomes

$$r_{2,t} = 2ed\hbar^{-1} E\ r_1$$
$$r_{1,t} = -\ 2ed\hbar^{-1} E\ r_2 \qquad (47)$$

with solution

$$r_2 = -\sin\sigma,\quad r_1 = -\cos\sigma$$
$$\sigma = 2ed\hbar^{-1}\int_{-\infty}^{t} E(x,t')\ dt' \ . \qquad (48)$$

Thus

$$\sigma_t = 2ed\hbar^{-1}E\ ,\ \text{or}\ \sigma_t = 2eV\hbar^{-1}$$

$$j_z = -\ 2\beta K\hbar^{-1}\ r_2\ ,\ \text{or}\ j_z = j_{zo}\sin\sigma \ . \qquad (49)$$

In their forms on the right these equations are Josephson's two equations: σ is the Josephson phase.

We have still to satisfy Maxwell's equation (46a). After one integration all through by time this becomes

$$\sigma_{xx} - \bar{c}^{-2}\sigma_{tt} = \lambda_0^{-2} \sin \sigma \quad . \tag{50}$$

This is the s-G scaled by the natural length

$$\lambda_0 = \{k\hbar\bar{c}^2/8\pi\beta K\}^{\frac{1}{2}} \quad . \tag{51}$$

It is also possible to obtain the s-G from the optical problem. We look for plane wave envelope solutions modulating resonant carrier waves by setting

$$E(x,t) = \hbar p^{-1}\, \varepsilon(x,t) \cos\{\omega_s(t - c^{-1}x)\}$$

$$r_1(x,t) = Q(x,t) \cos\{\omega_s(t - c^{-1}x)\} + P(x,t) \sin\{\omega_s(t - c^{-1}x)\} \quad . \tag{52}$$

One finds by using the fact that P, Q, ε vary on a 10^{-9}sec. time scale and $\omega_s \sim 10^{15}$ that $Q \approx 0$ and

$$P_{,t} = \varepsilon N$$

$$N_{,t} = - \varepsilon P \tag{53}$$

(where $N \equiv r_3(x,t)$) whilst

$$\varepsilon_x + c^{-1}\, \varepsilon_t = \alpha\, P \tag{54}$$

$(\alpha = 2\pi\, p^2 n\, \omega_s\, \hbar^{-1}\, c^{-1})$.

An "attenuator" is an initially unexcited medium. For this (53) is solved by

$$P = - \sin \sigma$$

$$N = - \cos \sigma$$

$$\sigma(x,t) = \int_{-\infty}^{t} \varepsilon(x,t')\, dt' \quad . \tag{55}$$

From (55), (54) is

$$\sigma_{xt} + c^{-1}\, \sigma_{tt} = - \alpha \sin \sigma \quad . \tag{56}$$

This is the s-G in unusual independent variables. Set

$$\sqrt{c}\, \xi = \sqrt{\alpha}\, (ct - 2x)\ , \quad \eta = \sqrt{\alpha c}\; t \quad . \tag{57}$$

Then

$$\sigma_{\xi\xi} - \sigma_{\eta\eta} = \sin \sigma \quad . \tag{58}$$

The crucial idea here is to exploit the resonance condition, and the approximation is very different from the assumption $r_3 \approx 0$ for the junction problem (r_3 is of course r_1 in the optical problem and this goes over to the out of phase component P of the dipole).

Both the s-G (58) for optical pulses and the s-G (50) for the large area junction have the multisoliton solutions (16) - (20). In the optical problem an arbitrary intensity envelope ($\propto \varepsilon^2$) breaks up in general into a train of sech^2 pulses. Each of these is the time derivative in the x,t coordinate system of a 2π-kink. The sech^2

pulses are solitons: they persist and the medium is transparent to them. The phenomena is therefore called self-induced transparency (SIT) which explains this phenomena introduced in §4. In the (x,t) system the 1-soliton is

$$\varepsilon(x,t) = p\, E_0\, \hbar^{-1} \operatorname{sech} \tfrac{1}{2} p\, E_0\, \hbar^{-1}(t - xv^{-1}) \,. \tag{59}$$

It is a soliton solution of the so called SIT equations [22]. The "area" $\Theta(x,t) \equiv \sigma(x,\infty)$ is independent of x:

$$\Theta(x,t) = \int_{-\infty}^{\infty} \varepsilon(x,t')dt' = [4 \tan^{-1} e^{\xi}]_{-\infty}^{\infty} = 2\pi \,. \tag{60}$$

A pulse of arbitrary area Θ entering the resonant medium undergoes a jump in Θ at the boundary of the medium to the value $2\nu\pi$ where ν is an integer. The pulse then reshapes in the medium to ν_1 2π-sech pulses (ν_1 kinks in σ), ν_2 2π-sechs of opposite sign (ν_2 antikinks in σ) and ν_3 breathers. We have $\nu_1 - \nu_2 = \nu$. Details are given in [22,6] and the references in [6].

The theory of the Josephson junction shows similar features in a slightly different way. In the rest frame the kink solution of (50) is

$$\sigma = 4 \tan^{-1} \exp[(x - x_0)/\lambda_0] \quad . \tag{61}$$

Use $\sigma_{tx} = \sigma_{xt} = 2e\hbar^{-1} V_{,x} = 2ed\hbar^{-1}c^{-1}B_{,t}$ to reach

$$\sigma_x = 2ed\, \hbar^{-1}c^{-1}\, B \tag{62}$$

(B is the y-component of the magnetic field). Then

$$2\pi = \sigma(\infty,t) - \sigma(-\infty,t) = 2ed\, \hbar^{-1}c^{-1} \int B\, dx = 2e\, \hbar^{-1}c^{-1} \int B d\, S \,. \tag{63}$$

Thus the kink carries one unit of flux hc/2e - the single "fluxon". The boundary conditions on a large junction will be $\sigma(\infty,t) = 2\nu_1\pi$, $\sigma(-\infty,t) = 2\nu_2\pi$ in general, and the total flux is $(\nu_2 - \nu_1)$ hc/2e. The sum of the numbers of kinks and antikinks is therefore $(\nu_2 - \nu_1)$. The breathers carry no flux.

The break up of optical pulses into their constituent solitons has been observed [28,6] Evidence of the existence of kinks in large area Josephson junctions has been obtained by looking at changes in the voltage/current characteristics [29]. Extra current spikes have appeared with particular voltage spacings [29]. In frequency units this spacing is just the fundamental even mode frequency of the Josephson equivalent cavity. Ref. [27] argues that a kink which reaches the end of an open ended cavity is reflected as an antikink. There is a natural cavity "mode" consisting of a kink with its antikink. There is therefore one node and the mode corresponds in that respect to a harmonic cavity mode with twice the fundamental frequency. Successive kink-antikink pairs induce current spikes (or current steps) just as radiation induces current steps in small area Josephson junctions. The observations [29] are in extraordinarily good agreement with this argument. However the s-G has not been solved for open-ended or close-ended boundary conditions on finite support $-\tfrac{1}{2}L \leq x \leq \tfrac{1}{2}L$. All the solutions we have described in this paper I are for the real line $-\infty < x < \infty$. Certain equations, notably the KdV (21) have been solved for periodic boundary conditions $u(0,t) = u(L,t)$, $u_x(0,t) = u_x(L,t)$, etc. [30,8].

We note finally a further application of the spin wave theory of this §5. This is to spin waves in the Fermi liquid ^{3}He below 2.6 mK. This liquid has two phases, the A-phase and the B-phase. The A-phase can be thought of as two interpenetrating superfluids carrying respectively 'up' spins and 'down' spins coupled by the very weak spin dipole interaction. Spin waves therefore satisfy the s-G. The B-phase admits the symmetric up-down states (the spin is unity not zero as in a superconductor). In consequence spin waves satisfy the double s-G

$$\sigma_{xx} - \sigma_{tt} = -(\sin\sigma + \tfrac{1}{2} \sin \tfrac{1}{2}\sigma) \,. \tag{64}$$

The evidence supports the view that this equation does not have true soliton solutions but its kink solutions have remarkable properties nevertheless. Equation (64) with positive sign on the right governs SIT with a hyperfine degeneracy quantum number $F = F' = 2$ on each of the two levels. The wobbling 4π-pulse solutions of (64) in this case have been observed [32]. The ^3He problem is treated in [33]. Both this and the degenerate SIT problem are studied in [6].

This completes our introduction to the soliton concept and its application to physical problems. The reader is referred to [6] and the references there for more details. In the lecture II which now follows we set up some considerably more powerful mathematical machinery. The list of references follows II.

Before we actually go on to II however it seems appropriate to offer the reader at least one illustrative diagram. The Fig. 1 which follows shows the break up of an optical pulse of area $\Theta = 8\pi$ (compare (60)) into two of the wobbling 4π-pulses. The optical medium is degenerate with the two hyperfine quantum numbers $F = F' = 2$ and the equation involved is precisely (64) with however the positive sign on the right side.

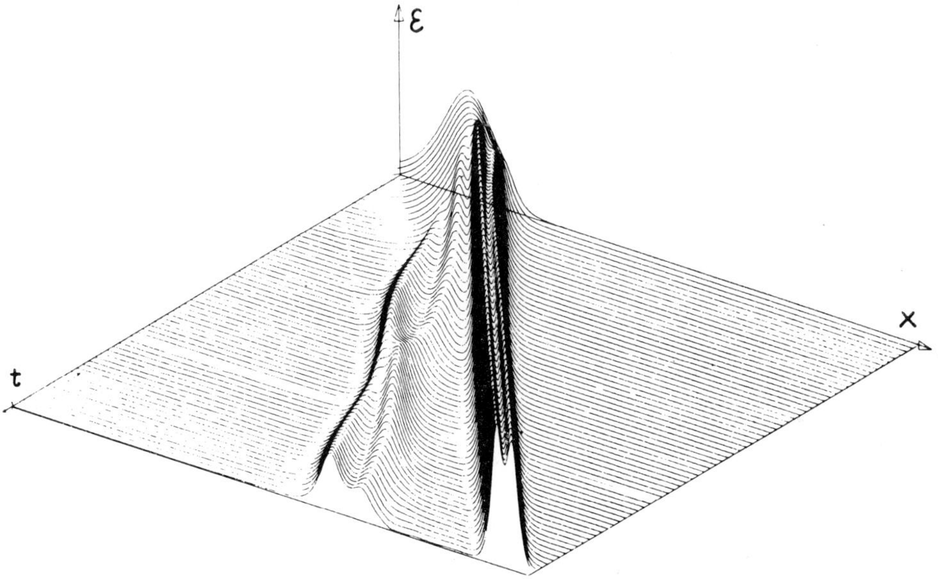

Fig.1 Break up of 8π optical pulse in a degenerate medium.

Solitons
II. Mathematical Structures

R. K. Bullough and R. K. Dodd

1. Introduction

The history of this subject is that GARDNER, GREENE, KRUSKAL and MIURA [34] suggested the inverse scattering scheme of solution for the KdV (22) in 1967, ZAKHAROV and SHABAT (ZS) [35] solved the NLS (24) in 1971, and WADATI and TODA [36] used the method of [34] to find the explicit solution of the KdV whilst WADATI [37] used the method of ZS [35] to solve the modified KdV (23) in 1972. Then HIROTA [38] and CAUDREY, GIBBON, EILBECK and BULLOUGH [39] found the multisoliton solutions of the s-G in 1972-73 and these were rapidly followed by the inverse scattering solution by ABLOWITZ, KAUP, NEWELL and SEGUR (AKNS) [40]. AKNS then gave a generalised ZS inverse scattering scheme [16]. Most of the work of this paper II is concerned with the AKNS-ZS scheme.

2. Bäcklund Transformations and Other Properties

The "Bäcklund transformation" for the s-G

$$\sigma_{xt} = \sin\sigma \tag{65}$$

was discovered in 1883 [41]. If σ is a solution of (65) then a second solution σ' is obtained from

$$\begin{aligned} \sigma'_x &= \sigma_x + 2k \sin \tfrac{1}{2}(\sigma' + \sigma) \\ \sigma'_t &= -\sigma_t + 2k^{-1} \sin \tfrac{1}{2}(\sigma' - \sigma) \end{aligned} \tag{66}$$

in which k is a real parameter. From (66)

$$\sigma'_{xt} = \sigma_{xt} - \sin\sigma' - \sin\sigma$$

so σ a solution of (65) means σ' a solution of (65). A particular solution is $\sigma = 0$ so that

$$\tfrac{1}{2}\sigma'_x = k \sin \tfrac{1}{2}\sigma' \;, \quad \tfrac{1}{2}\sigma'_t = k^{-1}\sin \tfrac{1}{2}\sigma'$$

$$\sigma' = 4 \tan^{-1} e^{(kx + k^{-1}t)} \;. \tag{67}$$

This is the 2π-kink for the s-G (65) in characteristic co-ordinates. By using this kink for σ one obtains the 4π-kink corresponding to (17) for σ'. In general (66) adds one soliton to σ.

We now know that the KdV (22), the modified KdV (23), the NLS (24), the s-G (25)(or (65)), and a number of other equations with soliton solutions have Bäcklund transformations (BTs). These are all auto-Bäcklund transformations (aBTs) transforming an equation to itself. There are also BTs which transform between two different equations. BTs do not imply soliton solutions. The Hopf-Cole transformation which linearises the Burgers equation (7) is a BT: the BT

$$u'_x = -(2\nu)^{-1}\, u\, u'$$

$$u'_t = -(4\nu)^{-1}(2\nu\, u_x - u^2)u' \tag{68}$$

maps solutions u of the Burgers equation

$$u_t + uu_x - \nu u_{xx} = 0 \qquad (\nu>0) \tag{69a}$$

into the heat equation

$$u'_t - \nu u'_{xx} = 0 \quad ; \tag{69b}$$

the transformation $u = -2\nu\{\ln u'\}_x$ is called the Hopf-Cole transformation and linearises (69a) as (69b). Since the initial value problem can be solved for (69b) it can be solved for (69a). The BT (66) "linearises" (65) in a much more subtle way: the linearisation is the linearisation of the inverse scattering method as we show below.

Four solutions of (65) are related by the consistency condition

$$\tan \tfrac{1}{4}(\sigma_3-\sigma_0) = \frac{k_2 + k_1}{k_2 - k_1} \tan\{\tfrac{1}{4}(\sigma_2-\sigma_1)\} \ . \tag{70}$$

It is therefore possible to use (70) as a superposition principle building up N-soliton solutions from 1-soliton solutions. Note that (70) is an algebraic relation and it is not necessary to solve even a first order differential equation like (66) once three properly related solutions are available (these three could be $\sigma_0 = 0, \sigma_1$ and σ_2 2π-kinks (67) with parameters k_1 and k_2,but in any case they must be σ_0 and its transform σ_1 with parameter k_1 and σ_2 with parameter k_2).

An example of a BT which is not an aBT is also

$$\sigma'_x = k^{-1} \sin(\sigma'-\sigma)$$

$$\sigma'_t = \sigma_t + k \sin \sigma' \tag{71}$$

between the s-G (65) and

$$\sigma'_{xt} = \{1 - k^2\sigma'^2_x\}^{\frac{1}{2}} \sin \sigma' \ . \tag{72}$$

From this BT we can derive an infinite set of conserved quantities for the s-G. All the other equations (22), (23), (24), etc. have infinite sets of conserved quantities and this is a feature of all the AKNS-ZS systems (see below).

Any relation $\{X(\sigma')\}_x + \{T(\sigma')\}_t = 0$ connecting the functionals $X(\sigma')$ and $T(\sigma')$ is a conservation low for σ' : $X(\sigma')$ is a flux and $T(\sigma')$ is a density. For the boundary conditions assumed here ($\sigma, \sigma_x, \sigma_{xx}$, etc. $\to 0$ as $|x| \to \infty$ or $\sigma \to 0$ (mod 2π) for the s-G, for example) $\int_{-\infty}^{\infty} \{X(\sigma')\}_x\, dx = [X(\sigma')]_{-\infty}^{+\infty} = 0$ so that

$$\frac{d}{dt}\int_{-\infty}^{\infty} T(\sigma')\, dx = 0$$

and $\int_{-\infty}^{\infty} T(\sigma')dx$ is a constant of the motion.

A conservation law for (72) is

$$\{(1 - k^2\sigma'^2_x)^{\frac{1}{2}}\}_t - k^2\{\cos \sigma'\}_x = 0 \ . \tag{73}$$

To find from this the conserved quantities of the s-G (65) write

$$\sigma' = \sum_{n=0}^{\infty} f_n(\sigma)\, k^n \ . \tag{74}$$

Then from (71)

$$\sigma' = \sigma + k\sigma_x + k^2\sigma_{xx} + k^3(\sigma_{xxx} - \frac{1}{3!}\sigma_x^3) + k^4(\sigma_{xxxx} + \sigma_x^2\sigma_{xx}) + \dots \quad . \tag{75}$$

If (75) is inserted into (73) we get, after dropping constants and constant factors, the expression (76) for $T(\sigma')$:

$$\begin{aligned} T(\sigma') = {} & \sigma_x^2 + k(2\sigma_x\sigma_{xx}) + k^2(\sigma_{xx}^2 + 2\,\sigma_x\,\sigma_{xxx} + \tfrac{1}{4}\,\sigma_x^4) \\ & + k^3(2\,\sigma_x\,\sigma_{xxxx} + 2\,\sigma_{xx}\,\sigma_{xxx} + 2\,\sigma_x^3\,\sigma_{xx}) \\ & + k^4(\sigma_{xxx}^2 + 2\,\sigma_{xx}\,\sigma_{xxxx} + 2\,\sigma_x\,\sigma_{xxxxx} + \tfrac{1}{8}\,\sigma_x^6 \\ & \qquad + 3\cdot\sigma_{xxx}\,\sigma_x^3 + \tfrac{13}{2}\,\sigma_x^2\,\sigma_{xx}^2) \\ & + \dots \end{aligned} \tag{76}$$

Since k is arbitrary the following quantities are conserved separately

$$\begin{aligned} & T^2 = \sigma_x^2\ , \quad T^4 = \sigma_{xx}^2 - \tfrac{1}{4}\,\sigma_x^4\ , \quad T^6 = \sigma_{xxx}^2 - \tfrac{5}{2}\,\sigma_{xx}^2\,\sigma_x^2 + \tfrac{1}{8}\,\sigma_x^6\ , \\ & T^8 = \sigma_{xxxx}^2 - \tfrac{7}{2}\,\sigma_{xxx}^2\,\sigma_x^2 + \tfrac{7}{4}\,\sigma_{xx}^4 + \tfrac{35}{8}\,\sigma_{xx}^2\,\sigma_x^4 + \tfrac{5}{64}\,\sigma_x^8\ , \\ & T^{10} \text{ etc.} \end{aligned} \tag{77}$$

The even rank quantities have been 'reduced' in (77) - see [42], the odd rank are perfect differentials and are trivially conserved.

The local conservation law for the density T^2 takes the explicit form

$$\tfrac{1}{4}\gamma^{-1}\{\sigma_x^2\}_t + \tfrac{1}{2}\gamma^{-1}\{\cos\,\sigma-1\}_x = 0 \tag{78}$$

where the constant γ is introduced as a dimensionless coupling constant to be used later. By comparison with the local conservation law $\mathcal{P}_t + \mathcal{H}_x = 0$ connecting momentum $\mathcal{P}$ and energy density $\mathcal{H}$ for the s-G (see however below (123)) we deduce that

$$\mathcal{P} = \tfrac{1}{4}\gamma^{-1}\,\sigma_x^2 \tag{79a}$$

$$\mathcal{H} = \tfrac{1}{2}\gamma^{-1}(\cos\sigma - 1) \quad . \tag{79b}$$

It is very relevant to later work to notice now that $\mathcal{H}$ is a Hamiltonian density: if $q = -\tfrac{1}{2}\sigma_x$ and $p = -2\gamma^{-1}\int_{-\infty}^{x} q\,dx' = \gamma^{-1}\sigma$, Hamilton's equation $q_t = \delta H/\delta p$ (in which $\delta H/\delta p$ is the Frechet or functional derivative of H) is the s-G equation (65). Further in terms of q, $\mathcal{H}$ is $\tfrac{1}{2}\gamma^{-1}\{\cos\text{-}2\int_{-\infty}^{x} q(x')dx' - 1\}$ and $p_t = -\delta H/\delta q = -\gamma^{-1}\int_{x}^{\infty}\{\sin\text{-}2\int_{-\infty}^{x'} q\,dx''\}\,dx'$ so that $\sigma_{tx} = \sin\,\sigma.$[2] Thus $-\tfrac{1}{2}\sigma_x$ and $\gamma^{-1}\sigma$ constitute a pair of canonically conjugate variables for the Hamiltonian density (79b).

This is an example of a fundamental canonical structure we shall return to later. Here we note that (79a) has the physical interpretation of a momentum density and (79b) of an energy density. However the physical interpretation of the higher rank densities T^4, T^6, ... and their associated fluxes is not clear. There is an infinity of the densities T^{2n} (for proof see [42]) each of which is a polynomial in σ_x, σ_{xx}, etc. The KdV, modified KdV, NLS, Hirota and other equations also have an infinity of polynomial conserved densities. The polynomial character may just be a feature of the AKNS-ZS systems however (it follows from the AKNS-ZS inverse scattering scheme - see (99) below):

---

2 The final form for $\delta H/\delta q$ is achieved by evaluating δH for variation δq and performing a change of order of the double integral involved.

the s-G in Lorentz invariant form (9) is not an AKNS-ZS system; it has an infinity of conserved quantities but these are not polynomials [42].

Intuitively the conserved quantities 'explain' the remarkable stability of soliton solutions. In practice it is not known whether their existence is even a necessary condition for solitons: certainly the distribution of the T^n may be neither every integer n nor every even integer 2n [43]. The condition is also not sufficient: for consider the simple wave (3) which is equivalent to

$$\{6(n+1)^{-1}\, u^{n+1}\}_x + \{n^{-1}\, u^n\}_t = 0 \quad \text{for} \quad n = 1, 2, \ldots .$$

We have however proved[3] two theorems [5,42] which in essence say that the generalised s-G $\sigma_{xt} = F(\sigma)$ has an infinity of polynomial conserved densities and has an aBT if and only if $F''(\sigma) + \alpha^2 F(\sigma) = 0$ for some complex constant α. This means that the s-G (65) has both whilst the double s-G (compare (64)) $\sigma_{xt} = \sin\sigma + \frac{1}{2}\sin\frac{1}{2}\sigma$ has neither. This does not necessarily mean the double s-G does not have soliton solutions. However our present conclusion is that it does not [33]. Two physical applications of the double s-G are referred to at the end of I. This seems to be a situation where the 'approximate' soliton solutions of a non-linear wave equation have important physical applications (compare the discussion of the ϕ-four eqn. in I §2).

An important consequence of the existence of an aBT is the following: in the aBT (66) set $\Gamma = \tan\frac{1}{4}(\sigma' + \sigma)$ to reach

$$\Gamma_{,x} = \tfrac{1}{2}\sigma_x(1 + \Gamma^2) + k\Gamma$$

$$\Gamma_{,t} = k^{-1}\Gamma\cos\sigma - (2k)^{-1}(1 - \Gamma^2)\sin\sigma . \tag{80}$$

By the Ricatti transformation $\Gamma = v_2/v_1$ one then obtains the linearised equations

$$v_{1,x} + \tfrac{1}{2}\,\sigma_{,x}\, v_2 = -\tfrac{1}{2}k\, v_1$$

$$-\tfrac{1}{2}\,\sigma_{,x} v_1 + v_{2,x} = \tfrac{1}{2}k\, v_2 \quad ; \tag{81a}$$

$$v_{1,t} = -\,(v_1/2k)\cos\sigma - (v_2/2k)\sin\sigma$$

$$v_{2,t} = -\,(v_1/2k)\sin\sigma + (v_2/2k)\cos\sigma . \tag{81b}$$

Eqn. (81a) is a linear scattering problem in which $\frac{1}{2}\sigma_x$ acts as a scattering potential and $-\frac{1}{2}i\,k$ is an eigenvalue. This can be seen as follows: the more general problem

$$\hat{L}v = \zeta v$$

$$\hat{L} = i\begin{bmatrix} \partial/\partial x & -q(x,t) \\ r(x,t) & -\partial/\partial x \end{bmatrix} , \quad v = \begin{bmatrix} v_1(x,t) \\ v_2(x,t) \end{bmatrix} \tag{82}$$

with ζ a complex eigenvalue maps into (81a) by $\zeta = -\frac{1}{2}i\,k, q = -\frac{1}{2}\sigma_x = -\,r$. On the other hand for $r \equiv -1$ in (82) v_2 satisfies the Schrödinger eigenvalue problem $-v_{2,xx} - q\,v_2 = \zeta^2\, v_2$ in which the scattering potential is $-q(x,t)$. Notice that this scattering potential depends on t so that in general the eigenvalues ζ^2 do so also. The same is true of the eigenvalues ζ of (82).

We generalise the pair of equations (81b) to

$$\hat{A}v = v_{,t} \quad , \quad \hat{A} = \begin{bmatrix} A(x,t) & B(x,t) \\ C(x,t) & -A(x,t) \end{bmatrix} \tag{83}$$

---

[3] The theorem on aBTs was first proved by McLaughlin and Scott as referenced in [5].

Under the condition ζ is independent of t the commutator relation

$$[\hat{L}, \hat{A}]v \equiv (\hat{L}\hat{A} - \hat{A}\hat{L})v = \hat{L}\, v_{,t} - \zeta\, v_{,t} = -\hat{L}_t\, v, \tag{84}$$

or

$$L_t = [\hat{A}, \hat{L}] , \tag{85}$$

holds. The operator equation (85) is actually a pair of non linear evolution equations (NEEs). In the case of (81) we find

$$i\begin{bmatrix} 0 & \tfrac{1}{2}\sigma_{xt} \\ \tfrac{1}{2}\sigma_{xt} & 0 \end{bmatrix} = \frac{-\sin\sigma}{2\zeta}\begin{bmatrix} 0 & 1 \\ 1 & 0 \end{bmatrix}\begin{bmatrix} \partial/\partial x & \tfrac{1}{2}\sigma_x \\ \tfrac{1}{2}\sigma_x & -\partial/\partial x \end{bmatrix} = i\begin{bmatrix} 0 & \tfrac{1}{2}\sin\sigma \\ \tfrac{1}{2}\sin\sigma & 0 \end{bmatrix} \tag{86}$$

and (85) is the original NEE. We now have an important result: the system (82) and (83) together with the condition $\zeta_t = 0$ is equivalent to the NEEs (85). The general form (85) as a single operator equation for the KdV equation was first given by Lax [44].

Of course the equivalence of (82) and (83) with $\zeta_t = 0$ to the NEEs (85) is useless unless we can solve these coupled systems. Ultimately this must require the discovery of the potentials q(x,t) and r(x,t). Thus (82) is to be solved inversely. This is the origin of "the inverse scattering method".

3. The Inverse Scattering Method

For given initial data q(x,0), r(x,0) we can solve the direct scattering problem (82). It is convenient to introduce the idea of scattering data. Consider the two Jost function solutions of (82) $\psi(x,t)$ and $\bar{\psi}(x,t)$. Set $\xi \equiv \mathrm{Re}\,\zeta$, $\eta \equiv \mathrm{Im}\,\zeta$. Since q, r $\to 0$ as $x \to \pm\infty$ (the boundary condition)

$$\psi \sim \begin{pmatrix} 0 \\ 1 \end{pmatrix} \exp i\,\xi\, x\,, \quad \bar{\psi} \sim \begin{pmatrix} 1 \\ 0 \end{pmatrix} \exp -i\,\xi\, x \tag{87}$$

as $x \to +\infty$. Correspondingly there is a pair of Jost functions $\bar{\phi}$, ϕ asymptotic to the same forms as $x \to -\infty$. Since there can only be one linearly independent pair we have $\phi = a\,\bar{\psi} + b\,\psi$, $\bar{\phi} = \bar{b}\,\bar{\psi} + \bar{a}\,\psi$ where a, b, $\bar{a}$, $\bar{b}$ depend only on ξ (and possibly t). Two solutions of (82) with row vectors $\tilde{u} = (u_1,u_2)$, $\tilde{v} = (v_1,v_2)$ have Wronskian $W(u,v) \equiv u_1v_2 - u_2v_1$ independent of x. It is easy to show $a = W(\phi,\psi)$, $b = -W(\phi,\bar{\psi})$, $\bar{a} = W(\bar{\phi},\bar{\psi})$, $\bar{b} = W(\bar{\phi},\psi)$ and $a\bar{a} - b\bar{b} = 1$.[4] The quantities a and b can be analytically continued into the upper half ζ-plane, $\bar{a}$ and $\bar{b}$ into the lower half ζ-plane. From the asymptotic behaviour of ϕ, ψ and $\bar{\psi}$, a^{-1} is the transmission coefficient and $ba^{-1} \equiv w$ is the reflexion coefficient for the wave starting as $\bar{\psi}$ at $x = +\infty$. The set $\{a,\bar{a},b,\bar{b}\}$ (of which one quantity is redundant) constitutes a part of the scattering data.

The Jost solutions of (82) are unbound scattering states. In addition there are bound states in general. These occur at the roots of $a(\zeta) = 0$ in the upper half of the complex ζ-plane: here $\phi = b(\zeta)\psi$, there is no 'input' wave, and the transmitted and reflected waves become a single bound state wave function. Typically the roots ζ_j of $a(\zeta) = 0$ have a finite imaginary part and both ϕ and ψ are exponentially damped as $|x| \to \infty$. Similar remarks apply to the roots of $\bar{a}(\zeta) = 0$ in the lower half ζ-plane.

It proves convenient to write $b(\zeta_j) = i\,k_j\,\dot{a}(\zeta_j)$; $\bar{b}(\bar{\zeta}_j) = -i\,\ell_j\,\dot{\bar{a}}(\zeta_j)$ $(\dot{f}(\zeta_j) \equiv (df/d\zeta)_{\zeta=\zeta_j})$. The set S of scattering data is then defined by

$$S = \{\zeta_j,\bar{\zeta}_s,k_j,\ell_s,a(\xi),\bar{a}(\xi),w(\xi),\bar{w}(\xi);\ \xi \in R;$$
$$j = 1, 2, \ldots, M;\quad s = 1, 2, \ldots, N\}.$$

[4] Our choice of signs in the asymptotic behaviour $\bar{\phi} \sim \begin{pmatrix} 0 \\ -1 \end{pmatrix} \exp i\zeta x$ as $x \to -\infty$ differs from that of [16] and [17] so that the minus sign multiplies $b\bar{b}$ in the normalisation condition. Note that $|a|$ and $|b|$ are unbounded.

Certain symmetries usually exist: for example if $r = -q$, $M = N$, $\bar{a} = a(-\zeta)$, $-\bar{b} = b(-\zeta)$, $\bar{\zeta}_k = -\zeta_k$, $\bar{b}(\bar{\zeta}_k) = -b(\zeta_k)$; if $r = -q^*$, $\bar{a}(\zeta) = a^*(\zeta^*)$, $\bar{b}(\zeta) = b^*(\zeta^*)$, $\bar{\zeta}_k = \zeta_k^*$, $\bar{b}(\bar{\zeta}_k) = \{b(\zeta_k)\}^*$. In the case of the s-G both conditions obtain since $q = -r$, and q is real. The bound state eigenvalues therefore arise on the imaginary ζ-axis ($\zeta_k = i\eta_k$, $\bar{\zeta}_k = -i|\bar{\eta}_k|$) or they arise as pairs $(\zeta_k, -\zeta_k^*)$ with $(\bar{\zeta}_k, -\bar{\zeta}_k^*) = (\zeta_k^*, -\zeta_k)$. The bound state eigenvalues $\zeta_k = i\eta_k$ determine the speeds and amplitudes of the 2π-kinks; the pairs $(\zeta, -\zeta^*)$ determine the breathers; the continuous part of the spectrum with eigenvalues ξ determines the background.

From the scattering data at $t = 0$ one can find the scattering data at the current time $t>0$. Conditions on A, B and C in (83) are $B,C \to 0$, $A \to \Omega(\zeta)$ as $|x| \to \infty$: for example $A = -\frac{1}{2k} = -\frac{1}{4i\zeta}$ and $B = C = 0$ when $\sigma = 0$ in (81b). Thus $\Omega(\zeta) = -\frac{1}{4i\zeta}$ for the s-G (65) and this is essentially the linearised dispersion relation. The result is generally true for AKNS systems [17,45]. For $x \to \infty$ we therefore have

$$(\hat{A} - \Omega(\zeta)I)\phi \sim \begin{bmatrix} 0 & 0 \\ 0 & -2\Omega(\xi) \end{bmatrix} \begin{bmatrix} a(\xi)\, e^{-i\xi x} \\ b(\xi)\, e^{i\xi x} \end{bmatrix} = \begin{bmatrix} a_{,t}(\xi)\, e^{-i\xi x} \\ b_{,t}(\xi)\, e^{i\xi x} \end{bmatrix} . \tag{88}$$

We can add $-\Omega(\zeta)I$ (I is the unit matrix) to $\hat{A}$ in (83) since this does not change the NEE (85) and indeed this choice is actually forced upon us by the need to satisfy the boundary conditions like (87) for the Jost functions. By this choice a is independent of t and $b(\xi,t) = b(\xi,0)e^{-2\Omega(\xi)t}$. A similar argument for the bound states yields $b(\zeta_k,t) = b(\zeta_k,0)e^{-2\Omega(\zeta_k)t}$. Thus for the s-G $b(i\eta_k,t) = b_0\, e^{-t/2\eta_k}$ (with $\eta_k > 0$). Notice, that equations (88) can be trivially integrated depends critically on the fact that $q,r \to 0$ as $x \to \infty$: in (82) q,r alone depends on t and the eigenvalues are by choice time invariant. The vanishing of q and r asymptotically underpins the whole of the inverse scattering method.

Now comes the crux of the argument: from the scattering data at time t we can infer the potentials q, r at time t. This finally solves the initial value problem for the coupled NEEs (85). The potentials are obtained from the generalised Gel'fand-Levitan-Marchenko equations

$$\bar{\underset{\sim}{K}}(x,y) + \begin{pmatrix} 0 \\ 1 \end{pmatrix} F(x+y) + \int_x^\infty \underset{\sim}{K}(x,z)\, F(z+y)dz = 0$$

$$\underset{\sim}{K}(x,y) + \begin{pmatrix} 1 \\ 0 \end{pmatrix} \bar{F}(x+y) + \int_x^\infty \bar{\underset{\sim}{K}}(x,z)\, \bar{F}(z+y)dz = 0 \tag{89a}$$

where

$$F(x) = (2\pi)^{-1} \int_{-\infty}^{\infty} w(\xi,t)\, e^{i\xi x}\, d\xi - i \sum_{j=1}^{M} k_j(t)\, e^{i\zeta_j x}$$

$$\bar{F}(x) = (2\pi)^{-1} \int_{-\infty}^{\infty} \bar{w}(\xi,t)\, e^{-i\xi x}\, d\xi + i \sum_{s=1}^{N} \ell_s(t)\, e^{-i\bar{\zeta}_s x} \tag{89b}$$

and

$$r(x,t) = -2\, \bar{K}_2(x,x,t) \;;\; q(x,t) = -2\, K_1(x,x,t) \tag{89c}$$

($\underset{\sim}{K}$ and $\bar{\underset{\sim}{K}}$ are column vectors with components K_1, K_2 and $\bar{K}_1$, $\bar{K}_2$). A number of inverse scattering problems are solved in [46]. The inverse scattering problem was already important in physics before its solution was applied to soliton theory since scattering potentials must usually be inferred from observed asymptotic scattering data. It is remarkable that the solution of the Schrödinger inverse scattering problem could be seized upon and exploited by GARDNER, GREENE, KRUSKAL and MIURA to solve the KdV [34].

As an example consider the single soliton problem which has $F(x) = - i\, k_0\, e^{-2\Omega(i\eta)t}\, e^{-\eta x} = \bar{F}^*$. We have to solve

$$K_1(x,y) + i\, k(t)\, e^{-\eta(x+y)} + i\, k(t) \int_x^\infty \bar{K}_1(x,z)\, e^{-\eta(z+y)}\, dz = 0$$

$$\bar{K}_1(x,y) - i\, k(t) \int_x^\infty K_1(x,z)\, e^{-\eta(z+y)}\, dz = 0 . \qquad (90)$$

The solution is obtained by setting $K_1(x,y) = f(x)e^{-\eta y}$ and to satisfy (90) we then need

$$f(x) = i\, k(t)\, e^{-\eta x}[1 - \tfrac{1}{4}\, k^2(t)\, \eta^{-2}\, e^{-4\eta x}]^{-1} . \qquad (91)$$

Thus

$$q(x,t) = -2\, K_1(x,x,t) = -2\, i\, k(t)\, e^{-2\eta x}\, [1 - \tfrac{1}{4}\, k^2(t)\, \eta^{-2}\, e^{-4\eta x}]^{-1}$$

$$= 2\eta\, \mathrm{sech}\, [2\Omega(i\eta)t + 2\eta x + \delta] \qquad (92)$$

in which η and δ can be treated as free parameters fixed by $q(x,0)$ ($\delta = \ln(-i\, k_0/2\eta)$ and $k_0 = k(i\eta, t=0)$ will be purely imaginary). The reader may care to find the 2-soliton solution similarly. Notice amazingly enough that the 1-soliton solution will always be a sech with speed $-\Omega(i\eta)/2\eta$ and only the function $\Omega(i\eta)$ relates the result to the original NEE. This leads us to expect that the most general equation soluble by the ISM in the AKNS-ZS formulation is determined by $\Omega(\zeta)$ alone. In fact one finds that a fairly general pair of NEEs solvable this way is [17,45]

$$\frac{\partial}{\partial t}\, \hat{\sigma}\, s(x,t) + 2\Omega(L^\dagger)\, s(x,t) = 0 \qquad (93a)$$

$$L^\dagger = \frac{1}{2i} \begin{bmatrix} \frac{\partial}{\partial x} - 2r \int_\infty^x dx'q & 2r \int_{-\infty}^x dx'r \\ -\, 2q \int_{-\infty}^x dx'q & -\frac{\partial}{\partial x} + 2q \int_\infty^x dx'r \end{bmatrix} , \quad s = \begin{pmatrix} r \\ q \end{pmatrix} , \quad \hat{\sigma} = \begin{bmatrix} 1 & 0 \\ 0 & -1 \end{bmatrix} \qquad (93b)$$

providing $\Omega(\zeta)$ is a simple ratio of entire functions. The reader may wish to express the s-G in the form (93).

In the case of the s-G, $\Omega(\zeta) = -\, 1/4i\zeta$ and $\Omega(i\eta) = (4\eta)^{-1}$. Thus (92) becomes

$$\sigma_x = 4\eta\, \mathrm{sech}\{(2\eta)^{-1}t + 2\eta + \delta\}$$

$$\sigma = 4\, \tan^{-1} \exp\{(2\eta^{-1})t + 2\eta x + \delta\} . \qquad (94)$$

This is the 2π-kink solution (67). The N-kink solution can be obtained similarly: the breathers come from pairs of eigenvalues $(\zeta, -\zeta^*)$.

4. Canonical structures

The argument develops from the fact that $a(\zeta)$ and $\ln a(\zeta)$ are constants of the motion, the result of (88). From $a(\zeta) = W(\phi,\psi)$, $a(\zeta) \sim \phi_1 e^{i\zeta x}$ as $x \to \infty$ (ϕ_1 is the 1-component of the Jost function ϕ). From (82) $\Phi \equiv \ln(\phi_1 \exp i\zeta x)$ satisfies [45,47]

$$2i\zeta\, \Phi_{,x} = \Phi_{,x}^2 - qr + q \frac{\partial}{\partial x} (q^{-1}\, \Phi_{,x}) . \qquad (95)$$

As $|\zeta| \to \infty$

$$\Phi_{,x} \sim \sum_{n=1}^{\infty} \mathcal{H}_n (2i\zeta)^{-n} \tag{96}$$

so from (95)

$$\mathcal{H}_{n+1} = q \frac{\partial}{\partial x}(q^{-1}\mathcal{H}_n) + \sum_{j+k=n} \mathcal{H}_j \mathcal{H}_k \, . \tag{97}$$

Also as $x \to \infty$ $\ln a(\zeta) \sim \Phi$ and since $a(\zeta) \to 1$ as $|\zeta| \to \infty$

$$\ln a(\zeta) \sim \sum_{n=1}^{\infty} \zeta^{-n} (2i)^{-n} \int_{-\infty}^{\infty} \mathcal{H}_n \, dx = \sum_{n=1}^{\infty} \zeta^{-n} H_n \quad . \tag{98}$$

Since a is a constant of the motion lna is also. Hence the $H_n \equiv (2i)^{-n} \int_{-\infty}^{\infty} \mathcal{H}_n \, dx$ are constants of the motion. The first few $\mathcal{H}_n$ prove to be

$$(2i)^{-1}\mathcal{H}_1 = -(2i)^{-1} qr \quad , \quad (2i)^{-2}\mathcal{H}_2 = (8)^{-1}(q\, r_x - q_x\, r)$$

$$(2i)^{-3}\mathcal{H}_3 = -(8i)^{-1}(r_x q_x + q^2 r^2) \, , \; (2i)^{-4}\mathcal{H}_4 = (32)^{-1}(q_x r_{xx} - q_{xx} r_x - 3r^2 q\, q_x + 3q^2 r\, r_x) \, . \tag{99}$$

If we set $q = -r = -\frac{1}{2}\sigma_x$, the $\mathcal{H}_{2n}$ vanish and the $\mathcal{H}_{2n+1}$ coincide with the T^{2n+2} in (77) (upto numerical factors).

Consider now the q,r as canonically conjugate variables (q = momentum, r = conjugate co-ordinate for definiteness). From the <u>Hamiltonian</u> $H = 8\, H_3$

$$q_t = -8\frac{\delta H_3}{\delta r} = i^{-1}(-q_{xx} + 2q^2 r) \tag{100a}$$

$$r_t = 8\frac{\delta H_3}{\delta q} = -i^{-1}(-r_{xx} + 2q\, r^2) \quad . \tag{100b}$$

If $q = -r^*$, (100a) is the NLS (24) and (100b) is its complex conjugate equation. From $H = -16\, H_4$

$$q_t = -\frac{\delta H}{\delta r} = -\{q_{xxx} - 6rq\, q_x\} \tag{101}$$

and if $q = -r$ this is the modified KdV (23). Formally we can also set $r = -1$ (although this does not satisfy the boundary conditions on r) to get the KdV (22). We can also choose $H = -16\sigma\, H_4 + 8\rho\, H_3$ (σ,ρ constants) and $r = -\gamma q^*$ to get

$$iq_t = i\sigma\, q_{xxx} + 6i\gamma\sigma|q|^2 q_x + \rho\, q_{xx} + 2\rho\, \gamma q|q|^2 \, . \tag{102}$$

This is the Hirota eqn. (25) with $2\gamma\sigma = \alpha$, $2\rho\gamma = \delta$ and $\alpha\beta = \sigma\delta$. Clearly we can choose any linear combination of the H_n to obtain interesting NEEs in q and r. In fact all these NEEs fall into the AKNS-ZS scheme [45,48]. And indeed all the coefficients of the H_n can depend on time without fundamentally changing this situation [45].

Notice also that for $q_t = -16\delta H_{2n+2}/\delta r$ ($n = 1,2,\ldots$) and $r = -q$ we obtain an infinite sequence of generalised modified KdV equations with (23) as first member; for $r = -q^*$, $q_t = -8\delta H_{2n+1}/\delta r$ ($n = 1,2,\ldots$) is an infinite sequence of NLS equations. The infinite series of KdV equations obtained formally by $q_t = -16\delta H_{2n+2}/\delta r$ and $r = -1$ was first discovered by LAX [44]. In any of the sequences each member has the same infinite set of conserved densities; each member proves to be soluble by the inverse method with scattering problem (82); and each member has the same spectrum. The flows described by these NEEs commute [44].

Perhaps the most striking result so far is the absence of the s-G (65). The $\mathcal{H}_{2n+1}$

coincide with the T^{2n+2} and for $q = -r = -\frac{1}{2}\sigma_x$ these are momentum densities for an infinite sequence of s-G equations (compare (78)). However it is clear that since $a(\zeta)$ is analytic in the upper half ζ-plane we can expand $\ln a(\zeta)$ about regular points μ (these points necessarily exclude the zeros of $a(\zeta)$). For a regular point

$$\ln a(\zeta) = \sum_{m=0}^{\infty} \bar{H}_m(\zeta-\mu)^m \ , \ \bar{H}_m = \frac{1}{m!}\left[\frac{\partial^m}{\partial\zeta^m} \ln a(\zeta)\right]_{\zeta=\mu} \qquad \mathrm{Im}\ \mu > 0 \ . \tag{103}$$

For the s-G we choose $r = -q$ and $\mu = 0$ (this is the choice which makes $\Omega(i\zeta)$ take its proper form - see (120) below). From (103) into (95)

$$\Phi_{,x} = q \tan(-\int_{-\infty}^{x} q\, dx') + \left[-(i \frac{\partial}{\partial x} \tan(-\int_{-\infty}^{x} q\, dx')) \cdot \int_{-\infty}^{x} \sin(-2\int_{-\infty}^{x'} q\, dx'')dx'\right]\zeta + \ldots \tag{104}$$

Then by using the boundary conditions $2\int_{-\infty}^{x} q\, dx' = -\sigma \to 0(\mathrm{mod}\ 2\pi)$ as $|x| \to \infty$ we find

$$\ln a(\zeta) = i\int_{-\infty}^{\infty} dx(1 - \cos(-2\int_{-\infty}^{x} q\, dx'))\zeta + \ldots \ . \tag{105}$$

From $q = -\frac{1}{2}\sigma_x$, $p = -\gamma^{-1}\int_{-\infty}^{x} q(x')dx' = \frac{1}{2}\gamma^{-1}\sigma$ we regain the s-G from

$$H = -\frac{1}{4\gamma}\int_{-\infty}^{\infty}\{2 - \cos(-2\int_{-\infty}^{x} q(x')dx') - \cos(2\gamma p)\}dx \tag{106}$$

for[5]

$$-q_t \ (= \tfrac{1}{2}\sigma_{xt}) = -\frac{\delta H}{\delta p} = \tfrac{1}{2}\sin\sigma$$
$$p_t \ (= \tfrac{1}{2}\gamma^{-1}\sigma_t) = -\frac{\delta H}{\delta q} = \frac{1}{2\gamma}\int_{-\infty}^{x}\sin\sigma\, dx' \ . \tag{107}$$

Notice in this example that the canonical coordinates are q and $-\gamma^{-1}\int_{-\infty}^{x} q(x')dx'$. With the coupling constant γ included such a pair typifies the canonical coordinates for the case where the potentials q and r are linearly dependent. In the previous examples (100)-(102), etc. q and r were linearly independent and were made dependent, as in (101) for example, only after finding the canonical equations. In the case of the modified KdV either formulation can be used [45].

The results of this section 4 show by demonstration only that the soluble equations discussed in this paper are all Hamiltonian systems and all such equations belong to an infinite sequence of similar equations. The key result however is that the mapping which transforms $q(x,0)$ and $r(x,0)$ into the scattering data is a canonical transformation (whether q and r are independent or not). Further the inverse mapping from scattering data at time t to $q(x,t)$ and $r(x,t)$ is an (inverse) canonical transformation. This means in particular that the scattering data define a new set of canonical co-ordinates.

---

5 From (104), with some difficulty, or more easily from (73), $\bar{H}_3 = -\frac{1}{2\gamma}\int_{-\infty}^{x}(\sigma_t^2\cos\sigma)dx$ and the next member in the infinite hierarchy of s-G's proves to be $\sigma_{xt} = 2\sigma_{tt}\cos\sigma - \sigma_t^2\sin\sigma$. We know of no physical application of this equation. On a different point, notice the factors of $\frac{1}{2}$ in the definition of p and H compared with the discussion following (79b). These factors are needed because (106) is symmetrised in p and q.

6. Canonical co-ordinates for the scattering data

We introduce [45,47] the symplectic form

$$\omega = \int_{-\infty}^{\infty} dx(dp(x) \wedge dq(x)) \equiv \int_{-\infty}^{\infty} dx[\delta p_1(x)\delta q_2(x) - \delta p_2(x)\delta q_1(x)] \ . \qquad (108)$$

$(\delta p_1,\delta q_1)$ and $(\delta p_2,\delta q_2)$ are independent variations of the pairs of canonical co-ordinates $(p(x),q(x))$. As a heuristic motivation for considering this quantity take Hamilton's principle for a single pair (p,q) of canonical co-ordinates

$$\delta \int (pdq - Hdt) = 0 \ . \qquad (109)$$

In the canonically transformed set $(\bar{p},\bar{q})$ with Hamiltonian $\bar{H}$

$$\delta \int (\bar{p}d\bar{q} - \bar{H}dt) = 0 \ , \qquad (110)$$

and so for some function F

$$\delta \int (pdq - Hdt - \bar{p}d\bar{q} + \bar{H}dt + dF) = 0 \ . \qquad (111)$$

Since for conservative systems the values of H and $\bar{H}$ are necessarily the same

$$\delta[pdq - \bar{p}d\bar{q} + dF] = 0$$

and

$$\delta pdq - \delta\bar{p}d\bar{q} + p\delta(dq) - \bar{p}\ \delta(d\bar{q}) + \delta(dF) = 0 \ ; \qquad (112)$$

but since δp and dq can be interpreted as independent arbitrary variations we also have

$$dp\delta q - d\bar{p}\delta\bar{q} + pd(\delta q) - \bar{p}\ d(\delta\bar{q}) + d(\delta F) = 0 \qquad (113)$$

so that

$$dp\delta q - \delta pdq = d\bar{p}\delta\bar{q} - \delta\bar{p}d\bar{q} \ . \qquad (114)$$

For many degrees of freedom we sum $dp_i\ \delta q_i - \delta p_i\ dq_i$. For continuously infinitely many degrees of freedom we integrate. This heuristic argument suggests that ω defined by (108) is invariant under a canonical transformation $(p(x),q(x)) \to \{\bar{p},\bar{q}\}$ (in which some at least of the $\bar{p},\bar{q}$ may be discrete). Conversely we find that if ω is invariant the transformation is canonical [45] . (An alternative route is to use the invariance of the Poisson brackets in the two systems of co-ordinates [48]).

We develop the theory only for the case when the potentials are linearly dependent. Although the independent case is in some ways simpler [45], the dependent case includes that of the s-G. It can be shown [45] in the case $r = -q$ and $q = q$, $p = -\gamma^{-1} \int_{-\infty}^{x} q\ dx'$ that under the scattering transform (82) the symplectic form (108) defined on the manifold with canonical coordinates $p(x)$, $q(x)$ transforms into

$$\rho = \sum_{j=1}^{M} dp_j \wedge dq_j + \int_{-\infty}^{\infty} d\xi\ (d\ P(\xi) \wedge d\ Q(\xi)) \qquad (115)$$

on the manifold with canonical coordinates

$$\{p_j,\ q_j,\ P(\xi),\ Q(\xi),\ j = 1,\ldots,M;\ \xi \in R\} \ .$$

The complex canonical coordinates are defined in terms of the scattering data by

$$p_j = \gamma^{-1} \ln \zeta_j \ , \quad q_j = 2 \ln b_j \quad j = 1,2,\ldots,M$$

$$P(\xi) = (2\pi\xi\gamma)^{-1} \ln \{a(\xi)\bar{a}(\xi)\} \ , \ Q(\xi) = \arg b(\xi) \ . \qquad (116)$$

The Hamiltonians and momenta corresponding to the $\bar{H}_m$ of (103) and the H_n in (98) are obtained by reconstructing $\ln a(\zeta)$ in terms of the scattering data and using (116). We find [45,48]

$$a(\zeta) = \sum_{j=1}^{M} \frac{(\zeta-\zeta_j)}{(\zeta+\zeta_j)} \exp - \frac{\zeta}{2\pi i} \int_{-\infty}^{\infty} \frac{d\xi}{(\xi^2-\zeta^2)} \ln(1 - w(\xi)\bar{w}(\xi)) \quad \operatorname{Im} \zeta > 0 \quad . \tag{117}$$

Thus the expansion of $\ln a(\zeta)$ in inverse powers of ζ as $|\zeta| \to \infty$ yields

$$H_{2m+1} = -2 \sum_{j=1}^{M} \frac{\exp\{-(2m+1)p_j\}}{(2m+1)} - i\gamma \int_{-\infty}^{\infty} d\xi \; \xi^{2m+1} \; P(\xi) \quad . \tag{118}$$

By expansion about an ordinary point μ

$$\bar{H}_m(\mu) = \sum_{j=1}^{M} - (m)^{-1}\{(\exp - p_j - \mu)^{-m} + (-1)^{m+1}(\exp - p_j + \mu)^{-m}\}$$
$$- \frac{1}{2i} \int_{-\infty}^{\infty} d\xi \; P(\xi)\xi\{(\xi-\mu)^{-(m+1)} + (-1)^{m+1}(\xi+\mu)^{-(m+1)}\} \quad . \tag{119}$$

Notice that the Hamiltonians (118) and (119) each depend only on the momenta. We can integrate the equations of motion therefore (all the momenta are constants) and connect the evolution of the $Q(\xi)$ in particular with the evolution of the scattering data $b(\xi,t)$ to see that $\Omega(\zeta) = 4i\;\zeta^{2n+1}$ from (118) and $\Omega(\zeta) = -\frac{1}{8i}\zeta\;\{(\zeta-\mu)^{-(m+1)} + (-1)^{m+1}(\zeta+\mu)^{-(m+1)}\}$ from (119). By considering the weak potential limit (q,r) both small) we can show [45]

$$\Omega(\zeta) = - \tfrac{1}{2}i\;\omega_q(-2\zeta) = \tfrac{1}{2}i\;\omega_r(2\zeta) \tag{120}$$

where ω_q and ω_r are <u>exactly</u> the linearised dispersion relations for the r and q equations.

We consider two particular cases:

$$H_s(p,q) = \frac{m^2}{4\gamma} \int_{-\infty}^{\infty} \{\cos(-2 \int_{-\infty}^{x} q\; dx') + \cos(-2\gamma p) - 2\}dx \tag{121}$$

which differs from (106) for the s-G only by the factor m^2, the square of a mass; and

$$H_k(p,q) = \frac{M}{2\gamma} \int_{-\infty}^{\infty} dx\{q^4 - q_x{}^2 + (\gamma p_x)^4 - (\gamma p_{xx})^2\} \tag{122}$$

which is essentially H_3 from (98) and (99). The co-ordinates are q and $p = -\gamma^{-1} \int_{-\infty}^{x} q dx'$ in each case. In (121) $q = -\frac{1}{2}\sigma_x$. Hamilton's equations yield $\sigma_{,xt} = m^2 \sin\sigma$ from (121) and $q_{,t} = M\,(6q^2 q_x + q_{xxx})$ from (122). (Notice that it is $\bar{H}_3$ which yields this result in the dependent case and not H_4 as in (101)).

The associated momenta are

$$P_s(p,q) = \frac{1}{2\gamma} \int_{-\infty}^{\infty} dx(q^2 + \gamma^2\; p_x{}^2)$$

$$P_k(p,q) = - \frac{1}{2\gamma} \int_{-\infty}^{\infty} dx(q^2 + \gamma^2\; p_x{}^2) \tag{123}$$

by appeal to (99); but notice the different structure between the pairs $(P_k,\, H_k)$ and $(P_s,\, H_s)$. The former are assigned by associating them with the first two conserved quantities H_1 and H_3 respectively derived from (99) (all the H_n are scaled by $-i\gamma^{-1}$ for a real coupling constant γ and each H_n then takes a convenient factor of 2^n: the factor M is then an energy scale). In this case the conservation laws for momentum density ($\propto q^2$) and energy density ($\propto (q_x{}^2-q^4)$) constitute two unrelated conservation laws.

The quantities P_S and H_S are obtained by associating $+2i\ \gamma^{-1}H_1$ with P_S (the sign does not matter as long as it is used consistently) and from this we find the conservation law $(4\gamma)^{-1}(\sigma_x{}^2)_t + \{(2\gamma)^{-1}(\cos\sigma - 1)\}_x = 0$ which is (78). Symmetry under interchange of x and t then induces the second conservation law $\{(2\gamma)^{-1}(\cos\sigma - 1)\}_t + (4\gamma)^{-1}(\sigma_t{}^2)_x = 0$ and H_S is obtained from the density $(2\gamma)^{-1}(\cos\sigma - 1)$ (which is actually the density corresponding to (105) and (106)). The quantity $8i\ \gamma^{-1}H_3$ is the momentum for the next member of the infinite set of s-Gs (see footnote 5). Notice that the particular conservation law $\mathcal{P}_t + \mathcal{H}_x = 0$ appealed to for eqns. (79) actually relies on the symmetry of the s-G under interchange of x and t.

We can now develop the argument. From (116) and (118) and (119)

$$P_S = \frac{2i}{\gamma}\left\{-2\sum_{j=1}^{K}(\zeta_j - \zeta_j{}^*) - 2i\sum_{j=1}^{N}\eta_j - i\gamma\int_{-\infty}^{\infty} d\xi\ \xi\ P(\xi)\right\}$$

$$H_S = \frac{-m^2 i}{\gamma}\left\{-2\sum_{j=1}^{K}(\zeta_i^{-1} - \zeta_j{}^{*-1}) + 2i\sum_{j=1}^{N}\eta_j{}^{-1} + i\gamma\int_{-\infty}^{\infty} d\xi\ \xi^{-1}\ P(\xi)\right\} \tag{124a}$$

$$P_k = -\frac{2i}{\gamma}\left\{-2\sum_{j=1}^{K}(\zeta_j - \zeta_j{}^*) - 2i\sum_{j=1}^{N}\eta_j - i\gamma\int_{-\infty}^{\infty} d\xi\ \xi\ P(\xi)\right\}$$

$$H_k = \frac{-8iM}{\gamma}\left\{-2\sum_{j=1}^{K}(\zeta_i{}^3 - \zeta_j{}^{*3}) - 2i\sum_{j=1}^{N}\eta_j{}^3 - i\gamma\int_{-\infty}^{\infty} d\xi\ \xi^3\ P(\xi)\right\}\ . \tag{124b}$$

We have used the fact that for $r = -q$ bound state eigenvalues arise as $i\eta_j$ ($\eta_j > 0$) or in pairs $(\zeta_j, -\zeta_j{}^*)$.

Equations (124) can be written in the forms

$$P = \sum_{j=1}^{K}\hat{p}_j + \sum_{j=1}^{N}p_j + \int_{-\infty}^{\infty} d\xi\ p(\xi)\ P(\xi)$$

$$H = \sum_{j=1}^{K}\hat{h}_j + \sum_{j=1}^{N}h_j + \int_{-\infty}^{\infty} d\xi\ h(\xi)\ P(\xi)\ . \tag{125}$$

(The p_j in (125) are <u>not</u> the p_j defined in (116). For the modified KdV with $\theta = \arg\zeta$

$$h_j\ p_j{}^{-3} = \tfrac{1}{4}\ M\ \gamma^2\ ,\quad \hat{h}_j\ \hat{p}_j{}^{-3} = \frac{\gamma^2 M}{16}\frac{\sin 3\theta}{\sin^3\theta} \tag{126a}$$

$$h(\xi)\{p(\xi)\}^{-3} = M\ . \tag{126b}$$

The dispersion relation (126b) is the linearised dispersion relation of the modified KdV equation. The relations (126a) define new 'particle' energy-momentum relations of a rather unusual kind. Relations (125) show that the 'particles' can be considered to form a collection of non-interacting entities. For the modified KdV we have been unable to extend the interpretation of the dynamics of these particles further. But everything we do below for the s-G including the quantisation of that equation and the discovery of its eigen spectrum can be done also for the modified KdV. The s-G has all the properties one might have hoped for it including a natural relativistic dynamics for each free particle.

From (124a)

$$h_j\ p_j = 4m^2\ \gamma^{-2}\ ,\quad \hat{h}_j\ \hat{p}_j = 16m^2\ \gamma^{-2}\ \sin^2\theta$$

$$h(\xi)\ p(\xi) = m^2\ . \tag{127}$$

If we define new energies and momenta by

$$h(\xi) = \tfrac{1}{2}\{h'(\xi) + p'(\xi)\}$$
$$p(\xi) = \tfrac{1}{2}\{h'(\xi) - p'(\xi)\} \qquad (128)$$

and define similar quantities for h_j, p_j and $\hat{h}_j$, $\hat{p}_j$, then if we define a new mass $m' = 2m$, we find, after dropping the primed notation, that

$$h_j = [4m^2\gamma^{-2} + p_j^2]^{\frac{1}{2}}\ ,\quad \hat{h}_j = [16m^2\gamma^{-2}\sin^{-2}\theta + \hat{p}_j^2]^{\frac{1}{2}}$$
$$h(\xi) = [m^2 + p(\xi)^2]^{\frac{1}{2}}\ . \qquad (129)$$

Since the expressions (125) go over to the same forms in the primed quantities the s-G is now described in terms of a collection of free relativistic particles of masses m, $2m\,\gamma^{-1}$ and $4m\,\gamma^{-1}\sin\theta$. In terms of a corrected coupling constant $\gamma' = 4\gamma$ these masses become m, $8m\,\gamma^{-1}$ and $16m\,\gamma^{-1}\sin\theta$. These results agree with those in §3 of I (see especially the remarks on the breather below (21).

For ease of reference we give the final expressions:

$$P_s = \sum_{j=1}^{M}\hat{p}_j + \sum_{j=1}^{L}p_j + \int_{-\infty}^{\infty} d\xi\ p(\xi)\ P(\xi)\ d\xi$$

$$H_s = \sum_{j=1}^{K}[256m^2\gamma^{-2}\sin^2\theta_j + \hat{p}_j^2]^{\frac{1}{2}} + \sum_{j=1}^{L}[64m^2\gamma^{-2} + p_j^2]^{\frac{1}{2}} + \int_{-\infty}^{\infty}[m^2 + p^2(\xi)]^{\frac{1}{2}}P(\xi)\ d\xi$$

(130)

The result in this form is due first of all to FADEEV [49]. It shows that the kinks, breathers, and background modes form a natural collection of independent elementary excitations. It would therefore seem that the statistical mechanics of a sine-Gordon system could be treated in these terms (compare [10]). The spectrum has three thresholds at m, $8m\,\gamma^{-1}$ and $16m\,\gamma^{-1}\sin\theta$. Since $0 \le \theta \le \pi/2$ only the energy gaps at m and $8m\,\gamma^{-1}$ will lead to exponentials in the specific heat. The situation is changed if the system is quantised. The quantisation of (130) is easy if $\gamma \ll 1$ [49]. One imposes the usual canonical commutation relations on the momenta p_j, $\hat{p}_j$ and $p(\xi)$. One discovers from (116) that the phase space available is $-\infty < p_j < \infty$, $-\infty < q_j < \infty$, etc. The invariance of the Poisson brackets guarantees that, except perhaps upto a renormalisation of the coupling constant [49], canonical quantisation of (130) is equivalent to the <u>operator</u> equation

$$\sigma_{xx} - \sigma_{tt} = \sin\sigma \qquad (131)$$

with[6]

$$[\sigma, \sigma_t] = i\,\gamma\,\delta(x-x')\ . \qquad (132)$$

However the phase space for the internal momentum θ_j of the breathers is compact: from (115) one finds the canonical pair $\{\gamma^{-1}\theta_j,\ 4\arg b_j\}$, so the available phase space has volume $\frac{1}{2}\pi\gamma^{-1} \times 8\pi = 4\pi^2\gamma^{-1}$. If the number of states is N this volume is also $2\pi N$ ($\hbar = 1$) so $N = 2\pi\gamma^{-1}$ and θ is quantised with discrete levels given approximately by $\theta_n = \frac{n\gamma}{4}$. In terms of γ' this becomes $\theta_n = \frac{n\gamma}{16}$. Thus [49] the continuous breather band of the classical s-G is replaced by the mass spectrum

---

6 From $[p,q] = [\gamma^{-1}\sigma, -\frac{1}{2}\sigma_x]$ and change of co-ordinates $x \to x + t$, $t \to x - t$ one finds $m' = 2m$ and $[\sigma,\sigma_t] = 4\,\gamma i\ \delta(x-x') = \gamma' i\ \delta(x-x')$. Then γ' is γ in (132).

$$M_n = \frac{16m}{\gamma} \sin \frac{n\gamma}{16} , \qquad n = 1,2,\ldots,N \tag{133}$$

(where $N \approx 2\pi\ \gamma^{-1}$).

DASHEN, HASSLACHER and NEVEU [11] obtained the same mass spectrum by a remarkable application of Feynman path integral techniques with the renormalisation $\gamma'' = \gamma[1 - (\gamma/8\pi)]^{-1}$. LUTHER [50] has shown that the BAXTER model [51] in statistical mechanics, for which the relevant part of the spectrum is known [52], maps into the massive THIRRING model [8] and the operator s-G (131). In this way he shows that the mass spectrum (133) can be interpreted as exact providing γ'' replaces γ. Notice that for $\gamma''<<1$, $M_n \approx nm$ (for $n<<N$) and the mass m of the field induces a tier of equally spaced levels corresponding to n "mesons".

For $\gamma'' = 8\pi$ the single remaining quantised level disappears as the breathers break up into pairs of heavy particles each of mass $8m\ \gamma''^{-1}$. We have shown [33] that, for spin waves in the A-phase of liquid ^{3}He there is a natural route to (131) with (132) with $\gamma'' \sim 10^5$. In this case the quantised breather spectrum is eliminated. Unfortunately the excitations at $8m\ \gamma''^{-1}$ become very low lying (corresponding frequencies are $\approx$ 1 Hz.) and the observation of this quantised spectrum will be complicated by other excitations (the so called orbital waves [33]).

The B-phase spin waves are governed by the double s-G (64) and we have not been able to quantise this. On the other hand both the ϕ-four equation (13) and the NLS equation (24) have been quantised: the ϕ-four is not an integrable system as is the s-G and the quantisation is not exact [53]; the NLS is integrable and the route quantising the s-G finds oscillator co-ordinates through the canonical formalism very much as in the work of this section. The independence of q and q^* means that the canonical formalism for the NLS must be developed for the independent case as in [45]. The result obtained is essentially exact [54].

References

1. D.J. Korteweg and G. de Vries, Phil.Mag. 39, 422 (1895).
2. J.S. Russell, Report on Waves, British Association Reports (1844).
3. G.B. Whitham, Linear and Non-linear Waves (John Wiley & Sons, New York, 1974) pp. 580-585.
4. J.D. Gibbon and J.C. Eilbeck, J.Phys.A : Gen.Phys. 5, L132 (1972).
5. R.K. Dodd and R.K. Bullough, Proc.Roy.Soc. A 351, 499 (1976).
6. R.K. Bullough, Solitons. Lectures at the Winter College on the Interaction of Radiation and Condensed Matter. I.C.T.P., Trieste, March 1976. To be published by the I.A.E.A., Vienna (1977).
7. J.M. Burgers, Adv.Appl.Mech. 1, 171 (1948).
8. S. Coleman, Classical lumps and their quantum descendants. Lectures at the 1975 International School of Subnuclear Physics "Ettore Majorana" (1975).
9. H. Haken, in Synergetics in Cooperative Phenomena, Edited by H. Haken (North Holland, Amsterdam, 1974).
10. J.A. Krumhansl and J.R. Schrieffer, Phys.Rev. B 11, 3535 (1975).
11. R.F. Dashen, B. Hasslacher, and A. Neveu, Phys.Rev. D 11, 3424 (1975).
12. P.J. Caudrey, J.C. Eilbeck and J.D. Gibbon, Il Nuovo Cimento 25, 497 (1975).

13. J.D. Gibbon and J.C. Eilbeck, J.Phys.A : Gen.Phys. 5, L122 (1972).

14. P.J. Caudrey, J.C. Eilbeck, and J.D. Gibbon, J.Inst.Math.Applics. 14, 375 (1975).

15. R. Hirota, Phys.Rev.Lett. 27, 1192 (1971).

16. M.J. Ablowitz, D.J. Kaup, A.C. Newell and H. Segur, Phys.Rev.Lett. 31, 125 (1973).

17. M.J. Ablowitz, D.J. Kaup, A.C. Newell and H. Segur, Studies in Appl.Math. 53, 249 (1974).

18. E. Fermi, J.R. Pasta and S.M. Ulam, Studies of nonlinear problems I, Los Alamos Rept. LA-1940 (May 1955) and Collected Works of E. Fermi Vol.II (Univ.of Chicago Press, 1965)pp.978-88.

19. J. Gibbons, S.G. Thornhill, M.J. Wardrop, and D. ter Haar, On the theory of Langmuir Solitons, preprint Univ. of Oxford, Dept. of Theoretical Phys. Ref.36/76 (1976).

20. V.E. Zakharov, Zh.Eksp.Teor.Fiz. 62, 1745 (1972) (Soviet Phys. J.E.T.P. 35, 908 (1975).

21. G.L. Lamb, Rev.Mod.Phys. 43, 99 (1971).

22. J.C. Eilbeck, P.J. Caudrey, J.D. Gibbon, and R.K. Bullough, 6, 1337 (1973).

23. B.B. Kadomtsev and V.I. Petviashvili, Dokl.Akad.Nauk.SSR, 192, 1337 (1973).

24. V.E. Zakharov and A.B. Shabat, Funkt.Anal.i Ego Prilozh. 8, 43 (1974).

25. H.C. Yuen and B.M. Lake, Phys.Fluids 18, 956 (1975).

26. R.P. Feynman, R.B. Leighton and M. Sands, The Feynman Lectures on Physics Vol.III (Addison-Wesley, Reading, Mass., 1965) pp.21.14-21.18.

27. J.D. Gibbon, P.J. Caudrey, R.K. Bullough and J.C. Eilbeck, Lett.al Nuovo Cimento 8, 775 (1973).

28. H.M. Gibbs and R.E. Slusher, Phys.Rev.A 6, 2326 (1972).

29. T.A. Fulton and R.C. Dynes, Solid State Comm. 12, 57 (1973).

30. S.P. Novikov, Funkt.Anal. i Ego Prilozh. 8, 54 (1974).

31. P.D. Lax, Comm. Pure and Appl.Maths. 28, 141 (1975).

32. R.K. Bullough, P.J. Caudrey, J.D. Gibbon, S. Duckworth, H.M. Gibbs, B. Bölger, and L. Baede, Optics Communications 18, 200 (1976).

33. R.K. Bullough and P.J. Caudrey, Bumping spin waves in the B-phase of liquid ^{3}He. Preprint (April, 1977).

34. C.S. Gardner, J.M. Greene, M.D. Kruskal and R.M. Miura, Phys.Rev.Lett. 19, 1095 (1967).

35. V.E. Zakharov and A.B. Shabat, Zh.Eksp.Teor.Fiz. 61, 118 (1971) (Soviet Phys. J.E.T.P. 34, 62 (1972)).

36. M. Wadati and M. Toda, J.Phys.Soc.Japan 32, 1403 (1972).

37. M. Wadati, J.Phys.Soc. Japan 32, 1681 (1972).

38. R. Hirota, J.Phys.Soc. Japan 33, 1459 (1972).

39. P.J. Caudrey, J.D. Gibbon, J.C. Eilbeck and R.K. Bullough, Phys.Rev.Lett. 30, 237 (1973).

40. M. Ablowitz, D.J. Kaup, A.C. Newell and H. Segur, Phys.Rev.Lett. 30, 1262 (1973).

41. P.P. Eisenhart, A treatise on the differential geometry of curves and surfaces (Dover, New York, 1960).

42. R.K. Dodd and R.K. Bullough, Proc.Roy.Soc. A 352, 481 (1977).

43. P.J. Caudrey, R.K. Dodd and J.D. Gibbon, Proc.Roy.Soc. A 351, 407 (1976).

44. P.D. Lax, Comm.Pure and Appl.Math. 21, 467 (1968).

45. R.K. Dodd and R.K. Bullough. To be published (April 1977).

46. M.A. Naimak, Linear differential operators Pt.II (Harrap, London, 1968).

47. V.E. Zakharov and L.D. Fadeev, Funkt.Anal.i Ego Prilozh. 5, 18 (1971).

48. H. Flaschka and A.C. Newell, "Integrable systems of non linear evolution equations" in Dynamical Systems Theory and Applications. Springer Lecture Notes in Physics. Edited by J. Moser (Springer Verlag, Heidelberg, 1975).

49. For example V.E. Korepin and L.D. Fadeev, Teoreticheskaya i Matematicheskaya Fiz. 25, 147 (1975).

50. A. Luther, Phys.Rev. B 14, 2153 (1976).

51. R.J. Baxter, Phys.Rev.Lett. 26, 832 (1971).

52. J.D. Johnson, S. Krinsky and B.M. McCoy, Phys.Rev. A 8, 2526 (1973).

53. R.F. Dashen, B. Hasslacher and A. Neveu, Phys.Rev. D 10, 4130 (1974).

54. D.J. Kaup, J.Math.Phys. 16, 2036 (1975).

Nonequilibrium Phase Transitions in Chemical Reactions

Chemical Instabilities as Critical Phenomena

A. Nitzan

With 1 Figure

1. Introduction

Much attention has been given in recent years to universal features characterizing instabilities and transition phenomena in equilibrium and non-equilibrium systems [1]. Laser transitions [2], hydrodynamical instabilities [3], cooperative phenomena in data processing [4] and various examples of chemical and biochemical instabilities [5] are only a few of the many non-equilibrium transition phenomena that were studied in analogy to equilibrium phase transitions and critical phenomena. It has been pointed out [2-5] that enhancement of fluctuations, long range order and critical slowing down are common to both equilibrium and non-equilibrium critical phenomena. All these features are interrelated and in the non-equilibrium case, where dynamical equations of motion (kinetic equations) are the starting point of all theoretical treatments, can be all traced to the existence of a long lifetime mode (or modes) near the transition point. Enhancement of the fluctuation amplitude follows from the non-equilibrium equivalent of the fluctuation dissipation theorem [6]. The possibility for the appearance of long range order is seen from the emergence of the quantity $\sqrt{\tau D}$, D being a characteristic diffusion coefficient and τ is the characteristic long lifetime, as a large scale length.

Unlike most equilibrium phase transitions, non-equilibrium transitions are often characterized by the appearance, beyond the instability point, of a state which is ordered either in space or in time or possibly even in both. The laser transitions are examples of instabilities leading to a limit cycle behavior; the Rayleigh-Bernard convection instability leads to spatial structure; the Prigogine Lefever model for chemical instability may exhibit both features. Even for such cases the analogy to equilibrium critical phenomena has been demonstrated [3d,3e,7]. The amplitude of the (spatially or temporally) ordered mode has been shown to play the role of an order parameter and to satisfy near the transition point an equation of motion similar to the time dependent Ginzburg Landau (TDGL) equation which characterizes order parameters near equilibrium critical points.

In the present work I focus attention on chemical instabilities in systems characterized by the occurrence of multiple homogeneous steady states. Such systems are more closely related to equilibrium phase transitions: The different homogeneous steady states play the role of different phases. Typical examples are the Schlögl model [5a]

$$\begin{aligned} A + X &\rightleftharpoons 2X \\ B + X &\rightarrow C \end{aligned} \qquad (1)$$

in which A, B and C are kept constant and play the role of externally controlled parameters, the Edelstein model [9]

$$\begin{aligned} A + X &\rightleftharpoons 2X \\ X + E &\rightleftharpoons C \\ C &\rightleftharpoons E + B \end{aligned} \qquad (2)$$

where A, B and E + C are externally controlled parameters and, finally, the photothermochemical instability [10] described schematically by

$$\begin{aligned} A + light &\rightleftharpoons A^* \\ A + M &\rightleftharpoons A + M + heat \\ A &\rightleftharpoons B + heat \end{aligned} \tag{3}$$

This reaction scheme occurs in a closed system which further exchanges heat with a thermostat. The light intensity and the external temperature are the controlled parameters. The concentration of A (or B) and the internal temperature are the state variable of this system.

It is interesting to note that the Schlögl model (1) which has been used by several workers for demonstrating properties of fluctuations near the transition point can never be realized because in the presence of the reverse reaction C ⟶ X + B no instability arises [5b]. The photothermochemical model represents a rare case where a multiple steady state situation has been predicted and then verified experimentally [11]. It should be noted that instabilities in illuminated systems are potentially easy to experimental observation: Maintaining a non-equilibrium state by energy flow is in principle much easier than achieving it by material flow. In fact, in most cases where quantitative studies have been made of the transition region (lasers, convection instability) the non-equilibrium state is maintained by energy flow.

The experimental observation of multiple steady states in a photothermochemical system is the main reason for focusing attention on these kind of instabilities. Here we have a system which potentially can be brought close to its critical point and may be studied there. From the theoretical point of view the following questions come to mind:

1) What is the order parameter characterizing the transition?
2) What is the equation of motion for this order parameter?
3) What type of critical exponents can be measured and what are the relations between them?
4) What is the limit of validity of a mean field theory? Can we expect "non-classical" critical exponents?
5) What is the dynamics which determines the transitions between steady states?

It should be pointed out that even without direct experimental measurements, quasi-experimental results bearing on these questions may be obtained from computer experiments [12].

2. Critical Conditions

My starting point is a set of equations of motion for a reacting diffusing system

$$\frac{\partial \underset{\sim}{x}}{\partial t} = \underset{\approx}{D}\, \nabla^2 \underset{\sim}{x} + \underset{\sim}{F}(\underset{\sim}{x}, \underset{\sim}{\lambda}) \tag{4}$$

where $\underset{\sim}{x}$ is a vector of variables describing the state of the system and $\underset{\sim}{\lambda}$ is a set of parameters which may be externally controlled. $\underset{\sim}{F}$ is a set of rate functions, and $\underset{\approx}{D}$ is the diffusion matrix for the variables $\underset{\sim}{x}$. In the case of the photothermochemical instability one of the components of $\underset{\sim}{x}$ is the temperature and then the matrix $\underset{\approx}{D}$ contains also heat conduction and thermal diffusion coefficients. The set of equations,

$$\underset{\sim}{F}(\underset{\sim}{x}, \underset{\sim}{\lambda}) = \underset{\sim}{0} \tag{5}$$

which correspond to the steady states of the system, will be called equations of state. In both experimental and theoretical studies we usually fix the values of all the parameters but one, and observe the behavior of the system when the remaining parameter, λ, is varied. For a certain range of values of λ the set (5) of equations of states has several solutions $\underset{\sim}{x}^i(\lambda)$, i=0,1,2... . Focusing on one of these, $\underset{\sim}{x}^0(\lambda)$ say, I define the steady state matrix and its corresponding determinant

$$\underset{\approx}{\Omega}_o(\lambda) = \underset{\approx}{\Omega}(X^o(\lambda), \lambda) \tag{6}$$

$$D_o(\lambda) = det(\underset{\approx}{\Omega}_o(\lambda)) = D(X^o(\lambda), \lambda) \tag{7}$$

where

$$\Omega_{ij}(\underset{\sim}{x}, \lambda) \equiv \left(\frac{\partial \underset{\sim}{F}(\underset{\sim}{x}, \lambda)}{\partial \underset{\sim}{x}}\right)_{ij} = \frac{\partial F_i(\underset{\sim}{x}, \lambda)}{\partial x_j} \tag{8}$$

and

$$D(\underset{\sim}{x}, \lambda) = det(\underset{\approx}{\Omega}(\underset{\sim}{x}, \lambda)) \tag{9}$$

As we are interested in transitions between homogeneous steady states I assume that at the marginal stability point, $\lambda = \lambda_{ms}$, $D_o(\lambda_{ms})$ vanishes. Furthermore, I assume that the matrix $\underset{\approx}{D}k^2 + \underset{\approx}{\Omega}_o(\lambda)$ does not become singular before this point, i.e., that the homogeneous mode is the first to become unstable.

A typical situation with multiple steady states is shown in Fig. 1a. The

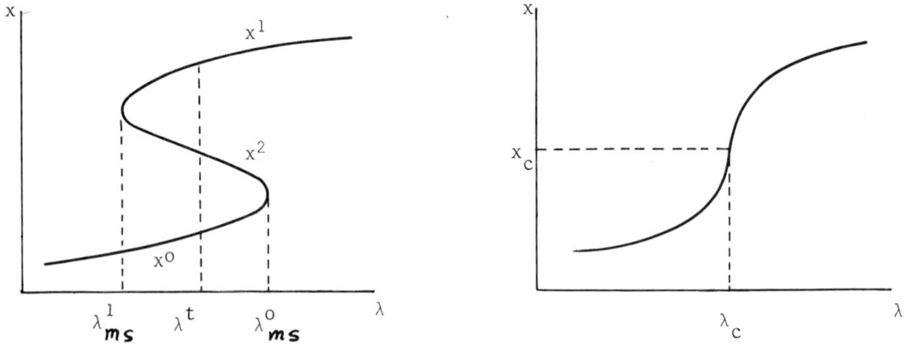

Fig. 1: Multiple steady states (a) and a critical point (b)

models (2) and (3) exhibit this kind of behavior. This situation gives rise to a hysteresis loop in the variation of x as a function of λ. As is well-known, the points corresponding to λ^o_{ms} and λ^1_{ms} cannot be realized: Finite amplitude fluctuations will cause a crossover from the less stable branch to the more stable branch much before these marginal stability points are reached. In fact, if λ is changed infinitely slowly we expect the transition to occur at a well-defined point λ_t and the hysteresis loop shrinks to zero. The situation is quite analogous to equilibrium phase transitions where λ_t corresponds to the Maxwell construction on the van der Waals equation and the extensions beyond it towards the marginal stability points correspond to the metastable branches. In studying critical behavior we thus have to focus on critical points like the one shown in Fig. 1b.

Mathematically a marginal stability point is characterized by the vanishing of $D_o(\lambda_{ms})$ Equivalently, as we see from Fig. 1a, it is characterized by the divergence of the response function $dx^o_j(\lambda)/d\lambda$ where x^o_j is any of the components of the vector $\underset{\sim}{x}^o$. The equivalence may be shown quite generally for the many variable case. In fact the following identity may be proven for a general n variable system

$$\frac{d\lambda}{dx_1}\left(\frac{\partial F_1}{\partial \lambda}\right)_{x_1, F_2 \dots F_n} = \frac{D}{D_1} \tag{10}$$

where

$$D_1 = \frac{\partial(F_2 \dots F_n)}{\partial(x_2 \dots x_n)} \tag{11}$$

Here all derivatives are evaluated at steady state, and the derivative $d\lambda/dx_1$ along a steady state line. Usually $(\partial F_1/\partial\lambda)_{x_1,F_2,\dots,F_n}$, as well as D_1 are finite at a marginal stability point whence the vanishing of D and of $d\lambda/dx_1$ are characterized by the same "critical exponent". Obviously, the same is true for all the variables x_j.

A critical point (marked by an index c) is defined as a point on the steady state line satisfying the conditions

$$\left(\frac{d\lambda}{dx_j}\right)_c = 0 \quad ; \quad \left(\frac{d^2\lambda}{dx_j^2}\right)_c = 0 \tag{12}$$

for at least one of the parameters in λ and at least one x_j. The derivatives are taken along the steady state line. When (12) holds for a particular x_j it holds for all x_k for which $(dx_k/dx_j)_c$ is finite. We saw that the first of the conditions (12) is equivalent to $D_c = 0$. It is convenient to transform also the second condition into a form that does not involve the explicit dependence of λ on x_j. It may indeed be shown that (12) written for j = 1 is equivalent to the pair of equations

$$D_c = 0 \tag{13a}$$

$$\left(\frac{\partial(D, F_2, \dots F_n)}{\partial(x_1, x_2, \dots x_n)}\right)_c = 0 \tag{13b}$$

Equation (13b) may be recast in another equivalent form. Choosing x_c as the origin for the state variables (i.e., $\underset{\sim}{x}_c = 0$), $\underset{\sim}{F}(\underset{\sim}{x},\lambda)$ may be expanded in the form

$$\underset{\sim}{F}(\underset{\sim}{x},\lambda) = \underset{\approx}{\Omega}(0,\lambda)\,\underset{\sim}{x} + \underset{\sim}{G}(\underset{\sim}{x},\lambda) \tag{14}$$

$\underset{\sim}{G}(\underset{\sim}{x},\lambda)$ contains only terms nonlinear in $\underset{\sim}{x}$. Now, the matrix $\underset{\approx}{\Omega}_c = \underset{\approx}{\Omega}(0,\lambda_c)$ has a zero eigenvalue. Define the corresponding right and left eigenvectors of $\underset{\approx}{\Omega}_c$ in the form

$$\begin{aligned} \underset{\approx}{\Omega}_c\, \underset{\sim}{u}_0 &= 0 \\ \tilde{\underset{\approx}{\Omega}}_c\, \underset{\sim}{u}_0^{(\ell)} &= 0 \end{aligned} \tag{15}$$

where $\tilde{\underset{\approx}{\Omega}}_c$ is the transpose of $\underset{\approx}{\Omega}_c$. Equation (13b) may be shown to be equivalent to

$$\sum_{i,j,k} (u_0^{(\ell)})_i \left(\frac{\partial^2 G_i}{\partial x_j\, \partial x_k}\right)_c (u_0)_j\, (u_0)_k = 0 \tag{16}$$

This critical point identity is important for the reduction of the equations of motion described in the next section.

3. Reduction of the Equations of Motion

I now apply reductive perturbation theory in an attempt to extract an equation of motion for the order parameter characterizing the transition near the critical point. I define the expansion parameter ε by

$$\lambda = \lambda_c + \varepsilon^2 \delta\lambda \tag{17}$$

and assume that λ was chosen as such that the root of $\underset{\approx}{\Omega}_0(\lambda)$, which vanishes for $\lambda = \lambda_c$ is proportional to $\lambda - \lambda_c$. I follow previous work [3d,7] in assuming that the time and length scale characterizing the order parameter near the critical point are of order ε^{-2} and ε^{-1}, respectively. As I am interested in cases where the transition involves a homogeneous non-oscillatory mode I can disregard the fast time and short length scales. I therefore take

$$\frac{\partial}{\partial t} \longrightarrow \varepsilon^2 \frac{\partial}{\partial \tau} \quad ; \quad \nabla_r \longrightarrow \varepsilon \nabla_R \tag{18}$$

and further assume that the deviation $\underset{\sim}{x}(\lambda)$ from the critical point $\underset{\sim}{x}_c = 0$ can be expanded in the form

$$\underset{\sim}{X}(\lambda) = \sum_{\ell=1}^{\infty} \varepsilon^{\ell} \underset{\sim}{X}_{\ell} \tag{19}$$

Finally, the matrix $\Omega_0(\lambda)$ is expanded near the critical point in the form

$$\underset{\approx}{\Omega}_0 = \underset{\approx}{\Omega}_c + \varepsilon\sqrt{\delta\lambda}\, \underset{\approx}{\Omega}_1 + \varepsilon^2 \delta\lambda\, \underset{\approx}{\Omega}_2 + \dots \tag{20}$$

Both $\underset{\approx}{\Omega}_1$ and $\underset{\approx}{\Omega}_2$ can be easily calculated from $\underset{\sim}{F}(\underset{\sim}{x},\lambda)$. In this calculation it is important to remember that $\underset{\sim}{x} = \underset{\sim}{x}_0(\lambda)$ is a function of λ on the steady state line. This functional dependence is responsible for the emergence of the $O(\varepsilon)$ term whose explicit form is given by

$$\underset{\approx}{\Omega}_1 = \alpha \left(\frac{\partial^2 \underset{\sim}{F}}{\partial \underset{\sim}{X}\, \partial \underset{\sim}{X}} \right)_c \cdot \underset{\sim}{U}_0 = \alpha \left(\frac{\partial^2 \underset{\sim}{G}}{\partial \underset{\sim}{X}\, \partial \underset{\sim}{X}} \right)_c \cdot \underset{\sim}{U}_0 \tag{21}$$

or

$$(\Omega_1)_{ij} = \alpha \sum_k \left(\frac{\partial^2 G_i}{\partial X_j\, \partial X_k} \right)_c (\underset{\sim}{U}_0)_k \tag{22}$$

where α is a constant.

Utilizing (18) and (20), the equations of motion (4) can be written in the form (correct to order ε^3)

$$\left[-\underset{\approx}{\Omega}_c - \varepsilon\sqrt{\delta\lambda}\, \underset{\approx}{\Omega}_1 + \varepsilon^2 \left(\frac{\partial}{\partial \tau} - \underset{\approx}{D} \nabla_R^2 - \delta\lambda\, \underset{\approx}{\Omega}_2 \right) \right] \underset{\sim}{X} = \underset{\sim}{G}(\underset{\sim}{X}) \tag{23}$$

Inserting (19) into (23), I proceed order by order in the conventional way. The first order equation

$$\underset{\approx}{\Omega}_c \underset{\sim}{X}_1 = 0 \tag{24}$$

yields

$$\underset{\sim}{X}_1 = W(\underset{\sim}{R}, \tau)\, \underset{\sim}{U}_0 \tag{25}$$

$W(\underset{\sim}{R},\tau)$ will be identified as the order parameter for this problem. It determines the behavior of the system on both sides of the critical point. To order ε^2 Eq. (23) gives

$$-\underset{\approx}{\Omega}_c \underset{\sim}{X}_2 - \sqrt{\delta\lambda}\,\underset{\approx}{\Omega}_1 \underset{\sim}{X}_1 = \left(\frac{\partial^2 \underset{\sim}{G}}{\partial \underset{\sim}{X}\,\partial \underset{\sim}{X}}\right)_c \underset{\sim}{X}_1 \underset{\sim}{X}_1 \tag{26}$$

The integrability condition for this equation

$$-\sqrt{\delta\lambda}\left(\underset{\sim}{u}_o^{(\ell)} \cdot \underset{\approx}{\Omega}_1 \underset{\sim}{X}_1\right) = \left(\underset{\sim}{u}_o^{(\ell)} \cdot \left(\frac{\partial^2 \underset{\sim}{G}}{\partial \underset{\sim}{X}\,\partial \underset{\sim}{X}}\right)_c \underset{\sim}{X}_1 \underset{\sim}{X}_1\right) \tag{27}$$

is automatically satisfied as both sides are identically zero. This follows from (16) and (21). Equation (26) may therefore be solved to yield

$$\underset{\sim}{X}_2 = -\left(\sqrt{\delta\lambda}\,W + W^2\right) \underset{\approx}{\Omega}_c^{-1} \left(\frac{\partial^2 \underset{\sim}{G}}{\partial \underset{\sim}{X}\,\partial \underset{\sim}{X}}\right)_c \underset{\sim}{u}_o \underset{\sim}{u}_o \tag{28}$$

In third order we have

$$\begin{aligned} -\underset{\approx}{\Omega}_c \underset{\sim}{X}_3 - \sqrt{\delta\lambda}\,\underset{\approx}{\Omega}_1 \underset{\sim}{X}_2 + \left(\frac{\partial}{\partial \tau} - \underset{\approx}{D} \nabla_R^2 - \delta\lambda \underset{\approx}{\Omega}_2\right) \underset{\sim}{X}_1 = \\ = \left(\frac{\partial^2 \underset{\sim}{G}}{\partial \underset{\sim}{X}\,\partial \underset{\sim}{X}}\right)_c \underset{\sim}{X}_1 \underset{\sim}{X}_2 + \left(\frac{\partial^3 \underset{\sim}{G}}{\partial \underset{\sim}{X}\,\partial \underset{\sim}{X}\,\partial \underset{\sim}{X}}\right)_c \underset{\sim}{X}_1 \underset{\sim}{X}_1 \underset{\sim}{X}_1 \end{aligned} \tag{29}$$

The integrability condition now yields an equation for the order parameter W

$$\left(\frac{\partial}{\partial \tau} - \bar{D} \nabla_R^2\right) W = \delta\lambda (a+b) W + \sqrt{\delta\lambda}\, b W^2 + (b+c) W^3 \tag{30}$$

where

$$\begin{aligned} a &\equiv \left(\underset{\sim}{u}_o^{(\ell)} \cdot \underset{\approx}{\Omega}_2 \underset{\sim}{u}_o\right) \quad ; \quad b \equiv -\left(\underset{\sim}{u}_o^{(\ell)} \cdot \underset{\approx}{\Omega}_1 \underset{\approx}{\Omega}_c^{-1} \left(\frac{\partial^2 \underset{\sim}{G}}{\partial \underset{\sim}{X}\,\partial \underset{\sim}{X}}\right)_c \underset{\sim}{u}_o \underset{\sim}{u}_o\right) \\ c &\equiv \left(\underset{\sim}{u}_o^{(\ell)} \cdot \left(\frac{\partial^3 \underset{\sim}{G}}{\partial \underset{\sim}{X}\,\partial \underset{\sim}{X}\,\partial \underset{\sim}{X}}\right)_c \underset{\sim}{u}_o \underset{\sim}{u}_o \underset{\sim}{u}_o\right) \quad ; \quad \bar{D} \equiv \left(\underset{\sim}{u}_o^{(\ell)} \cdot \underset{\approx}{D}\, \underset{\sim}{u}_o\right) \end{aligned} \tag{31}$$

Equation (30) is of the Ginzburg Landau type. In conclusion I wish to stress again the importance of the critical conditions, presented in the previous section, in obtaining this result.

4. Fluctuations

Fluctuations may be introduced phenomenologically into the present formalism by adding Langevin "forces" $\underset{\sim}{f}$ into the r.h.s. of (4).

$$\frac{\partial \underset{\sim}{X}}{\partial t} = \underset{\approx}{D} \nabla^2 \underset{\sim}{X} + \underset{\sim}{F}(\underset{\sim}{X}) + \underset{\sim}{f} \tag{32}$$

In a reacting diffusing system, where $\underset{\sim}{x}$ represents concentrations of chemical components, there are contributions to $\underset{\sim}{f}$ from the reaction and the diffusion processes [13]. In the vicinity of the critical point, where we focus on long wavelength phenomena, the diffusional contributions are expected to be insignificant. Indeed, the diffusional correlation function contains terms like $\nabla_r \nabla_{r'} \delta(r-r')\delta(t-t')$ and becomes proportional to ε^{d+4} under the scaling (18). The diffusional random force therefore does not contribute to order ε^3 for $d = 3$. The reaction random force correlation function has the form

$$\langle f_j(\underset{\sim}{r},t)\, f_{j'}(\underset{\sim}{r}'t')\rangle = Q_{jj'}\,\delta(\underset{\sim}{r}-\underset{\sim}{r}')\,\delta(t-t') \qquad (33)$$

where

$$Q_{jj'} = \sum_{\kappa} Q_{jj'}^{(\kappa)} \qquad (34)$$

and

$$Q_{jj'}^{(\kappa)} = \nu_{j\kappa}\,\nu_{j'\kappa'}\,\frac{\vec{r}_{\kappa}+\overleftarrow{r}_{\kappa}}{A_o} \qquad (35)$$

Here j and j´ are indices denoting chemical components, κ denotes a particular chemical reaction, $\nu_{j\kappa}$ is the stochiometric coefficient of component j in the reaction κ, A_0 is Avogadro's number and, finally, $\vec{r}_{\kappa}$ and $\overleftarrow{r}_{\kappa}$ are forward and backward reaction rates. Equation (35) follows from a second order expansion of a master equation, and may fail too close to the critical point [15]. I assume here that this procedure is adequate for a rough estimate of the Ginzburg's critical region, as described in the next section.

Following Graham I now assert that for distances not too close to the critical point the random force terms will contribute not earlier than in the third order. Introducing these terms as $0(\varepsilon^3)$ terms results in a Langevin equation equivalent of (30)

$$\left(\frac{\partial}{\partial\tau} - \bar{D}\nabla_R^2\right)W = \delta\lambda(a+b)W + \sqrt{\delta\lambda}\, bW^2 + (b+c)W^3 + f_w(\underset{\sim}{R},\tau) \qquad (36)$$

where

$$f_w = \left(\underset{\sim}{U}_o^{(\ell)}\cdot \underset{\sim}{f}\right) \qquad (37)$$

and a scaling $(\underset{\sim}{r}t) \longrightarrow (\underset{\sim}{R}\tau)$ has been done on $\underset{\sim}{f}$ according to (18). Equation (36) is a time dependent Langevin Ginzburg Landau equation for the order parameter W. It can be converted to Fokker Planck equation for the probability distribution $P(W(R,\tau))$ [14] and solved for the steady state distribution in terms of a potential function $U(W(R))$, as is shown in the next section. Within the range of validity of Eq. (36) we may find analogs to all the critical exponents encountered in equilibrium critical phenomena.

5. The Ginzburg Criterion

A Langevin equation of the form

$$\frac{d}{dt}W = \sigma\nabla^2 W + F(w) + f(\underset{\sim}{r},t) \qquad (38)$$

with f being a Gaussian random variable satisfying

$$\begin{aligned} &\langle f(\underset{\sim}{r},t)\rangle = 0 \\ &\langle f(\underset{\sim}{r},t),\, f(\underset{\sim}{r}'t')\rangle = \Phi\,\delta(\underset{\sim}{r}-\underset{\sim}{r}')\,\delta(t-t') \end{aligned} \qquad (39)$$

is known to yield a steady state probability distribution

$$P[W(r)] = \frac{1}{Z} \exp\left[-\frac{1}{\Phi}\int dr\left(U[W(r)] + \frac{1}{2}\sigma(\nabla W(r))^2\right)\right] \tag{40}$$

where Z is a normalizing factor and U is related to F by

$$F(x) = -\partial U(x)/\partial x \tag{41}$$

For the case

$$U(w) = \mu w^2 + \frac{1}{2}\nu w^4 \tag{42}$$

Ginzburg [16] has provided a criterion for the validity of mean field theory or the Gaussian approximation (which neglects the quartic term). This criterion is

$$\frac{\Phi K_d \nu N}{(2\pi)^d(\frac{1}{2}\sigma)^2}\left(\frac{\frac{1}{2}\sigma}{\mu}\right)^{\frac{1}{2}(4-d)} \ll 1 \tag{43}$$

where d is the dimensionality of the system, K_d is the surface area of a d dimensional

$$N = \int_0^\infty dx \frac{x^{d'}}{1+x^2} \tag{44}$$

with d′ being the non-integer part of d. Equation (43) provides a measure for the size of the critical region inside which the mean field theory or the Gaussian approximation fail. Equations (31), (35) and (36) may now be used in translating the quantities ϕ, ν, σ and μ to chemical and diffusional terms. The result is

$$\frac{1}{A_o}\frac{(D\theta)^{-d/2}}{\gamma} \ll (\widetilde{\delta\lambda})^{\frac{1}{2}(4-d)} \tag{45}$$

where D, θ and γ are typical diffusion coefficient, time (inverse rate) and concentration in the systems and $\widetilde{\delta\lambda}$ is a (dimensionless) measure for the distance from the critical point. In an open chemical system λ may represent a feeding rate for one of the reactants and is measured in mole/(cm^{3}sec). The dimensionless $\widetilde{\delta\lambda}$ is related to $\delta\lambda$ by

$$\widetilde{\delta\lambda} = \delta\lambda\frac{\theta}{\gamma} \tag{46}$$

For a typical liquid chemical system in three dimensions we have $D \cong 10^{-5}$ cm^2/sec and we get from (45) and (46)

$$\delta\lambda \gg \frac{10^{-16}}{\gamma^{1/2}\,\theta^2} \tag{47}$$

where θ should now be measured in seconds and γ in moles/cm^3 . The r.h.s. of (47) is a measure for the size of the critical region, expressed in terms of a feeding rate as an externally controlled parameter. It is seen that for very fast reactions ($\theta < 10^{-5}$ sec) and low enough concentrations this region may be experimentally accessible. Within it, critical exponents should deviate from their mean field values. We should remember, however, that due to the limitations imposed on the derivation of section 4, the TDGL equation (38) itself may not be valid there.

6. First Order Transitions

So far we were interested in systems near their critical points. Far from the critical point the system may have several steady states. A typical situation with three steady states is shown schematically in Fig. 1a, where x^0 and x^1 represent stable branches and x^2 is an unstable branch. In the analogous case of equilibrium first order phase transitions we expect the transition between the two stable branches to occur at a point, λ_t , where the free energies of these branches become equal. Similar reasoning may be made in the present non-equilibrium case where instead of free energy we now have the potential function $U(x(r))$ defining the steady state probability distribution [15]. This approach is limited to the vicinity of the critical point or to very simple, usually unrealistic, models where such a potential function may be found. An alternative way is based on the macroscopic deterministic kinetics and may be understood from the following gedanken experiment: Suppose we prepare the system initially in a long tube, such as to the right we have it in the state x^0, and to the left in x^1. At time zero we let the two phases diffuse into each other. Depending on λ one of the phases will take over the other. λ_t is defined as this point where both remain unchanged.

To study the dynamics underlying this process I take a simple example [17]. Consider a system represented by the equation

$$\frac{\partial x}{\partial t} = D\frac{\partial^2 x}{\partial r^2} - k\,(x-a)(x-b)(x-c) \qquad ; \qquad b \geq c \geq a \tag{48}$$

where r is now a one-dimensional length coordinate. The system is characterized by three homogeneous steady states: a and b are stable and c is unstable to small homogeneous perturbations. A solution of (48) satisfying the boundary conditions $x(r \longrightarrow \infty) = b$ and $x(r \longrightarrow -\infty) = a$ is given by [18]

$$x(\varphi) = a + (b-a)[1+\exp(-\beta\varphi)] \tag{49}$$

where

$$\varphi = r - v t \tag{50}$$

$$\beta = \left(\frac{k}{2D}\right)^{1/2}(b-a) \tag{51}$$

and where

$$v = \left(\frac{1}{2}kD\right)^{1/2}(a+b-2c) \tag{52}$$

This solution represents a soliton with constant velocity v moving in one direction or another, depending on the magnitudes of a+b and c. If there is a single parameter λ which determines the state of the system then a, b and c are functions of λ, and λ_t is the solution of

$$a(\lambda) + b(\lambda) = 2c(\lambda) \tag{53}$$

Numerical calculations were done also on two- and three-dimensional systems with initial conditions representing a drop of one phase immersed in a bulk of the other [17]. There, for $\lambda > \lambda_t$ say, we can identify a critical radius above which the drop of x^1 immersed in x^0 will grow and below which it will shrink. In case of growth the behavior again approaches a soliton solution as the front of the growing sphere becomes plane-like.

Numerical solutions for systems with more than one variable show essentially similar behavior [15,16]. In fact, close to the critical point the similarity between (48) and (30) indicates that a soliton-type behavior is typical for a general system. Far from the critical point analytical approach is much more limited though several perturbational methods have been recently developed for a variety of special cases [17,19].

7. Conclusion

As was pointed out by GRAHAM [14b] macroscopic instabilities, such as hydrodynamical and chemical instabilities, are different from equilibrium phase transitions; the former are of macroscopic origin while the latter are microscopic in nature. However, the analogy between these phenomena can be carried out quite far. With the advances made in designing controlled non-equilibrium systems, and in numerical simulations of reacting diffusing systems, implications of this analogy, such as discussed here, are expected to be further investigated.

ACKNOWLEDGMENT

This work is partially supported by the United States-Israel Binational Science Foundation, Jerusalem, Israel.

References

1a. Synergetics, Edited by H. Haken, B.G. Teubner, (Stuttgart, 1973).
1b. Cooperative Effects: Progress in Synergetics, Edited by H. Haken (North-Holland, Amsterdam, 1974).
1c. H. Haken, Rev.Mod.Phys. 47, 67 (1975).

2a. R. Graham and H. Haken, Z.Physik. 237, 31 (1970).
2b. V. DeGiorgio and M.O. Scully, Phys.Rev. A2, 1170 (1970).
2c. H. Haken, Phys.Letters 53A, 77 (1975).
2d. R. Graham, Phys.Letters 58A, 440 (1976).

3a. V.M. Zaitsev and M.I. Shliomis, Sov.Phys.JETP 32, 866 (1971).
3b. J.P. Boon, J.Phys.Chem.Liquids 3, 157 (1972).
3c. H.N.W. Lekkerkerker and J.P. Boon, Phys.Rev. A10, 1355 (1974).
3d. R. Graham, Phys.Rev. A10, 1762 (1974).
3e. H. Haken, Phys.Letters A46, 193 (1973).

4. R. Landauer, Ferroelectrics 2, 47 (1971).

5a. F. Schlögl, Z.Physik 253, 147 (1972).
5b. A. Nitzan, P. Ortoleva, J. Deutch and J. Ross, J.Chem.Phys. 61, 1056 (1974).
5c. C.W. Gardiner, K.J. McNeil, D.F. Walls and I.S. Matheson, J.Stat.Phys. 14, 307 (1976).

6. M. Lax, Rev.Mod.Phys. 32, 25 (1960).

7. Y. Kuramoto and T. Tsuzuki, Prog.Theor.Phys. 52, 1399 (1974); 54, 687 (1975).

8. H. Haken, Phys.Letters 51A, 125 (1975); Z.Physik B20, 413 (1975); B21, 105 (1975).

9. B.B. Edelstein, J.Theor.Biol. 29, 57 (1970).

10. A. Nitzan and J. Ross, J.Chem.Phys. 59, 241 (1973).

11. C.L. Creel and J. Ross, J.Chem.Phys. 65, 3779 (1976).

12. P. Ortoleva and S. Yip, J.Chem.Phys. 65, 2045 (1976).

13a. S. Grossmann, J.Chem.Phys. 65, 2007 (1976).
13b. C.W. Gardiner, J.Stat.Phys. 15, 451 (1976).

14a. R. Graham, Springer Tracts in Modern Physics No. 66 (Springer, Berlin, 1973).
14b. R. Graham, in Fluctuations, Instabilities and Phase Transitions, Edited by T. Riste (Plenum, New York, 1976).

15. N.G. van Kampen, Adv.Chem.Phys. 34, 245 (1976).

16. V.L. Ginzburg, Fizika Tverdogo Tela 2, 2031 (1960).

17. A. Nitzan, P. Ortoleva and J. Ross, Far.Symp. 9, 241 (1974).

18. E.W. Montroll, in Statistical Mechanics, Edited by S.A. Rice, K.F. Freed and J.C. Light (Univ. of Chicago Press, 1972), p. 69.

19. P. Ortoleva and J. Ross, J.Chem.Phys. 60, 5090, (1974); 63, 3398 (1975).

Chemical Evolution far from Equilibrium

A. Pacault

With 28 Figures

Lecture given at the workshop of Elmau summarizing the most important results obtained in this field by J. Boissonade, P. Hanusse, P. De Kepper, A. Pacault, A. Rossi, J.C. Roux, C. Vidal

1. Introduction

Chemical instabilities are able to give rise to strange phenomena such as oscillating reaction in homogeneous media. The reaction of hydrogen peroxide and potassium iodate in dilute aqueous sulfuric acid solution, published in 1921 by BRAY (1), is the first experimental evidence of oscillating behavior in an homogeneous chemical reaction. More recently, BELOUSOV (2)(1959) reported sustained oscillations in the oxidation of citric acid by potassium bromate catalyzed by ceric sulfate in dilute aqueous sulfuric acid. This reaction and its variants were extensively investigated by ZHABOTINSKII (3) in the early 1960's. The interest in this field becomes more and more intensive (fig. 1). Moreover the experiments are few in which are fulfilled the thermodynamical conditions required to obtain sustained oscillations in chemical reactions : existence of fluxes between the system and its surroundings (table I).

This condition can be provided by a suitable continuous flow stirred-tank reactor (fig. 2) in which were made all the experiments described below. We have investigated the BELOUSOV-ZHABOTINSKII reaction (B.Z.) and the H_2O_2 - KIO_3 - $CH_2(CO_2H)$ - $MnSO_4$ - $HClO_4$ reaction, proposed by BRIGGS and RAUSCHER (4), which ressembles the reaction of BRAY but operates at room temperature with greater amplitude in oscillations (B.R.). Here is the most complete study of homogeneous reactions performed in a well fitted reactor and under conditions which exhibit many interesting phenomena such as : sustained oscillations, multiple stationary states, excitation, composite oscillations, peculiar regulation, chemical turbulence (5 to 23).

Because of unusual features of the chemical reactions treated in this lecture it is necessary to introduce a suitable definition of the terms employed (see table I).

Table 1 Definition of terms

Constraints	Variables that might be controlled by the experimenter *(for example concentrations or flows)*
Surroundings or external medium	The whole set of constraints
Responses	Variables measurable by the experimenter ; all the responses are very rarely measured.
System	The whole set of responses
Object of investigation	Conjunction of the system and its surrounding
State of the object	A set of values of the responses for a given set of constraints

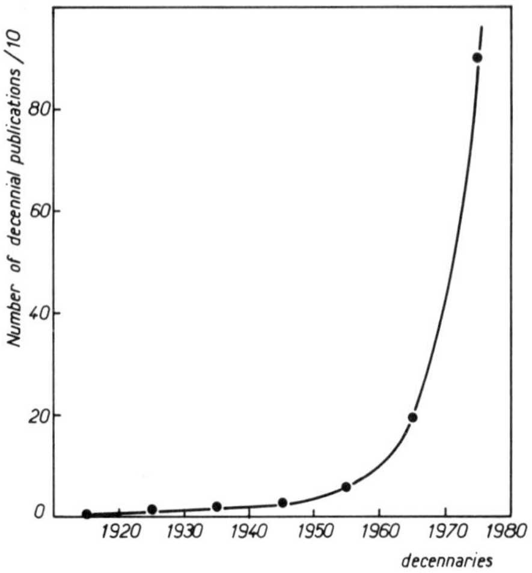

Fig. 1.
Approximate number of papers published on chemical evolution far from equilibrium as a function of time.

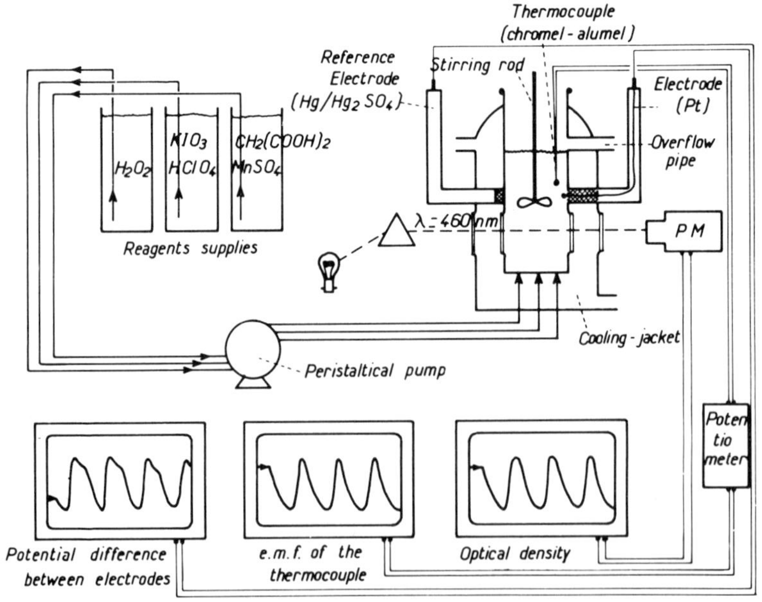

Fig. 2. Scheme of the continuous-flow stirred-tank reactor.

In these experiments, the constraints are stirring, light intensity, temperature T, pressure P, residence time τ (ratio of the reactor volume to the total flow rate) and the concentrations $|A_i|_0$ of the inlet species A_i after mixing and before any reaction. The reponses (Table I^0) measured are the temperature, the oxidation-reduction potential E, and the absorption of light by an intermediate species.

2. Bench experiments

2.1 *State characteristics* : Qualitatively one may distinguish steady non oscillating states from sustained oscillating states which have a very stable frequency with a relative occuracy within 5.10^{-3}. Figure 3 shows three oscillating responses which can last as long as reactants are provided ; for example some experiments were carried over 10^4 periods. It must be compared with the chemical oscillations generally published since the Bray's first result (fig. 4).

The fast addition of a small amount of an intermediate species in the reactor disturbs the oscillations (fig. 5 : A,B,C) but they recover their initial form after a while. This stability to perturbation is characteristic of a limit cycle behaviour. Its representation in the responses space (fig. 6) shows very well the evolution of the reaction during one period (points on the cycle are time equidistant).

Indeed, such oscillating reactions give a good illustration of a real thermodynamical clock besides atomic clock and astronomical clocks. Thus, we have built a clock driven by a chemical reaction (12).

The existence of these stable oscillating states suggests an improvement of the experimental procedure in order to reach the nature of chemical species and their production (or consumption) rates. Indeed periodicity allows accumulation of any signal over several periods, thus increasing signal to noise ratio. An apparatus coupling the reactor to a computer has been built (17)(fig. 7). A variable, for example the absorption of light at a given wavelength, is measured during a suitable number of periods, paying attention to trigger the recording at the same time. The oxido-reduction potential is used as a trigger for each cycle. The efficiency of such a method can be seen comparing the curves a and b of figure 8.

It is noteworthy that stable oscillating states may also be found in living systems : the whole field of ecology provides examples of population oscillations which suggest the same kind of mathematical models as oscillating chemical systems. Microbial populations are a convenient material to perform laboratory experiments exhibiting oscillatory behaviour of some simple living systems. As in the case of chemical reactions a system open to exchanges with the surroundings is necessary. A continuous flow-stirred-tank-reactor contains a nutrient medium which is inoculated with a bacterial prey (Escherichia COLI) and a bacterial predator (Bdellovibrio Bacteriovorus): the prey and the predator populations undergo large periodic oscillations closely in phase opposition (fig. 9)(30)

2.2 *State diagram* : Depending on the values of the constraints, sustained oscillations or steady states are observed in the reactor ; it is then possible to establish, in the constraints space, the critical surface separating these two classes of states.

Figure 10 shows examples of sections in the ($|CH_2(COOH)_2|_0$, $|KIO_3|_0$) plane of the state diagram of B.R. reaction. Periods of oscillating states are largely dependent on the values of the constraints ; they are given in seconds. Figure 11, section in the ($[CH_2(COOH)_2]_0$, $[KIO_3]_0$) for another residence time and figure 12 (T, $|CH_2(COOH)_2|_0$) are other examples of sections of the state diagram.

In this way one is able to build up state diagrams in the constraints space (H_2O_2 , KIO_3 , $CH_2(COOH)_2$) at every temperature and at every residence time all other constraints remaining constant ; figure 13 is an example of this.

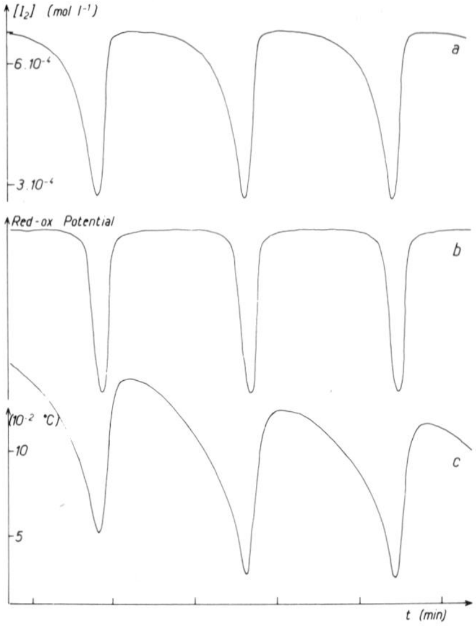

Fig. 3. Three responses of the B.R. reaction versus time.
a/ optical density of iodine
b/ oxido-reduction potential
c/ temperature

Values of the constraints : $|KIO_3|_0 = 0{,}068$ M
$|H_2O_2|_0 = 0{,}344$ M
$|CH_2(CO_2H)_2|_0 = 0{,}505$ M
$|HClO_4|_0 = 0{,}0515$ M
$|MnSO_4|_0 = 0{,}0941$ M
$\tau = 2{,}7$ min
$P = 1$ atm.
$T = 25°C$

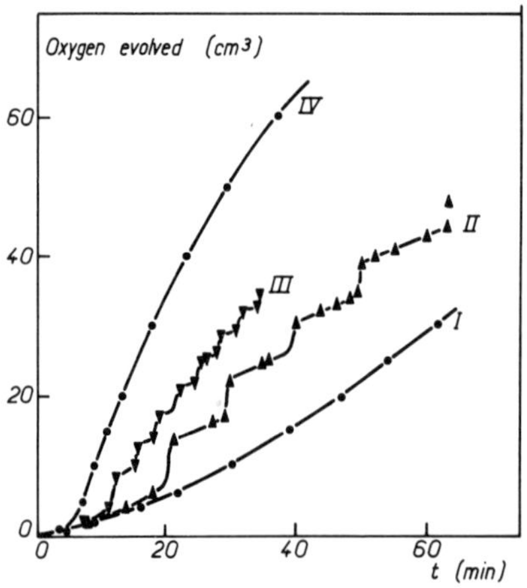

Fig. 4.
Amount of oxygen evolved versus time published by Bray (1)

I	SO_4H_2	0,055 N	
II		0,073 N	P = 1 atm.
III		0,0916 N	T = 60°C
IV		0,110 N	

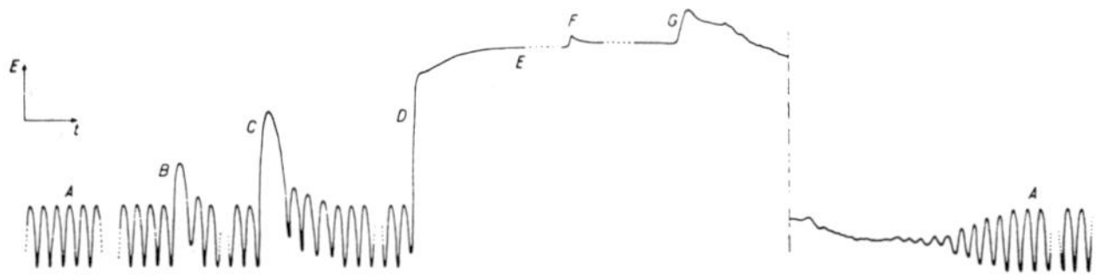

Fig. 5 .
Oxido-reduction potential versus time of an oscillating state A of a B.R. reaction $|KIO_3|_o$ 0,024 M, $|CH_2(CO_2H)_2|_o$ 0,056 M, $|H_2O_2|$ 1,2 M, $|HClO_4|_o$ 0,058 M, $|MnSO_4|_o$ 0,004 M ; T = 25°C, P = 1 atm., τ = 6,4 min perturbed by

B addition of 10^{-4} mole l^{-1} I^- leading to recover the limit cycle.

C addition of 3.10^{-4} mole l^{-1} I^- leading to recover the limit cycle.

D addition of 5.10^{-4} mole l^{-1} I^- ; over this threshold transition towards a non oscillating stable state E occurs.

F addition of 1,12 10^{-2} mole l^{-1} $CH_2(CO_2H)_2$ leading to recover the state E.

G addition of 4,5 10^{-2} mole l^{-1} $CH_2(CO_2H)_2$ leading to the initial limit cycle state.

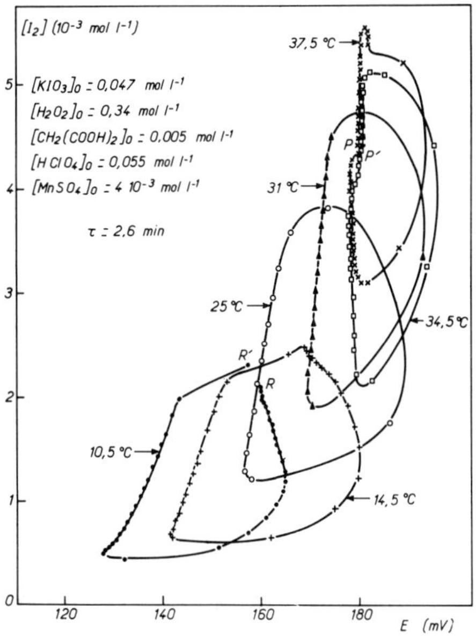

Fig. 6. Trajectories of the B.R. reaction in the responses space (iodine concentration - oxido-reduction potential) at different temperatures.

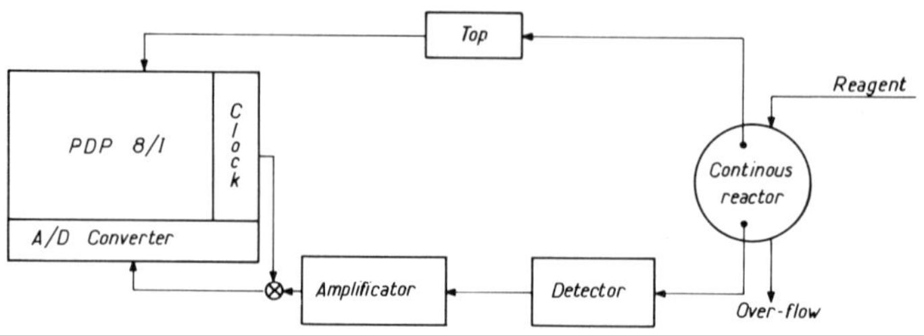

Fig. 7.
Scheme of apparatus allowing the accumulation of results obtained during a large number of periods.

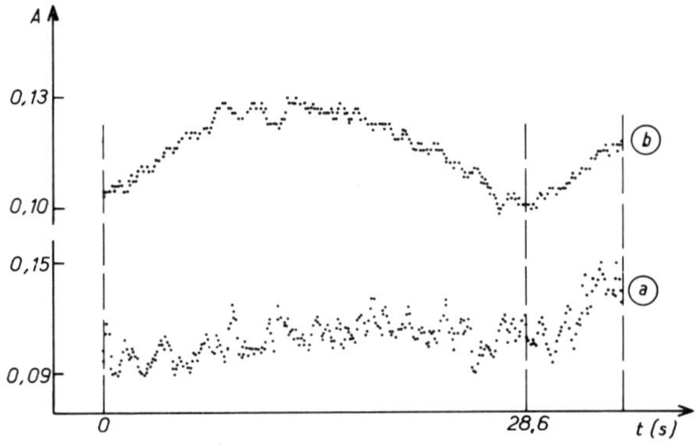

Fig. 8. Absorption at 700 nm versus time
a) initial spectrum
b) average after a hundred measurement.

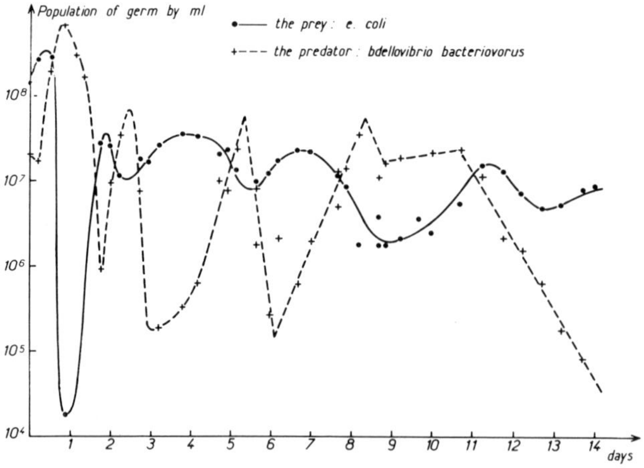

Fig. 9. Population of prey (o) and predator (△) versus time.

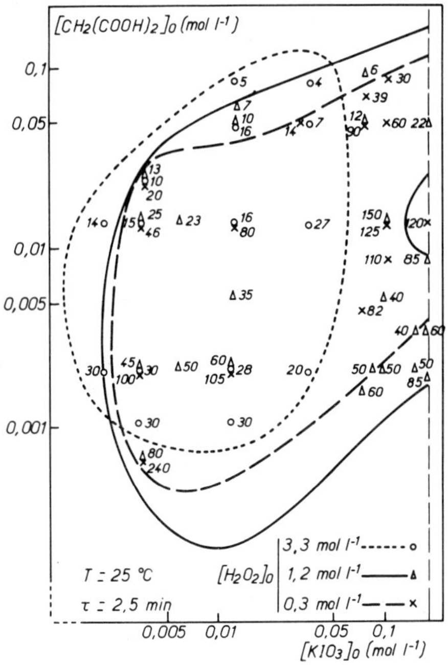

Fig. 10.
Section of state diagram of B.R. reaction.

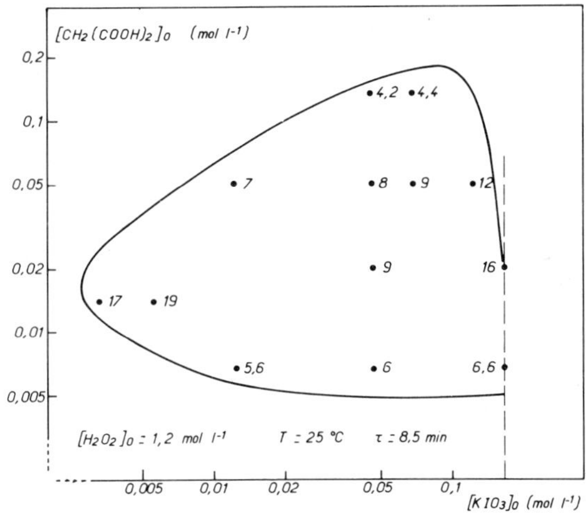

Fig. 11. Section of the state diagram of B.R. reaction.

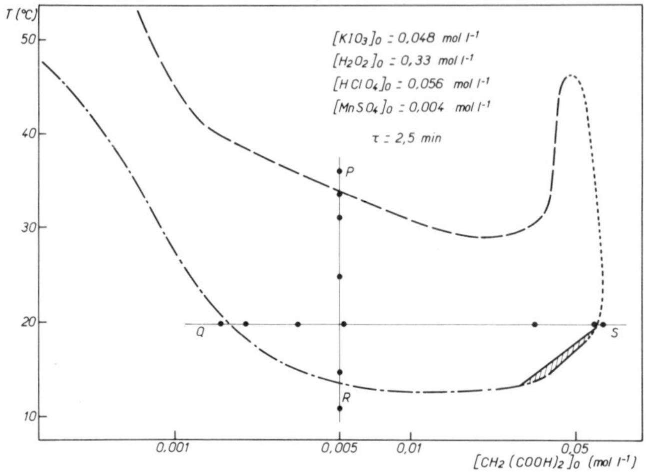

Fig. 12.
Section of the state diagram of B.R. reaction. Different kinds of transitions are observed at the boundary of the oscillating domain.

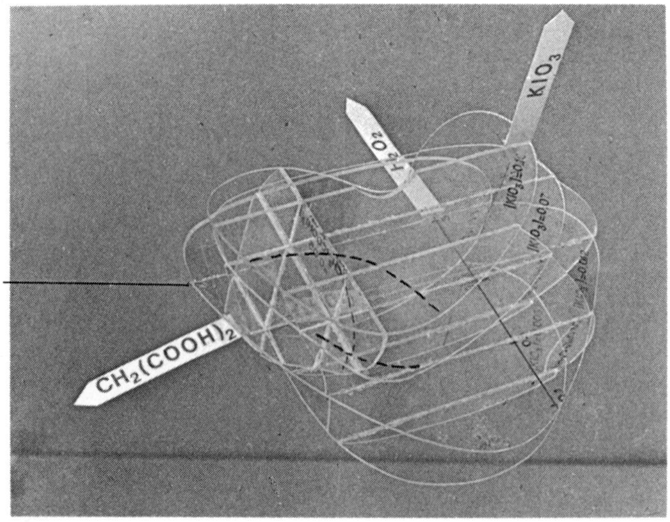

Fig. 13. State diagram in the constraints space KIO_3 , H_2O_2 , $CH_2(CO_2H)_2$.

In the same way we have set up the state diagram of B.Z. reaction at 40°C in the constraints space ($NaBrO_3$, $Ce_2(SO_4)$, $CH_2(CO_2H)_2$). Fig. 14 and 15 show two sections of this diagram. Depending on the region of the constraints space, the transitions from one class of states to the other by change of constraints exhibit very different characteristics. The transition may either show smooth changes in the amplitude of the limit cycle which fades at the transition point or, in contrast, present a jump in the average value of the responses. In this last case the limit cycle disappears with a finite amplitude. Sometimes, for the same set of constraints, there may exist several states, steady state or oscillating. This happens in the shaded regions of figure 10 where a sustained oscillatory state coexists with a steady state, the system being on either of them depending on the initial conditions of the experiment. In these regions, transitions may also occur for a fixed set of constraints, by a perturbation : i.e., the fast addition by the experimenter of a small but sufficient amount of an intermediate chemical species in the reactor (fig. 5 : D and G). The shapes and amplitudes of oscillations are constraints dependent.

2.3 *Multiple Stationary States* : Let us suppose the B.R. reaction to be in an oscillating state, B(fig. 16). The experimenter decreases constraint C on a time scale larger than the relaxation time of the system. For a critical value, C_s, of the constraint, the system regime changes abruptly to a nonoscillating state, inducing a steep variation of the average value in the iodine concentration. Now, if we reverse the evolution of the constraint C, the system remains in this steady state until the constraint reaches another critical value C_0($C_0 \neq C_s$) where oscillatory process is recovered with a discontinuity in iodine average concentration. If one decreases again constraint C, the system will remain in this oscillating state until $C = C_s$.

The spontaneous transition between oscillating and steady states only happens for two characteristic values C_0 and C_s of the constraint. Then, for any value of the constraint between C_0 and C_s, the system may be on either state A or state B : it is a bistable system. Transition from one state to the other is not *"invertible"** since there is a hysteresis phenomenon. Now, for a given value of the constraint between C_0 and C_s, one can trigger transitions from one state to the other by a perturbation. This has been done by producing a steep variation in the iodine concentration in the reactor.If after this perturbation the system, temporarily drawn from its stationary state, goes back to its initial state, the perturbation is said to be regressive and quoted "V" ; if, for a greater perturbation, the system shifts to the other stationary state, the perturbation is said to be nonregressive and quoted "Λ" (fig. 16). Between these two cases, for different values of the constraint, a threshold line of perturbation (dashed line) points out the frontier between regressive and nonregressive perturbations.

Fig. 16 illustrates the evolution of a bistable system as a function of the inverted ratio of the residence time τ(this quantity is proportional to the total flow rate through the reactor). Here $C_0 = 1/\tau_0 = 5.1 \times 10^{-3}\ s^{-1}$ and $C_s = 0.3 \times 10^{-3}\ s^{-1}$. Stationary steady state corresponds to high concentration in iodine (curve A). It is interrupted for values of iodine concentration greater than 20×10^{-4} mol $l.^{-1}$, the value above which solid iodine appears in the mixture. Curve B corresponds to the second state ; as a function of the constraint $C = 1/\tau$ this state shows sustained oscillations (vertical segments indicate their amplitude) until the value of $C = 1/\tau = 1.1 \times 10^{-3}\ s^{-1}$, below which it gives place to a damped oscillatory state (14). Between these two branches the threshold line S gives the minimum perturbation in the iodine concentration which triggers transitions from B to A states. This threshold line is not necessarily the locus of unstable states predicted by theory (14).

It is possible to have an opposite triggered transition by extracting iodine from the solution, dropping in the reactor a few milliliters of benzene, which rapidly floats on the reagent surface and is removed by an overflow pipe with a certain amount

---

* *We distinguish, with DUHEM, the "invertible" feature of a chemical reaction (as they are represented by $\Sigma_{\nu_i} A_i \rightleftarrows \Sigma_{\nu_i'} A_i'$) from a "reversible" process in which a chemical reaction may be involved. Indeed, invertible chemical reactions proceed very often far from thermodynamical reversibility.*

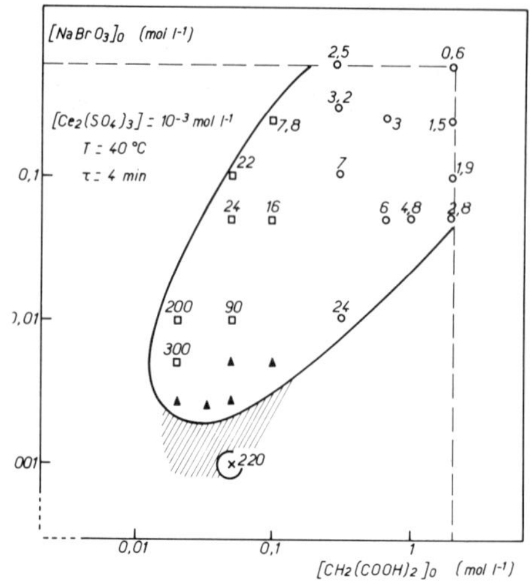

Fig. 14. Section of the state diagram of B.Z. reaction.

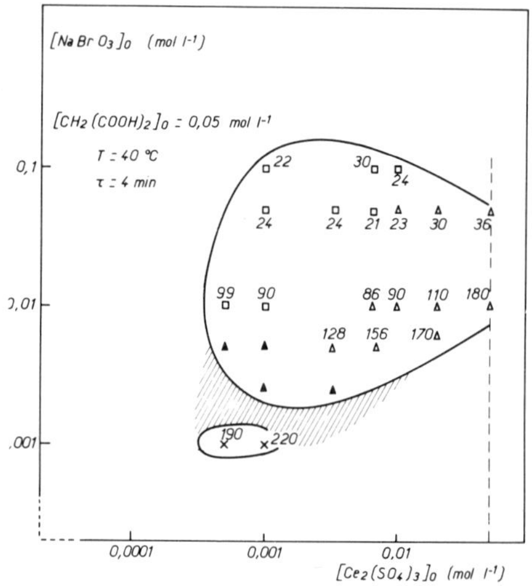

Fig. 15. Section of the state diagram of B.Z. reaction.

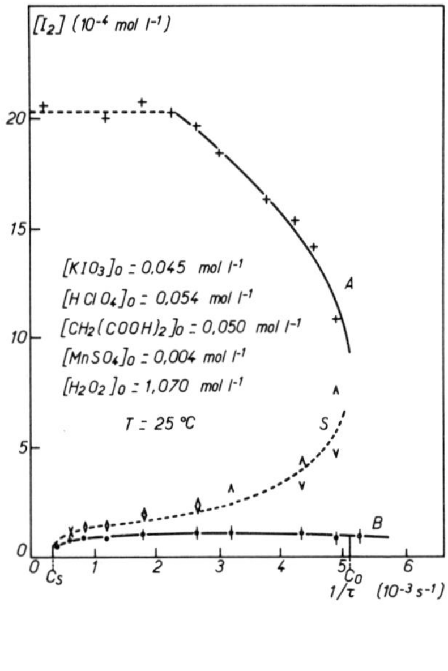

Fig. 16 Bistationary in the B.R. reaction showed in a response-constraint space

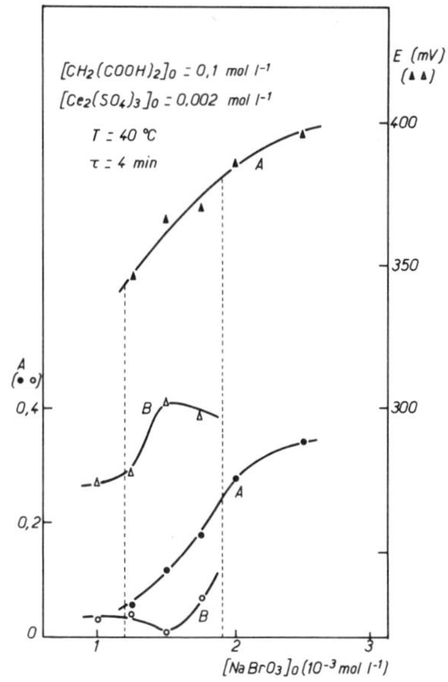

Fig. 17
Bistationary in the B.Z. reaction seen by two responses, optical density at 365 nm and oxido-reduction potential versus the constraint $NaBrO_{3\ 0}$.

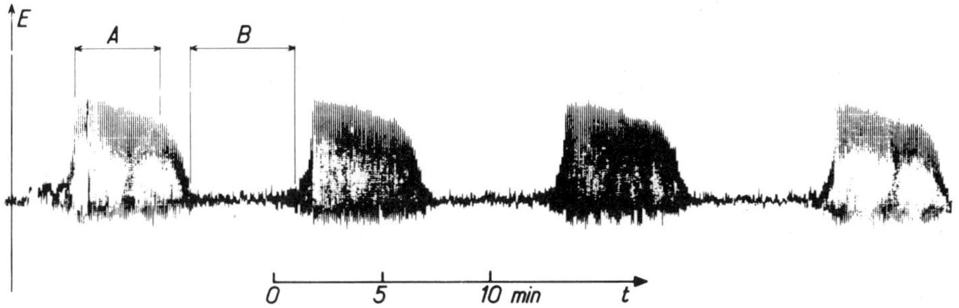

Fig. 18 Composite double oscillation in the B.R. reaction : oxido-reduction potential versus time

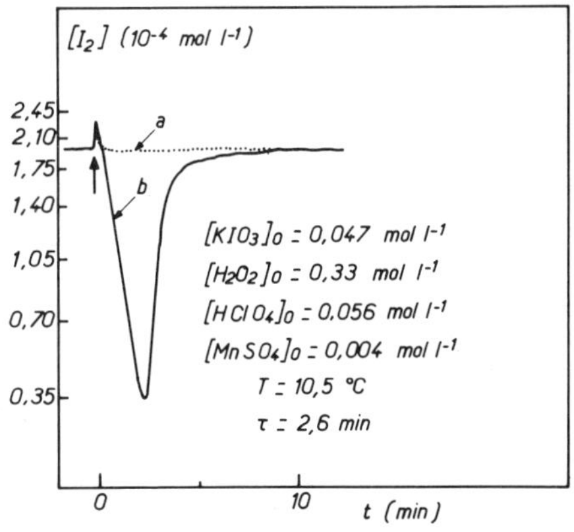

Fig. 19
Excitability. Iodine concentration versus time after an iodine perturbation of
a) o.o8 lo^{-4} mole l^{-1}
b) o,lo lo^{-4} mole l^{-1}

of iodine ; if enough iodine is extracted, the oscillatory state is restored.

For the first time a bistability was discovered in the state diagram of B.Z. reaction where two non oscillating steady states A and B exist (fig. 17). These two states can be shown when measuring the two following responses : oxido-reduction potential and optical density at 365 nm. versus the constraint $|NaBrO_3|_0$.

The theory does not predict a limit to the number of multiple steady states. Till now only bistability has been reported in chemical systems ; but furthermore an intensive study on B.R. reaction allowed us to find a region in the constraints diagram where three non oscillating steady states exist.

2.4 *Composite Double Oscillation* : Another striking behavior of this system is the composite double oscillation such as those observed in figure 18 where for a given set of constraints there is a periodical burst and disappearance of higher frequency oscillations.

This can be understood as a periodic transition between two pseudostationary states, one of them having an intrinsic oscillatory character. A group of intermediate species of the system evolving on a larger time scale than the remaining one can then be considered as pseudoconstraints for the latter intermediate species ; drifts of these pseudoconstraints are such that they flip-flop the system from one pseudostationary state to the other. A reaction scheme explaining this behavior has recently been developed (VIII, table II).

2.5 *Excitability* : Let us have a stationary steady state ; for perturbation (in $|I_2|$, for example) under a threshold value, the system relaxes rapidly to its initial steady state (a, in figure 19). Perturbation over a threshold will induce a decrease in the iodine concentration, having no correlation with the intensivity of the perturbation ; then, after a while, the system gets back to its initial steady state (b, in figure 19).

If the response trajectories for perturbations under and over the threshold are very different, beyond this threshold the response trajectory is nearly independent of the intensity of the perturbation. This phenomenon of chemical amplification of a perturbation is known as excitability. It is often observed in the neighborhood of a sustained oscillatory state, and the response trajectory after excitation is very similar to that exhibited by the neighboring limit cycle.

2.6 *Peculiar regulation* : Suppose that be B.R. reaction evolve in a domaine of the state diagram corresponding to an excitable steady state. We change the reaction adding a continuous flow of an iodine aqueous solution. Iodine which was only a response becomes now a constraint species. This new system is able to show a very peculiar type of regulation. As shown in fig. 20, when there is no iodine input, the system has a high concentration iodine steady state. A small iodine flux leads to a small increase in the iodine concentration of the reactor but, for a critical value of this flux, there is a sudden drop in the reactor iodine concentration. Its inversion with respect to the iodine flux presents an hysteresis phenomenon. That means that if the iodine flux decreases the low iodine concentration holds untill the iodine flux reaches another critical value.

2.7 *Chemical turbulence* : A computer experiment on chemical turbulence will be shown in the film exhibited at the end of this talk. Moreover, we found experimentally many different kinds of waves probably due to a chemical turbulence. A mathematical analysis will allow to precise if this regime is really chaotic (29).

2.8 *Kinetic Analysis of These Phenomena* : A better understanding of these phenomena, qualitatively described, requires a more quantitative approach wich becomes possible using the accumulating apparatus described before.

In a first step, the optical density is recorded as a function of time at different wavelenghs (fig. 21).

Table II Reaction Schemes

I	1920, A.J. Lotka (24) $A = B$	$A + X \rightarrow 2X$ $X + Y \rightarrow 2Y$ $Y \rightarrow B$	First theoretical model to predict sustained chemical oscillations, but oscillations are marginally stable.
II	1952, A.M. Turing (25) $A + 2B = D + E$	$A \rightarrow X$ $X + Y \rightleftarrows C$ $C \rightarrow D$ $B + C \rightarrow W$ $W \rightarrow Y + C$ $Y \rightarrow E$ $Y + V \rightarrow V'$ $V' \rightarrow E + V$	Spatial structure
III	1968, I. Prigogine and R. Lefever (26) $A + B = D + E$	$A \rightarrow X$ $2X + Y \rightarrow 3X$ $B + X \rightarrow Y + D$ $X \rightarrow E$	Limit cycle ; chemical waves ; localized and nonlocalized spatial structures
IV	1970, B. Edelstein (27) $C = B$	$C + X \rightleftarrows 2X$ $X + Y \rightleftarrows Z$ $Z \rightleftarrows Y + B$	Multiple steady states ; hysteresis
V	1972, C. Vidal (6,7) $A + 2B = D + E$	$A \rightleftarrows Y$ $B + Y \rightleftarrows X + Y$ $2X \rightleftarrows D + X$ $Y + X \rightleftarrows E$	Multiple steady states that can generate spatial organization
VI	1974, R.M. Noyes and R.J. Field (28) $fA + 2B = fP + Q$	$A + Y \rightarrow X$ $X + Y \rightarrow P$ $B + X \rightarrow 2X + Z$ $EX \rightarrow Q$ $Z \rightarrow fY$	Limit cycle ; excitability ; was suggested by the mechanism of the Belousov-Zhabotinskii reaction
VII	1974, P. Hanusse (13) $2A = B$	$A \rightarrow X$ $2X \rightarrow 2Y$ $Y + Z \rightarrow 2Z$ $X + Z \rightarrow B$	Limit cycle ; spatial structures and spatiotemporal structures
VIII	1975, J. Boissonade (20) $A + 2B + 3C + D = P + P' + X'$	$A \rightarrow X$ $B + X \rightarrow 2X$ $D + X \rightarrow P$ $X \rightarrow X'$ $B + X' \rightarrow Y$ $Y \rightarrow Z + X'$ $C \rightarrow \alpha$ $2\alpha \rightarrow 2\beta$ $\beta + Z \rightarrow 2Z$ $\alpha + Z \rightarrow P'$	Composite double oscillations in flow reactor conditions

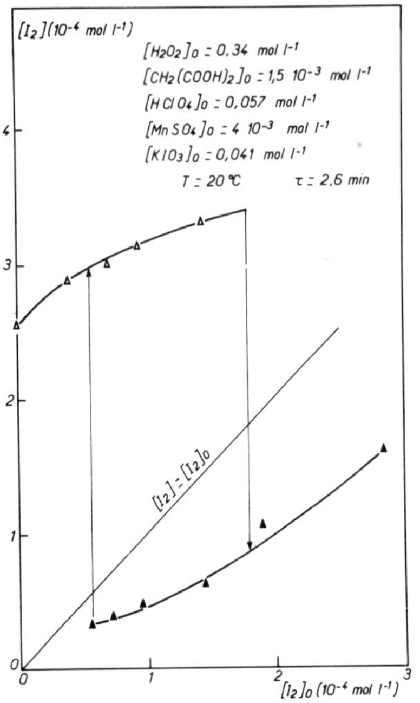

Fig.2o Peculiar regulation. Iodine concentration in the reactor $|I_2|$ versus the iodine flux $|I_2|_o$

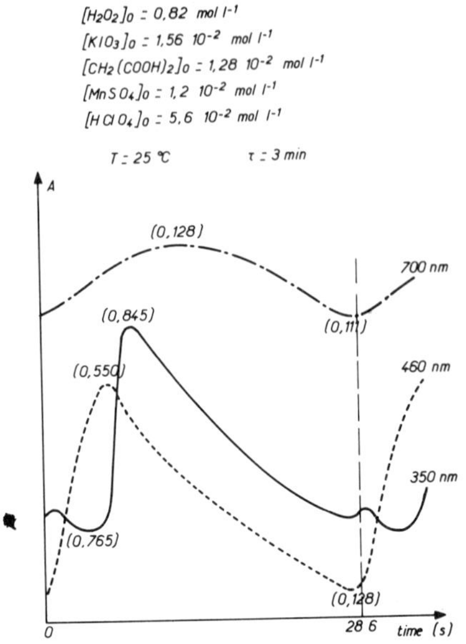

Fig.21 Optical density versus time at several wavelengths. A = f(t).

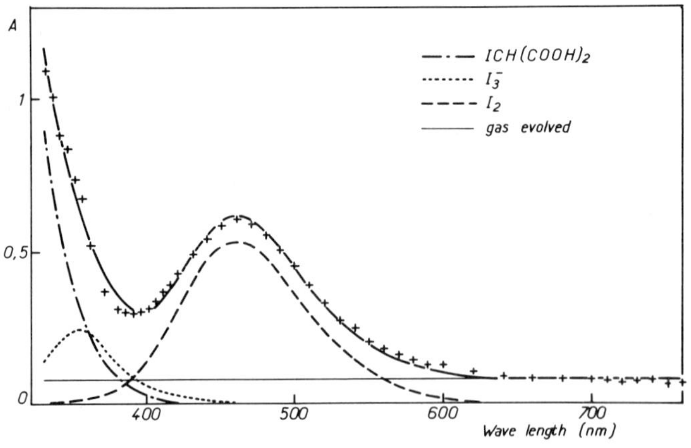

Fig. 22 Spectra of different species 3,9 s after triggering by the signal given by the oxido-reduction potential

Secondly these experimental results are sorted by a computer in order to give the spectra, i.e. optical density versus wavelength for every instant.

Thirdly, we attempt to recover these experimental spectra by a linear combination of spectra of the species assumed to be present in the medium (fig. 22). Fit is carried out by means of a least-squares analysis which leads to coefficients of the linear combination for each spectrum. As far as molecular extinctions are known, one obtains in this way the concentration of every species as a function of time (22)(fig. 23).

From these data we determine the reaction rate and phase for each species during a cycle (23)(fig. 24).

3. Computer experiments

A possible explanation of all these results is found using imaginary reaction schemes (table II) from which one can derive a set of differential equations.

Many mathematical methods were used to solve them. Moreover, it may also be fruitful to use a computer to do some experiments using suitable reaction schemes.

This film shows some interesing results obtained in this way.

3.1 *Explanations for Understanding this Film* : This text is meant to be a presentation of the super-8 film titled "Simulation des systèmes dissipatifs chimiques par une méthode de Monte-Carlo"(Monte-Carlo Simulation of Chemical Dissipative Systems) by Patrick HANUSSE.

3.1.1 Method : in this film we present some examples of the behavior of Chemical Dissipative Systems (31) studied by a Monte-Carlo simulation method. Chemical systems evolving far from equilibrium are interesting for their deterministic behavior as well as for their stochastic properties, that is when fluctuations are taken into account. The simulation method that we have used enables us to study the deterministic and stochastic dynamical behavior of chemical systems in time and space. The simulation process can be optimized for each case, deterministic or stochastic, separately. We will only describe here the formulation used to study stochastic properties, which can deal with the deterministic properties as well (9)(32).

The formulation of the simulation process is similar to the stochastic theory of chemical reaction (33) in which one considers the probability of finding in the system a given number of particles of each chemical species. From the description of each elementary process one derives a Master Equation which describes the evolution of this probability density. Thus, basically, the method we use gives a solution of the Master Equation for which an analytic solution is very rarely available.

Let us consider a system containing X "particles" of species X, the reaction $X \rightarrow Y$, and a constant k such that kX is less than unity. Then, a random number generator gives us a number between zero and one. If that number is greater than kX we only increment a time counter ; if lower, we decrease X by a certain amount n and increase Y by the same amount and then increment the time counter. The same sequence of operation is repeated again and again.

If X is large (a few hundred times n), then the evolution of the number of particles will be related in a straightforward way to the solution of the deterministic kinetic equation for this reaction, here, an exponential decay.

If X is small (a hundred times n or less), fluctuations around that solution will appear, and their characteristics (moments, correlations in time and space) can be measured (calculated) by averaging over time, mainly for systems at stationary state, or by averaging over a large number of identical systems, typically a hundred. The method can be generalized to any reaction step, to any complex reaction scheme and to the description of diffusion in space, having several identical systems or "cells", interact by a linear process (9)(32).

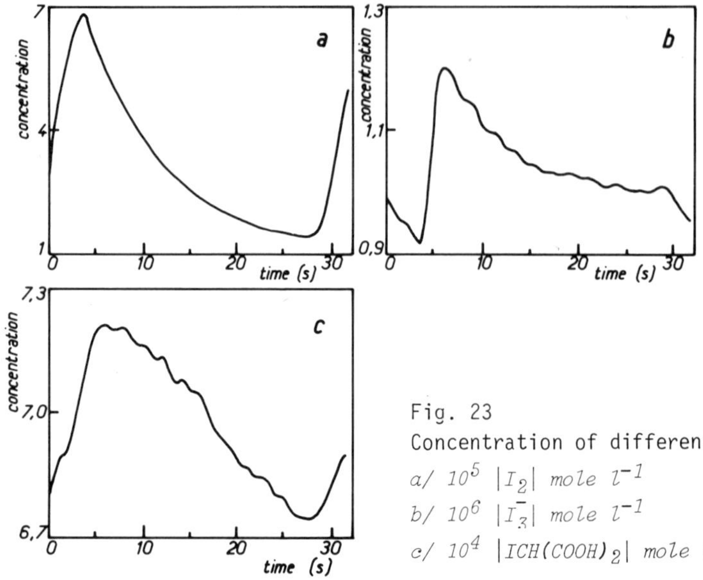

Fig. 23
Concentration of different species versus time
a/ $10^5\ |I_2|$ *mole* l^{-1}
b/ $10^6\ |I_3^-|$ *mole* l^{-1}
c/ $10^4\ |ICH(COOH)_2|$ *mole* l^{-1}

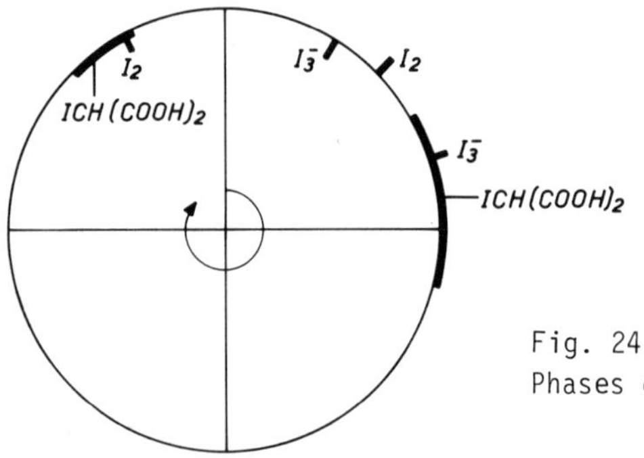

Fig. 24
Phases of extremums : maximum outside the circle
minimum inside the circle

3.1.2 Applications

First model : a first group of sequences describe some aspects of the behavior of the reaction scheme VII (table II) (13)(32).

- Limit Cycle and Inhomogeneities

For the following value of the parameters, $A = k_1 = k_3 = k_4 = 1$, $k_2 = 0.1$, there exists a Limit Cycle, i.e. an asymptotic closed trajectory in the X,Y,Z space. Therefore, in a uniform system (e.g. well stirred), the concentrations of X, Y and Z will oscillate in time. The stability analysis of the system shows (13)(32) that the stationary state of the system is unstable to uniform perturbations and to long wavelength perturbations only. This situation is the most common one for a Limit Cycle. Next, we allow for diffusion in a one dimensional space with the following value of the diffusion coefficients, $D_X = D_Y = 1$, $D_Z = 0.5$*. The flux of species X, Y and Z is zero at the boundary. The film shows the concentration of species X as a function of space in a system of 50 length units, starting on the Limit Cycle with uniform concentrations.

---

* *the length unit is the cell.*

Therefore, the main phenomenon observed is an overall oscillation of the concentration. Clearly, the uniformity of the concentration is not maintained. Two kinds of inhomogeneities can be observed on the film and confirmed by a more careful analysis of the fluctuations by calculating the spatial correlation function and the spectral density (32). First, long wavelength inhomogeneities that appear as a non-uniformity of the phase of the oscillation. This is expected since on a time scale τ, diffusion will only be able to maintain uniformity over a length of the order $(D\tau)^{1/2}$. Second, we observe short wavelength disturbances that are particularly visible during some stages of the overall oscillation. It is clear that a coherence appears on a short length scale and that some wavelength are particularly favored. This is not only a non-uniformity of the phase but a coherent periodic modulation of the phase in space. Of course, fluctuations are present, making the picture less clear in the real space. This phenomenon is related to the stability of the uniform motion around the Limit Cycle with respect to non uniform perturbations. This simple example gives and idea of the kind of spatial phenomena which can appear in an unstirred oscillating system ; such systems may also exhibit chemical waves.

- Oscillating Spatial Structures.

Chemical turbulence : in the second sequence we consider the case when the uniform stationary state is stable to uniform perturbations but unstable to non-uniform perturbations within a finite domain of wavelengths. The stability is such that the divergence from the uniform stationary state occurs in an oscillating fashion**. At the boundaries, the concentrations of X, Y and Z are kept constant at their stationary value. The value of the parameters are : $A = k_1 = 1$; $k_2 = 0.005$; $k_3 = 1.8$; $k_4 = 0.6$; $D_X = 2$; $D_Y = D_Z = 0.1$.

A first important observation is that the system does not seem to attain any asymptotic periodic behavior. The complex oscillation in time and space has nevertheless a number of interesting features. First, at some positions, fixed points appear : at these points the concentrations does not oscillate at all or very slightly. Between two fixed points or between a fixed point and a boundary, one can observe either oscillations with propagation from one point to the other or in phase oscillation with no propagation (fig. 25). Both cases may be observed at a given time and at different

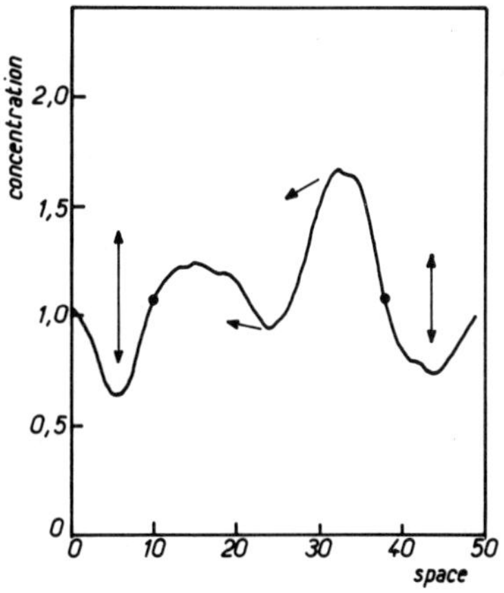

Fig. 25 Oscillating Spatial Structure : concentration profile of species X as a function of space. The dots are nodes of vibration. In the center of the figure a wave emitted by the right node is propagating towards the left node and is absorbed by it. On each sides the oscillation occurs without propagation. The arrows indicate the direction of motion.

** *i.e. there exists two complex conjugate unstable stability eigenvalues (13).*

positions. Such a pattern does not last and a fixed point, particularly one "absorbing" a wave emitted by a second one, may disappear, with formation of another fixed point elsewhere in the system. Thus, a fixed point can appear disappear or move in the system. Moreover, the portion of space between to fixed points seem to form an isolated subsystem with little influence on the surrounding and little influenced by it.

We have used the term "chemical turbulence" to qualify such a complex behavior. By turbulence we mean that a coherent behavior can appear only for a finite time and over a finite portion of space. It seems that in such a case the system has no "attractor" stable enough for a global coherent motion to build up in time. Of course, the fluctuations that result from the stochastic processes involved, even when small, entrance that feature. Other kinds of chaotic behavior have been predicted for homogeneous reaction models (34). Although different, they all exhibit weak stability of motion which is probably the main macroscopic characteristic of turbulence. Nevertheless a more accurate definition of this concept in chemical reactive systems is still needed, particularly for experimental purposes.

Vibrating String : in some cases oscillation without propagation can extend over the entire system. For the following value of the parameters, $D_X = 2$, $D_Y = D_Z = 0$ (others as before), we have observed an oscillating structure that looks like a stationary vibrating string (13) with six nodes of vibration (fig. 26). Such a structure is not stable and small perturbations will usually make it evolue to a chaotic behavior as before. In the film, time is progressively accelerated to show at the end the average profile of the concentration.

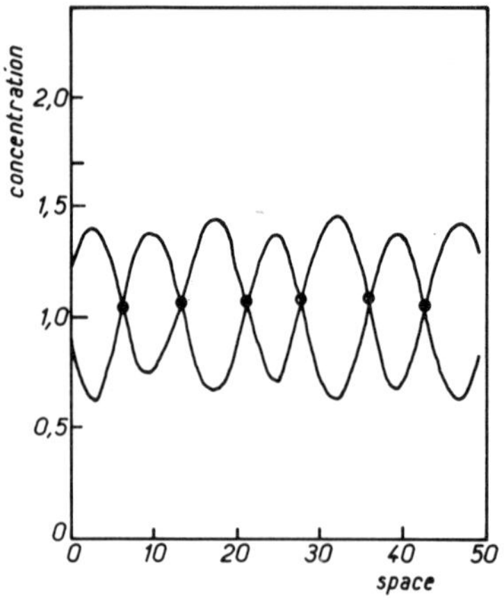

Fig. 26 Vibrating string : two out of phase profiles in space during the course of an oscillation without propagation.

- Fixed Spatial Structures

For the same model there exists stability conditions such that the uniform stationary state is unstable to non-uniform perturbations as before but with a monotonous divergence from the uniformity*. In that case, fixed spatial structures are obtained.

The film shows the formation of various structures for a system length of 50, 100 and 200 and for the following value of the parameters : $A = 1$; $k_1 = 2$; $k_2 = 0.1$;

* *i.e. there exists one real positive stability eigen value.*

$k_3 = 1$; $k_4 = 1/1.4$, $D_X = 0$, $D_Y = 1000$, $D_Z = 20$. Boundary conditions are of the "zero flux" type. The system is initially in the uniform stationary state. Fluctuations grow and a new stable non-uniform structure appear (see fig. 27). For a given length several structures may be achieved : symmetric structures or asymmetric or "polar" ones. When a polar structure appears the symmetry of the system is broken since the boundary conditions are symmetric and all processes are locally isotropic. This is an important feature of the evolution of chemical systems far from equilibrium (31). The wavelength of the final structure is not necessarily in the range of wavelength with respect to which the uniform stationary state is unstable. In fact the stability analysis is only valid in the vicinity of the stationary state and does not give any information about the characteristics of other non-uniform stationary states. It is interesting to look at the dynamics of fluctuations during the formation of the structure by Fourier analysis of the equal time spatial correlation function. One sees then that during the early stages of the formation of the structure, even before it appears macroscopically, Fourier components of fluctuations in the domain of instability are dominent. Thereafter the component of the final structure builds up progressively.

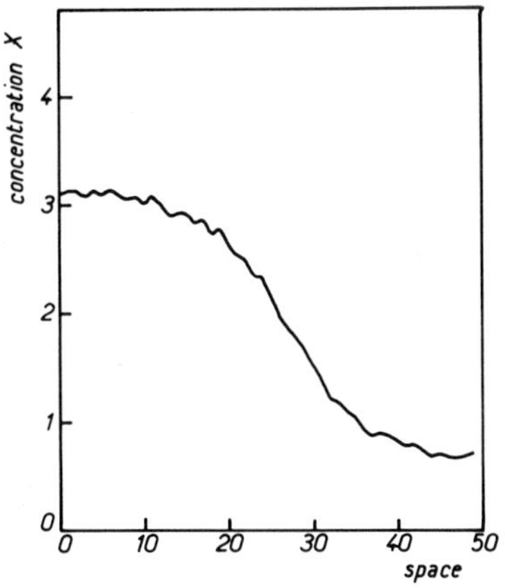

Fig. 27 Fixed Spatial Structure : this asymmetric structure has been formed after a symmetry breaking transition from an unstable uniform stationary state. System length 50.

Edelstein Model : eventhough fluctuations were present and played an important role in what we have seen before, the deterministic behavior of the system was the essential aspect under consideration. We now term to situations where fluctuations are larger, to such an extend that we can no longer talk about deterministic behavior but must consider stochastic properties. Fluctuations are studied for themselves and their effect on the macroscopic behavior of the system. Several interesting problems can be investigated in this field using Monte-Carlo simulation (35). The film presents one application. We consider the Edelstein model (IV table II) with the following value of the parameters : $k_i = 1$, $k_{-i} = 0.1$, $B = 0.1$, $Y + Z = 5$.

- Homogeneous Fluctuations and Bistability

The deterministic kinetic equations for this model indicate that there are three stationary states (solid line in fig. 28), two stable, one unstable for A between A_1 and A_2 in fig. 28 : however, stochastic treatment (36) predicts only one stationary distribution function, hence only one value of the macroscopic concentration for a given value of A. The stochastic simulation allows us to measure easily the distribution function and its moments. This sequence of the film shows the number of particles X (ordinate) in an ensemble of 200 identical independent systems (abscissa). At the beginning all the systems are set on branch E_1 in fig. 28 for A a greater than A_T. At first the number of particles fluctuate around branch E_1. Then, in some systems the number of particles X increases sharply untill a stationary state is obtained for the ensemble, in which the picture shows two "clouds" of points with a continuous exchange between them. We then have the stationary bimodal distribution predicted by the stochastic theory. Successive sequences of the film show the same process for increasing values of A. The moments of the distribution can be calculated (see fig. 28) as well as the correlation time of fluctuations in each branch and the transition rate from one branch to the other. The first moment (dashed line in fig. 28) has a single value for each value of A so that no hysteresis is observed when sweeping A back and forth, as it is when the deterministic equations are used (solid line in fig. 28). In this example we see that the average of the stochastic behavior can differ from the deterministic prediction.

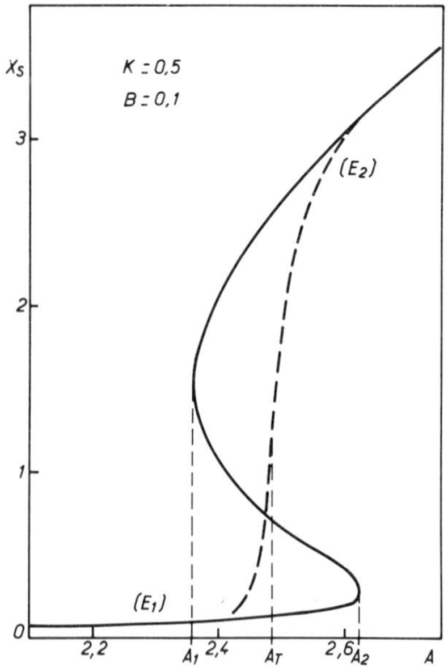

Fig. 28 Bistability : stationary state versus parameter A for the Edelstein model as predicted by the deterministic kinetic equations (solid line) and average number of particles (dashed line) obtain by stochastic simulation. On concentration unit is equivalent to 20 "particles"(35).

- Transition by Nucleation

We now consider diffusion between those 200 systems in a one dimensional space as before. The system is started in branch E_1 for A = 2.56. Macroscopic uniformity is maintained for a while due to diffusion and in spite of fluctuations. But after an interval there appears on the left of the system a small portion of space with higher concentration. The size of this "nucleus" increases due to diffusion again. Finally it fills up the system which is then on branch E_2. That state is said to be more stable than the metastable initial one E_1. Here, the transition between two macroscopically uniform states is achieved more easily by a non-uniform transient state (nucleation process) rather than by a most unlikely uniform transition. The evolution of such a chemical system is obviously analogous to a first order phase transition.

3.1.3 Conclusion

This film was meant to give a few examples of the fascinating behavior of chemical systems evolving far from equilibrium and at the same time to show the effectiveness of the Monte-Carlo simulation method in the study of those systems. Of course, this method is particularly well suited to studing stochastic properties and can be used with benefit to investigate problems that cannot be handled analytically at all or only by making questionnable approximations. The validity of such approximations can be checked by such calculations. On the other hand this method does not seem to present the problems of accuracy encountered in some numerical calculations (37) and is probably easier to use than molecular dynamics (38) in the same range of application.

In the future, several interesting problems should be investigated, in particular those related to the analogy to phase transitions such that the behavior of fluctuations and correlations near critical points.

4. General Conclusion

There is a common feeling that complexity can give rise to new things. Nevertheless the strange phenomena discussed here (multiple stable states, excitability, temporal and spatial structuration) are still amazing in view of the quite ordinary chemistry involved. Will it then be possible, as we have done, to study in detail these new phenomena and thus have solid bases for their qualitative development in other fields (biology, sociology, linguistics) in which the great number of parameters make quantitative explanation difficult ?

REFERENCES

1 W.C. Bray, *J. Am. Chem. Soc. 43, 1262 (1921).*

2 B.P. Belousov, *Sb. Ref. Radiats. Med. Moscow, 1958, 145 (1959).*

3 *"Oscillatory Processes in Biological and Chemical Systems", vol. I.,* G.M. Franck, ed., *Nauka, Moscow, 1967 ; vol. II,* E.E. Sel'kov, ed., *Puschino-na-Oka, Moscow, 1971.*

4 T.S. Briggs and W.C. Rauscher, *J. Chem. Educ. 50, 7, 496 (1973).*

5 P. Hanusse, *C.R. Acad. Sc. Paris, C.274, p. 1245 (1972).*

6 C. Vidal, *C.R. Acad. Sc. Paris, C.274, p. 1713 (1972).*

7 C. Vidal, *C.R. Acad. Sc. Paris, C.275, p. 523 (1972).*

8 P. Hanusse, *C.R. Acad. Sc. Paris, C.277, p. 263 (1973).*

9 P. Hanusse, *C.R. Acad. Sc. Paris, C.277, p. 93 (1973).*

10 A. Pacault, P. De Kepper et P. Hanusse, *C.R. Acad. Sc. Paris, C.280, p. 73 (1975).*

11 A. Pacault, P. De Kepper, P. Hanusse et A. Rossi, *C.R. Acad. Sc. Paris, C.281, p. 215 (1975).*

12 A. Pacault, P. De Kepper et P. Hanusse, *Proceedings of the 25th International Meeting of the Société de Chimie-Physique (1975).*

13 P. Hanusse et A. Pacault, *Proceedings of the 25th International Meeting of the Société de Chimie-Physique, 1974, Elsevier, Amsterdam, 1975, p. 50.*

14 P. De Kepper, A. Pacault et A. Rossi, *C.R. Acad. Sc. Paris, C.282, p. 199 (1976).*

15 A. Pacault, P. Hanusse, P. De Kepper, C. Vidal et J. Boissonade, *Fundamenta Scientiae n° 52, Université Louis Pasteur (1976).*

16 P. De Kepper, *C.R. Acad. Sc. Paris, C. 283, p. 25 (1976).*

17 J.C. Roux, S. Sanchez et C. Vidal, *C.R. Acad. Sc. Paris, B.282, p. 451 (1976).*

18 P. De Kepper, A. Rossi et A. Pacault, *C.R. Acad. Sc. Paris, C.283, p. 371 (1976).*

19 A. Pacault, P. Hanusse, P. De Kepper, C. Vidal et J. Boissonade, *Accounts of Chemical Research, vol. 9, p. 438 (1976).*

20 J. Boissonade, *J. Chim. Phys. n° 5, p. 540 (1976).*

21 P. Hanusse, *Phys. Letters, 59A, p. 421 (1977).*

22 J.C. Roux et C. Vidal, *C.R. Acad. Sc. Paris, C. 284, p. 293 (1977).*

23 C. Vidal, J.C. Roux, A. Rossi, *C.R. Acad. Sc. Paris, C.284, p. 585 (1977).*

24 A. Lotka, *J. Phys. Chem., 14, 271 (1910) ; Proc. Natl. Acad. Sci. USA, 6, 410 (1920) ; J. Am. Chem. Soc., 42, 1595 (1920).*

25 A.M. Turing, *Philos. Trans. R. Soc. London, Ser. B, 237, 37 (1952).*

26 I. Prigogine and R. Lefever, *J. Chem. Phys., 48, 795 (1968) ;* M. Herschkowitz and G. Nicolis, *ibid., 56, 1890 (1972).*

27 B.B. Edelstein, *J. Theor. Biol., 29, 57 (1970).*

28 R.J. Field and R.M. Noyes, *J. Chem. Phys., 60, 1877 (1974) ;* R.J. Field, *ibid., 63, 2289 (1975) ;* R.M. Noyes, R.J. Field and E. Körös, *J. Am. Chem. Soc. 94, 1394 (1972) ;* R.J. Field, R.M. Noyes and E. Körös, *ibid., 94, 8649 (1972).*

29 T.Y. LI and J.A. Yorke, *Am. Math. Monthly, 82, 985 (1975).*

30 A. Marchand and E. Dulos, *Appl. Microbiol. 30, n° 6, p. 994 (1975) ;* A. Marchand and E. Dulos, *private communication.*

31 *"Structure, Stabilité et Fluctuations",* P. Glansdorff and I. Prigogine, *Masson, Paris (1971).*

32 P. Hanusse, *Ph. D. Dissertation, Bordeaux, France (1976).*

33 McQuarrie, *Adv. Chem. Phys. XV, 149 (1969).*

34 O. Rössler, *Z. Naturforsch 31a, 259-64 (1976) ; 31a, 1168-1172 (1976) ; 31a, 1664-1670 (1976).*

35 P. Hanusse, *J. Chem. Phys. (1977) in the press.*

36 I. Matheson, D.F. Walls and C.W. Gardiner, *J. Stat. Phys. 12, 21 (1975).*

37 J.S. Turner, *Adv. Chem. Phys. XXIX, 63 (1975).*

38 A. Nitzan, P. Ortoleva, J. Deutsch and J. Ross, *J. Chem. Phys. 61, 1056 (1974).* P. Ortoleva and S. Yip, *J. Chem. Phys. 65, 2045 (1976).*

Entropy and Critical Fluctuations in a Stochastic Model of Second Order Nonequilibrium Phase Transition

G. Czajkowski

With 1 Figure

1. Introduction

A large class of phenomena in the field of dissipative structures shows closed analogy of certain steady state changes to phase transitions. These phenomena therefore are called "nonequilibrium phase transitions" [1-7]. In particular, certain chemical reaction models were discussed showing phenomena of this kind of phase transition [6-12].

In the following we shall discuss a chemical reaction model showing a nonequilibrium phase transition of second order. We show that a stationary solution of an appropriate Markovian master equation is equivalent to a representative distribution obtained by information-theoretical methods. For this distribution a relative d fluctuation at the critical point appears as a universal constant. We show also that the entropy attains always its relative maximum. Finally, we show that a decrease of the absolute value of the entropy does not necessarily mean an increase of the order.

2. A Phase Transition of Second Order

The following chemical reactions between four chemical species A, B, C, X may be assumed:

$$\begin{aligned} A + X &\underset{k_2}{\overset{k_1}{\rightleftarrows}} 2X \\ A &\overset{k_3}{\rightarrow} X \\ X + B &\underset{\rightarrow}{k_4} C \end{aligned} \tag{1}$$

[6,13]. The concentrations of the species A, B, C (which we shall also denote by A, B, C) shall be held constant. Only the concentration X can vary with time. The derivative of X with respect to time is

$$dX/dt = (a - b)X - c\,X^2 + a \tag{2}$$

where we denoted

$$a: = k_1A\ , \quad b: = k_4B\ , \quad c: = k_2 \tag{3}$$

and put $k_3 = k_4$ (k_i's are reaction constants).

For the steady state we get

$$X^s = \begin{cases} 0 & a \leq b \\ (a - b)/c & a > b \end{cases} \tag{4}$$

if $X \gg 1$. This is the behaviour of a phase transition of second order. We can compare it either with the phase transition of a ferromagnetic substance or with a laser transition [6,13].

3. A Stochastic Model

In a stochastic model the following equation of motion for the probability function $P_n(t)$ that there are n molecules of species X at time t is required

$$\partial P_n(t)/\partial t = a[n\, P_{n-1}(t) - (n+1)\, P_n(t)] + b[(n+1)\, P_{n+1}(t) - n\, P_n(t)] + c[(n+1)n\, P_{n+1}(t) - n(n-1)\, P_n(t)] \quad , \tag{5}$$

cf. [13]. Its stationary solution has the form

$$P_n = P_0 \exp\left[n \ln y - \sum_{k=1}^{n} \ln(1 + uk)\right] \tag{6}$$

with $y{:}= a/b$ and $u{:}= c/b$. P_0 is obtained by the normalization of P_n:

$$P_0^{-1} = {}_1f_1(1\ ;\ 1 + u^{-1}\ ;\ y/u) \tag{7}$$

where ${}_1f_1$ is the hypergeometric function. The distribution P_n shows a threshold behaviour which is typical for phase transitions of the second order. For $y < y_c = 1 + u$ it is approximately a geometric distribution

$$P_n = (1 - y)\, y^n \quad . \tag{8}$$

The mean and standard deviation of this distribution are

$$\langle n\rangle = y\,(1 - y)^{-1} \tag{9}$$

$$\sigma = \sqrt{\langle n^2\rangle - \langle n\rangle^2} = y\,(1 - y)^{-1} \quad . \tag{10}$$

The relative fluctuation

$$\frac{\sigma^2}{\langle n\rangle^2} = 1 + \frac{1}{\langle n\rangle} \tag{11}$$

equals approximately 1 for macroscopic values of $\langle n\rangle$. Above threshold $y > y_c$ the distribution (6) is approximately a Poisson distribution

$$P_n = e^{-n_0}\,\frac{(n_0)^n}{n!} \quad , \quad n_0{:}= a/c \tag{12}$$

for which the mean and standard deviation are:

$$\langle n\rangle = n_0 \quad , \quad \sigma = (n_0)^{1/2} \quad . \tag{13}$$

The relative fluctuation

$$\frac{\sigma^2}{\langle n\rangle^2} = \frac{1}{n_0} \tag{14}$$

tends to zero for a distribution peaked well away from zero.

4. An Information-Theoretical Approach

From the point of view of information thermodynamics the macroscopic situation of a system is characterized by a set of mean values of certain macroscopic observables. In the case considered these are first two statistical moments of the variable n. Following the scheme of information thermodynamics, we consider a set of probability distributions P_n which realize a given macroscopic situation, namely have the same first two moments. The set is called a macrostate and its elements are microstates (for the presentation of information thermodynamics cf., e.g., [14]). For the system under consideration a macrostate W is defined as

$$W = \left(P_n : P_n \geq 0 , \sum_n P_n = 1 , \sum_n n^r P_n = m_r , r = 1,2\right) . \tag{15}$$

If the mean values m_r fulfill some conditions, then there exists a unique representative microstate $P^0 \in W$ which maximizes the entropy, i.e., such that

$$S(W) := \sup_{P \in W} s(P) = s(P^0) \tag{16}$$

where

$$s(P) = -\sum_n P_n \ln P_n \quad . \tag{17}$$

The representative microstate has the form

$$P_n^0 = Z(\lambda_1,\lambda_2)^{-1} \exp(-\lambda_1 n - \lambda_2 n^2) \tag{18}$$

where

$$Z(\lambda_1,\lambda_2) = \sum_n \exp(-\lambda_1 - \lambda_2 n^2) \tag{19}$$

and the equations

$$m_r = - \frac{\partial \ln Z(\lambda_1,\lambda_2)}{\partial \lambda_r} \tag{20}$$

have unique solutions for λ_1,λ_2. We show that the representative microstate (18) has the same properties as the stationary distribution (6). The threshold condition reads now $\lambda_1 = 0$.

a) Below threshold $\lambda_1 \gg \lambda_2 > 0$. Then the representative distribution is approximately a geometric distribution

$$P_n^0 = \frac{1}{m_1 + 1}\left(\frac{m_1}{m_1 + 1}\right)^n \tag{21}$$

with the relative fluctuation

$$\frac{\;^2}{m_1^2} = 1 + m_1^{-1} \tag{22}$$

which corresponds to the situation defined by (8-11).

b) Above the threshold $\lambda_1 < 0$ and the representative distribution may be approximated by a Gaussian distribution (and thus by a Poisson distribution)

$$P_n^o = \tilde{Z}^{-1} \exp\left[-\lambda_2\left(n - \frac{\lambda_1}{2\lambda_2}\right)^2\right] \tag{23}$$

$$\tilde{Z} = \exp(\lambda_1^2/4\lambda_2) Z(\lambda_1,\lambda_2) \quad . \tag{24}$$

For $(-\lambda_1/2\lambda_2) >> 1$ we have

$$\lambda_2 = (2\sigma^2)^{-1} \ , \quad \lambda_1 = -m_1/\sigma^2 \ , \quad Z = \sigma\sqrt{2\pi} \tag{25}$$

cf. also [15]. Exact formulae for λ_1,λ_2 as functions of m_1,m_2 were given in the continuous case, i.e., for the distribution

$$P^o(x) = Z(\lambda_1,\lambda_2)^{-1} \exp(-\lambda_1 x - \lambda_2 x^2) \tag{26}$$

with

$$Z(\lambda_1,\lambda_2) = \int_0^\infty dx \exp(-\lambda_1 x - \lambda_2 x^2) \tag{27}$$

and

$$m_r = \int_0^\infty dx\, x^r\, P^o(x) \tag{28}$$

[15,16]. It was shown that for the above distribution

$$0 < \frac{\sigma^2}{m_1^2} \leq 1 \quad . \tag{29}$$

The threshold condition $\lambda_1 = 0$ gives the critical value of the relative fluctuation:

$$\left(\frac{\sigma^2}{m_1^2}\right)_{cr} = \frac{\pi}{2} - 1 \simeq 0.57 \quad , \tag{30}$$

i.e., for the statistical moments m_1,m_2

$$\frac{m_2}{m_1^2} = \frac{\pi}{2} \quad . \tag{31}$$

Below the threshold

$$1 \geq \frac{\sigma^2}{m_1^2} > \frac{\pi}{2} - 1 \tag{32}$$

and the distribution (26) is approximately an exponential distribution

$$P^o(x) = \lambda\, e^{-\lambda x} \ , \quad \lambda = 1/m_1 \quad . \tag{33}$$

Above the threshold

$$0 < \frac{\sigma^2}{m_1^2} < \frac{\pi}{2} - 1 \quad . \tag{34}$$

For $(\sigma^2/m_1^2) \to 0$ and $m_1 >> 1$ the distribution (26) is practically a Gaussian

$$P^0(x) = \frac{1}{\sqrt{2\pi\sigma^2}} \exp\left[-\frac{(x-m_1)^2}{2\sigma^2}\right] . \tag{35}$$

For the stationary distribution (8) the critical relative fluctuation is given by:

$$\left(\frac{\sigma^2}{m_1^2}\right)_{cr} + 1 = V(N,u) = \frac{\left(\sum_{n=0}^{N} n^2 e^{f_n}\right)\left(\sum_{n=0}^{N} e^{f_n}\right)}{\left(\sum_{n=0}^{N} n e^{f_n}\right)^2} \tag{36}$$

where

$$f_n = f(n,u) = n \ln(1+u) - \sum_{k=1}^{n} \ln(1+uk) . \tag{37}$$

Some numerical values for the function V(N,u) for N = 300 are given in Table 1.

$u.10^3$	V	m_1
10	1.519	6.273
9	1.522	6.634
8	1.524	7.069
7	1.528	7.608
6	1.531	8.303
5	1.535	9.244
4	1.539	10.623
3	1.545	12.937
2	1.552	18.162
1	1.553	19.127

Thus for sufficiently small values of u the critical behaviour of the stationary distribution (6) is equivalent to that of the representative distribution. It follows also from the expansion of f(n,u). For $0 < u << 1$ we have $f(n,u) \simeq -un^2/2$.

5. The Entropy

With respect to the relations (16-20) the entropy S(W) of the macrostate (15) is given by

$$S = S(\lambda_1,\lambda_2) = \ln Z(\lambda_1,\lambda_2) + \lambda_1 m_1 + \lambda_2 m_2 . \tag{38}$$

Since $\lambda_r = \lambda_r(m_1,m_2)$ by the relations (20), the entropy is a function of the moments:

$$S(\lambda_1,\lambda_2) = F(m_1,m_2) = F_1\left(m_1,\frac{\sigma^2}{m_1^2}\right) . \tag{39}$$

The above relation defines a certain surface (the entropy surface) in 3-dimensional space $(m_1, m_1/\sigma^2, S)$.

a) In a plane m_1 = const S is a monotonic increasing function of σ^2/m_1^2. It increases from the value

$$S_{min} = (1/2)\ \ln 2\pi\ e\sigma \tag{40}$$

for $(\sigma^2/m_1^2) \to 0$ (far above the threshold) to the maximal value

$$S_{max} = \ln m_1 + 1 \quad . \tag{41}$$

The critical value of the entropy obtained for continuous distribution by setting $\lambda_1 = 0$ is given by

$$S_c = \ln m_1 + \ln(\sqrt{e}\ \pi/2) < S_{max} \quad . \tag{42}$$

b) In a plane σ^2/m_1^2 = const the entropy is an increasing function of m_1 (Fig.1). Thus the system may evolve from a less ordered (chaotic) state I characterized by $(\sigma^2/m_1^2) > (\sigma^2/m_1^2)_{cr}$ to an ordered state II with $(\sigma^2/m_1^2) < (\sigma^2/m_1^2)_{cr}$ by different ways on the entropy surface. It may happen that $S_I > S_{II}$, or $S_I < S_{II}$, or $S_I = S_{II}$.

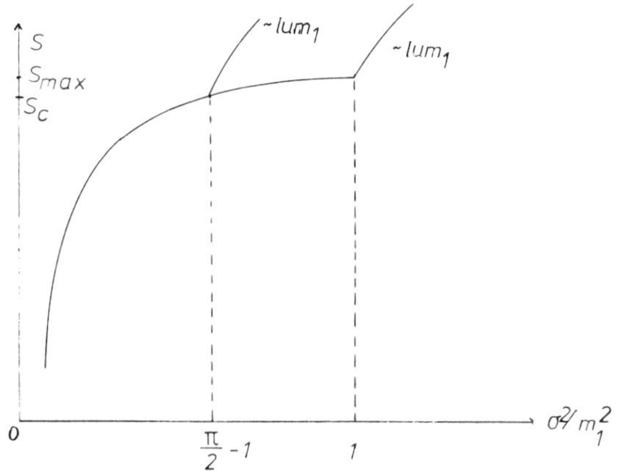

References

1. Haken, H.: In Cooperative Effects, Progress in Synergetics, ed. H. Haken (North-Holland, Amsterdam 1974)
2. Pytte, E., Thomas, H.: Phys. Rev. A179, 431 (1969)
3. DeGiorgio, V., Scully, M.O.: Phys. Rev. A2, 1170 (1970)
4. Grossmann, S., Richter, P.H.: Z. Physik 242, 458 (1971)
5. Woo, J.W.F., Landauer, R.: IEEE J. QE-7, 435 (1971)
6. Schlögl, F.: Z. Physik 253, 147 (1972)
7. Haken, H.: Rev. Mod. Phys. 47, 67 (1975)
8. Glansdorff, P., Prigogine, I.: Thermodynamic Theory of Structure, Stability, and Fluctuations (Wiley, New York 1971)
9. Czajkowski, G.: Z. Physik 270, 25 (1974)
10. Nitzan, A., Ortoleva, P., Deutch, J., Ross, J.: J. Chem. Phys. 61, 1056 (1974)
11. Ortoleva, P., Ross, J.: J. Chem. Phys. 63, 3398 (1975)
12. Czajkowski, G., Ebeling, W.: J. Non-Eq. Thermodyn. 2, 1 (1977)

13. McNeil, K.J., Walls, D.: J. Stat. Phys. 10, 439 (1974)

14. Ingarden, R.S.: Acta Phys. Polon. A43, 3 (1973)

15. Czajkowski, G.: Acta Phys. Polon. A44, 747 (1973)

16. Czajkowski, G.: Bull. Acad. Polon. Sci.: Ser. Sci. Math. Astron. Phys. 21, 759 (1973)

Chemical Waves and Turbulence

Chemical Waves and Chemical Turbulence

Y. Kuramoto

With 7 Figures

1. Introduction

The purpose of this talk is to present a unified viewpoint about the pattern formation in an oscillating reaction-diffusion system. A general equation describing it, is:

$$\dot{X} = R(X)+D\nabla^2 X \tag{1}$$

Here X is a vector composed of concentration variables, R(X) represents the reaction part, and D is a diagonal diffusion matrix. Without going into the details of the properties of the subsystem $\dot{X}=R(X)$, we assume here simply that this subsystem has a stable time-periodic solution $X_0(t)$ of limit cycle type. Thus we are concerned with the pattern formation taking place in a system composed of many local nonlinear oscillators coupled with each other through diffusion.

The next section is a brief sketch of how two types of contracted kinetic equations may be derived from (1). These equations are universal in the sense, that their forms are model-independent. Each equation can be obtained for a certain extreme situation, which is attained either by changing a system parameter or restricting our observation to such a class of phenomena, that has a long characteristic time or spatial scale. As critical phenomena tell us, a universal nature of a system normally latent, is likely to emerge in some extreme situations. In our case the resulting kinetic equation is a compact dynamical representation of a usually hidden universality.

The first extreme situation is realized in the vicinity near the bifurcation point of a certain kind, and in the second one a class of phenomena with slow spatial variation of phase will be considered. Technically speaking, the contraction of (1) is possible by means of certain perturbation methods, since some small parameters are definable in the above situations. Our perturbation methods are different from the bifurcation theory |1|, because we do not directly look for solutions, but look for a generator of solution via kinetic equations by using perturbation theory. Though this peculiarity of our approach yields on the one hand some ambiguity of our theory from a mathematical viewpoint, at least our theory has undoubtedly a great utility. Also, a similar idea like ours has long been one of the most important viewpoints in the field of nonequilibrium statistical mechanics, and can be found e.g. in the very successful works of CHAPMAN and ENSKOG on the problem of deriving hydrodynamic equations from the Boltzmann equation |2|.

The application of our contracted kinetic equations to pattern formation will be discussed in sections 3~5. Three types of patterns are considered there: circular waves, spiral or scroll waves, and a chaotic pattern, which one may call a type of chemical turbulence.

2. Two Universal Equations

Let us begin with the first case, in which the system is near a bifurcation point. The bifurcation we are now concerned with, is of the following type: Below some threshold value of a system parameter the equation $\dot{X}=R$

has a stable steady state, and above that value it bifurcates into an unstable steady state and a stable limit cycle as a result of the unstable growth of a couple of eigenmodes with complex conjugate eigenvalues. In the vicinity of this instability point one may expect that the dynamics governed by $\ddot{X}=R$ is essentially confined in the 2-dimensional subspace spanned by the above-mentioned critical eigenmodes. This implies, that a suitably defined complex variable $w=x+iy$ is sufficient to describe the dynamics near the bifurcation. A conclusion from the reductive perturbation scheme |3||4| is that if one chooses the directions and the scales of x and y axes in the concentration space in a suitable way, the equation governing w takes the form

$$\dot{w}=(\lambda-g|w|^2)w+d\nabla^2 w \tag{2}$$

Due to its similarity to the time-dependent Ginzburg-Landau equation for superconductivity, we shall retain the same name for (2) briefly TDGL. Eq.(2) however, has a nontrivial difference from the ordinary TDGL, namely, the parameters λ, d and g in (2) are all complex numbers. The real parts of λ and d are generally positive above the bifurcation point, and Re g is also positive, because normal bifurcation is being considered. This means, that by scaling the amplitude w, the time t and the spatial coordinate r, we can normalize all these real parts to 1, and we get

$$\dot{w}=(1+ic_0)w+(1+ic_1)\nabla^2 w-(1+ic_2)|w|^2 w \tag{3}$$

For $t\to\infty$ the solution of (3) in the absence of the $\nabla^2 w$ term is a limit cycle oscillation along a unit circle in the x-y space with the frequency c_0-c_2. Though this is a great simplification of the general nonlinear oscillation $X_0(t)$, we now have the complex coupling constant $1+ic_1$ among these idealized oscillators. The nonvanishing imaginary part c_1 comes from the fact, that in general there is a difference in magnitude among the diffusion constants of the original concentration variables.

Let us next consider the second case for which another universal equation can be obtained |5|. Although in this case the system state has not to be near a bifurcation point, we must restrict our concern now to the class of phenomena with a slow spatial dependence of the phase.

If the concentrations form a spatially uniform oscillation, we have

$$X=X_0(\phi) \tag{4}$$

Here ϕ is the phase of the oscillation, its convenient definition being

$$\phi=\omega t+\phi_1, \qquad \omega=2\pi/T, \tag{5}$$

where T is the period of oscillation, and ϕ_1 is an arbitrary constant, or

$$\dot{\phi}_1=0 \tag{6}$$

Eq. (6) states that the phase has a neutral stability, and that in a sense the excess phase ϕ_1 may be looked upon as a constant of motion. Let us next consider a slightly more general case in which the state point of each local nonlinear oscillator still lies approximately on the limit cycle orbit, but its phase varies slowly in space. Since $\phi_1(r,t)$ is a kind of conserved quantity, its slow spatial variation means the appearance of a characteristic long time scale. Thus in such a situation a reaction-diffusion equation will be greatly simplified just as the Boltzmann equation when extremely contracted yields hydrodynamic equations which are expressed in terms of five slowly varying conserved quantities. The state (4) may be compared to the complete equilibrium of a gas, while $X_0(\phi(r,t))$ to the local equilibrium state. It is a well-known fact, however, that the transport phenomena are the result from the deviations from the local equilibrium state. Thus, in the spirit of ENSKOG, we assume

$$X=X(\phi,\nabla\phi,(\nabla\phi)^2, \ldots) \tag{7a}$$

$$\dot{\phi}=\Omega(\phi,\nabla\phi,(\nabla\phi)^2, \ldots) \tag{7b}$$

We replace ∇ by $\varepsilon\nabla$ and expand X and Ω in powers of ε:

$$X=X_0+\varepsilon^2 X_1+\varepsilon^4 X_2+\ldots \tag{8a}$$

$$\Omega=\Omega_0+\varepsilon^2\Omega_1+\varepsilon^4\Omega_2+\ldots \tag{8b}$$

It is evident that the above expansion terms must have the form

$$X_0=X_0(\phi),\quad X_1=X_1^{(1)}(\phi)\nabla^2\phi+X_1^{(2)}(\phi)(\nabla\phi)^2,\ldots \tag{9a}$$

$$\Omega_0=\omega,\qquad \Omega_1=\Omega_1^{(1)}(\phi)\nabla^2\phi+\Omega_1^{(2)}(\phi)(\nabla\phi)^2,\ldots \tag{9b}$$

so that the nontrivial lowest order approximation means

$$X=X_0(\phi(r,t)), \tag{10a}$$

$$\dot{\phi}_1=\Omega_1^{(1)}(\phi)\varepsilon^2\nabla^2\phi_1+\Omega_1^{(2)}(\phi)\varepsilon^2(\nabla\phi_1)^2 \tag{10b}$$

The unknown quantities are the coefficients $\{X_n^{(\nu)}(\phi),\ \Omega_n^{(\nu)}(\phi)\}$. These are solved iteratively by putting (9a) and (9b) into (1) and making a balance equation in each (n,ν); it turns out that all the above coefficients become 2π-periodic in ϕ in the limit $\phi\to\infty$, or equivalently, $t\to\infty$. Since ϕ_1 has a slow time-variation, the coefficients $\Omega_1^{(1)}(\phi)$ and $\Omega_1^{(2)}(\phi)$ in (10b) may be replaced by their time-average over one period of oscillation. Thus (10b) is reduced to

$$\dot{\phi}_1=\nu\nabla^2\phi_1+\mu(\nabla\phi_1)^2 \tag{11}$$

where we have put $\varepsilon=1$, and

$$\nu=\frac{1}{2\pi}\int_0^{2\pi}\Omega_1^{(1)}(\phi)\,d\phi,\qquad \mu=\frac{1}{2\pi}\int_0^{2\pi}\Omega_1^{(2)}(\phi)\,d\phi \tag{12}$$

The description of the dynamics in the form (10a) and (11) is sometimes very convenient in discussing pattern formation. This is because the skeleton of a concentration pattern is essentially formed by the equi-concentration contour which is given by the condition

$$\phi(r,t)=2n\pi \qquad (n=0,\pm1,\pm2,\ldots) \tag{13}$$

according to (10a). Thus the essential part of a pattern can be obtained without the knowledge of the function $X_0(\phi)$; one has only to find ϕ_1 from (11) and put it into (13). Eqs. (3) and (11) are the two universal equations, which we shall use now to discuss patterns.

3. Circular Waves

The first pattern we want to study is a circular wave pattern or a concentric ring pattern |6| first observed by ZAIKIN and ZHABOTINSKY |7| for the BELOUSOV-ZHABOTINSKY reaction |8|. Some important features which must be theoretically accounted for are the following.

A. During the spontaneous bulk oscillation, leading centers (pacemakers)

appear which have higher frequency than the frequency of the bulk oscillation. The phase difference thus produced between the bulk medium and a leading center propagates outward as a chemical wave in the form of concentric rings or a target pattern. It is known that one has always an impurity at the pacemaker position.
B. It is evident that the outermost ring of a target should merge into the background once during a period of the bulk oscillation. But the fact that it never reappears needs explanation.
C. When the waves from different pacemakers collide, they become annihilated, such that a cusp structure appears in an oblique collision.
D. If two target patterns collide, the pattern controlled by the faster pacemakter expands at the expense of the pattern of the slower one. In this way only one target pattern, the one with the highest frequency, eventually dominates the whole space.

To explain these properties we only need the nonlinear phase diffusion equation (11) because that equation is somewhat simpler than TDGL (3). However, a difficulty arises from the fact that (11) was originally derived for a homogeneous system (1) while now we must consider a heterogeneous system including impurities. This is not a very serious problem. The experimental fact that chemical waves never diffuse in any distant regions far from heterogeneous nuclei implies that an undamped wave solution may exist for the homogeneous system (1) or its reduced form (11). The ro le of heterogeneous nuclei is considered to form such a kind of boundary condition at the nuclei, that would realize the above-mentioned wave solution for the homogeneous system (11). We shall find later, that a particular solution of (11) corresponding to an undamped wave train is $\nabla\phi_1=c$, where c is a constant vector. It is important to notice that $\nabla\phi_1=c$ means the change of the oscillation frequency by μc^2 as compared with the case $\nabla\phi_1=0$. In other words the system has an ability of changing its own frequency by assuming a phase distribution of constant slope, and the undamped propagation of waves is closely related to such a self-synchronizing ability. The above argument suggests the reason why ORTOLEVA and ROSS' phase wave theory |9| which in the homogeneous limit gives only a spatially uniform limit cycle solution, did not succeed to explain a <u>steadily</u> expanding target pattern. From such a class of solutions which tend to uniform limit cycle solutions in the homogeneous limit, a family of self-synchronizing solution as mentioned above will be excluded. Furhtermore, the works by KOPELL and HOWARD |10| are in accord with our viewpoint. The solution $\nabla\phi_1=c$ is nothing but the long-wavelength limit of a one-parameter family of periodic solutions as has been stated in their works; and they believe, quite reasonably, that such a family of solutions essentially corresponds to the wave train of rings observed byZAIKIN and ZHABOTINSKY.

After having obtained now some feeling about the role of the heterogeneity, we shall content ourselves by introducing the heterogeneity phenomenologically by adding an inhomogeneous term to (11):

$$\dot{\phi}_1=\nu\nabla^2\phi_1+\mu(\nabla\phi_1)^2+\sum_j s_j(r-r_j). \qquad (14)$$

Here s_j is assumed to be essentially zero except in the vicinity of the j-th impurity position r_j. One may notice that the above inhomogeneous term represents nothing but the heterogeneity in the natural frequency. We assume further $\nu>0$, and take a length unit in which $\nu=1$. This assumption is very important. In fact it is quite possible that ν becomes negative, and then we are led to chemical turbulence (see section 5).

Under the transformation $\phi_1=\mu^{-1}\ln q$, (14) reduces to a linear differential equation

$$\dot{q}=(\nabla^2+\sum_j v_j(r))q, \qquad (15)$$

$$v_j = \mu s_j \tag{16}$$

We write the time-dependence of an eigenstate of (15) like $e^{-\lambda t}$, and express a general solution q as a sum of the zero-eigenvalue state $q_0(r)$ and the other contribution $\psi(r,t)$:

$$q = q_0 + \psi \tag{17}$$

If (15) admits no negative λ, then ψ is damped in time to zero and we have a trivial result $q=q_0$, which corresponds to no interesting patterns. In fact in this case ϕ_1 is obviously constant in time, and this means that the oscillation frequency is everywhere equal to ω and therefore the heterogeneities do not play the role of pacemakers; the heterogeneitis are entrained by the background oscillation. In the other case where we have negative λ, ψ grows in time and eventually dominates q_0, so that we may expect the appearance of a certain pattern developing without limit.

One may notice that if one introduces an imaginary time $\tau=-it$, (15) written in terms of ψ and τ takes the form of a Schrödinger equation for a particle moving in the potential field Σv_j. It is clear that the condition for the existence of a negative λ is nothing but the condition for the existence of a bound state for the above mentioned potential system. In the following we shall be concerned with the asymptotic behavior of ψ for $t\to\infty$ and $|r-r_j|\to\infty$. Then one may retain only the ground state contribution which gives the maximum growth rate of ψ in terms of the real time t. Since each v_j is assumed to be sufficiently localized near r_j, the contribution from the ground state may well be approximated by the sum of the ground state contributions coming from the respective potentials. Thus ψ may essentially be expressed as

$$\psi = \sum_j{}' a_j \exp\{\omega_j t - \sqrt{\omega_j}\,|r-r_j|\} \tag{18}$$

where ω_j is the absolute value of the ground state eigenfrequency of the potential v_j, a_j is a constant, Σ' means the summation only over the potentials which allow for bound states, and also we have used the fact that the long distance behavior of the ground state eigenfunction shows exponential decay as a function of $|r-r_j|$ with the decay constant $\sqrt{\omega_j}$. Putting (18) into (17), and approximating $q_0(r)$ by a constant, namely, neglecting its slow spatial dependence as compared with the exponential behavior of ψ, we obtain

$$q(r,t) \simeq q_0\left[1+\sum_j{}'\exp(f_j)\right] \tag{19}$$

where

$$f_j = \omega_j(t-t_j) - \sqrt{\omega_j}\,|r-r_j|$$

with $t_j=-\ln(a_j/q_0)$. We shall now retein the most dominant term of the right-hand side of (19). This approximation may be permitted because the relative magnitude among the above terms behaves exponentially so that in most region in space one of the terms will be outstandingly large. If we use the terminology of KOPELL and HOWARD'S quite recent paper (10), the above approximation amounts to neglecting a <u>shock</u> structure and approximating a shock as a discontinuity between two plane waves. It is interesting to notice that the shock under consideration is literally the shock wave of Burgers equation into which (11) can be transformed under $\nabla\phi_1=v$. With the above approximation we obtain

$$\phi_1 = \mu^{-1}\,\mathrm{MAX}\,[0, f_1, f_2, \ldots.] \tag{20}$$

where a trivial constant shift of phase, $\mu^{-1}\ln q_0$, has been omitted.

Consider a 2-dimensional system with positive μ, and suppose there are two pacemakers. Then ϕ_1 expressed by (20) has the form of joined cones as shown in Fig. 1(a). Inside the cones ϕ_1 is increasing

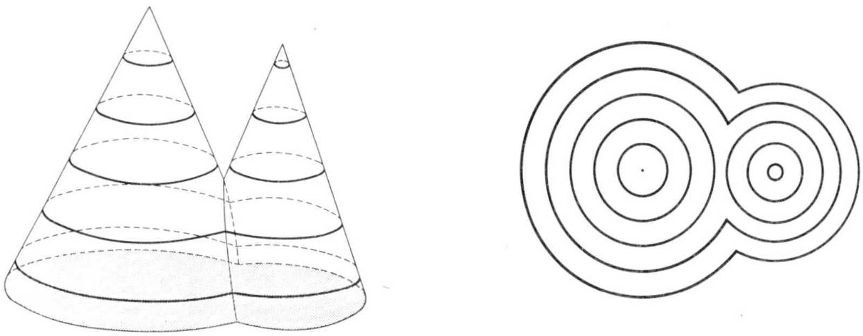

(a) Fig. 1 (b)

steadily in time, which means that the frequency there is larger than the frequency of the outside region. Since the cones are growing, such an entrained region by pacemakers will expand without limit. The concentration pattern must be determined from the condition (13) or $\phi_1=2n\pi-\omega t$, and is clearly given by joined target patterns like Fig.1(b). From the way of constructing the pattern implied by Fig. 1(a) and (b), one easily confirms that the previously stated properties A∿D are all reproduced.

4. Spiral Waves

A spiral pattern |11| is known to appear without heterogeneity. In contrast to circular waves, this pattern cannot be understood consistently in terms of the phase function only |6|, |12|, but an amplitude effect is essential in the core region of a spiral as we shall find out later. The TDGL equation will therefore be a suitable equation to explain this pattern.

Let us first look for such a 2-dimensional particular solution of TDGL that would describe a fully developed spiral pattern |12|. We introduce an amplitude P and a phase ϕ by $w=P\exp(i\phi)$, and assume a solution from (3) in polar coordinates (r,θ) in the form

$$\phi=\omega t\pm\theta+S(r), \tag{21}$$

$$P=R(r) \tag{22}$$

where ω is an unspecified constant and $S(r)$ and $R(r)$ are unspecified functions of r. Substituting (21) and (22) into (3) we obtain coupled nonlinear ordinary differential equations for S and R. Since S and R must be time-independent, the time t must not appear in those equations. This condition requires that ω should essentially be an eigenvalue of a certain Schrödinger equation similar to (15), the wave function being $\psi=R\exp(\nu^{-1}\mu S)$ with $\nu=1+c_1c_2$ and $\mu=c_2-c_1$. Though we have here no heterogeneity, an effective potential arises from the angular dependence of ϕ and R, and this potential looks like a kind of interatomic potential with a strong repulsive core and a long range attractive force. Again the ground state contribution becomes dominant, and in this way ω, $S(r)$ and $R(r)$ may uniquely be determined. The behavior of $S(r)$ and $R(r)$ are given in Fig. 2 with the parameter values $c_1=-1.0$ and $c_2=0.6$. We see that R is exactly zero at r=0, and for large r

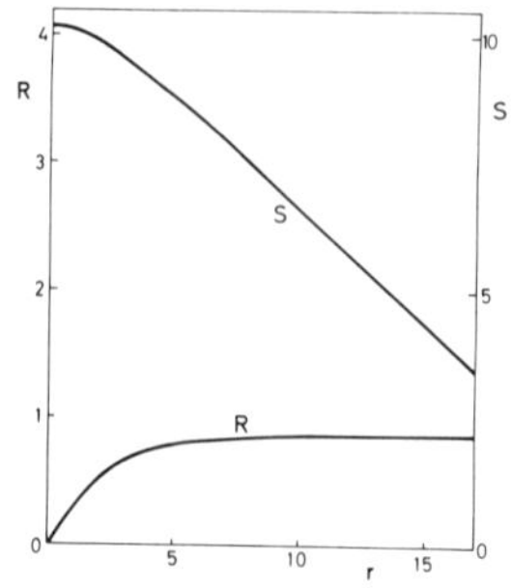

Fig. 2

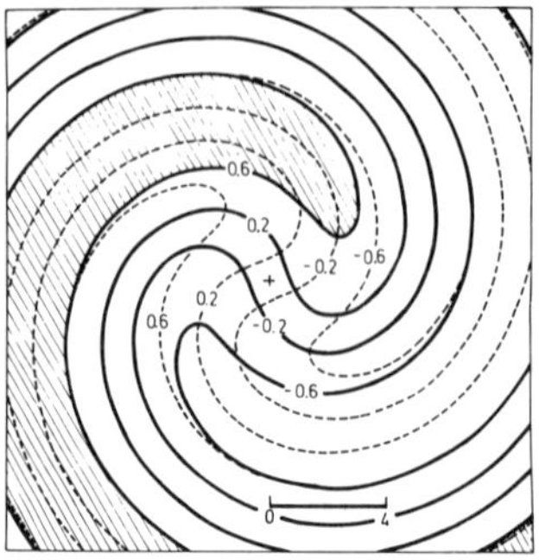

Fig. 3

it tends to a constant and S is essentially linear in r. The fact that R=P is constant means that the equiconcentration contour for the concentrations x=Re w and y=I_mw may again be given by the condition $\phi=2n\pi$. Thus if we use an approximate expression $S=\alpha r$ in (21) we have

$$\omega t\pm\theta+\alpha r=2n\pi \tag{23}$$

which represents an Archimedean spiral rotating with the frequency ω. A spiral pattern obtained from the exact S(r) and R(r) is shown in Fig.3, where some equiconcentration contours for x and y are given by the solid lines and broken lines, respectively, and the cross indicates the phaseless point at which P=o. It is clear that the above particular solution may also be regarded as a 3-dimensional solution of (3) in cylindrical coordinates (r,θ,z) and represents a scroll pattern with a straight scroll axis on which P=o.

A particular solution like the above one does not tell us the origin and the process of forming this pattern. Furthermore there is a question why experiments in a 3-dimensional system often show a scroll ring pattern instead of simple scroll waves with a straight axis. We shall now consider this problem for our TDGL system. Suppose that we have initially a large disturbance in the concentrations x and y. Generally such a disturbance will produce in a 3-dimensional system a curved surface on which x=o and likewise another surface y=o. It is quite possible that these surfaces cross each other thus produce a ring on which x=y=o. This ring will play the role of a phaseless axis from which spiral waves will be generated, and eventually a scroll ring will be formed. This speculation was confirmed by computer calculation. The initial distribution of x and y was assumed in such a way that the surfaces x=o and y=o are spherical with equal radii. In one of these spheres the concentration x takes a certain positive constant value and abruptly chandes into a certain negative constant value at the spherical surface, and similarly for y. Thus the phaseless axis is a perfect circle and any pattern produced under this initial condition must be of rotational symmetry, so that instead of considering a full 3-dimensional pattern we may only consider the section obtained by an intersection with a plane on which the axis of symmetry lies. Figure 4 shows the development in such a section of the pattern, where

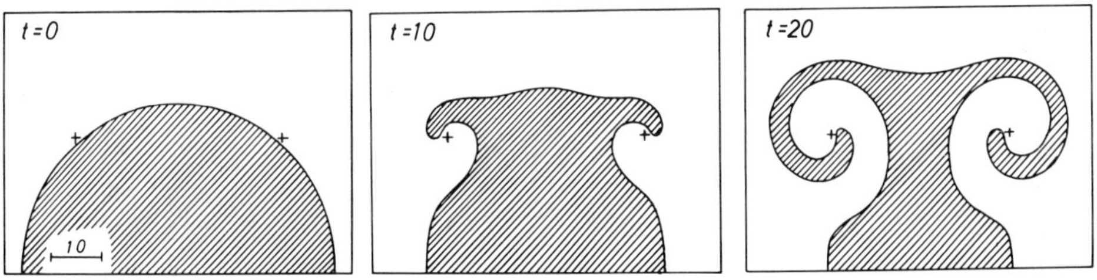

Fig. 4

in the hatched region x is greater than 0.4 and the crosses are the phaseless points. The parameter values are c_o=o.o, c_1=-1.o and c_2=o.6. The results clearly support our conjecture mentioned above.

5. Chemical Turbulence

The last pattern to be discussed is a chaotic spatio-temporal pattern. This pattern appears if ν is negative. Negative ν means the instability of a spatially uniform limit cycle oscillation with respect to a sufficiently long wavelength disturbance of phase. We first explain

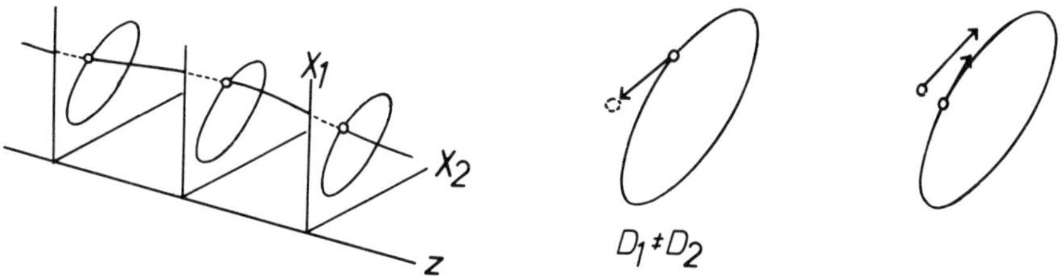

(a) Fig. 5 (b) (c)

qualitatively why such an instability is possible in a system with positive diffusion constants for the original concentration variables. In Fig. 5(a) a system of many oscillators is shown schematically. At each point in the system represented by the 1-dimensional coordinate z, we can define a concentration space which is shown in that figure by the 2-dimensional space (x_1, x_2). The points of the local state which are indicated by the open circles are on the local limit cycle trajectories and these points form a slow spatial variation of phase. The disturbance of phase will at first tend to diffuse due to the diffusion of the concentrations x_1 and x_2. If the diffusion constants D_1 and D_2 are equal, the disturbance produced along the orbit will simply be dampled along the orbit, and nothing interesting will happen. If $D_1 \neq D_2$, however, the disturbance along the orbit will not retrace its way, but in its decay process the state points will inevitably be pulled apart from the orbit (Fig. 5(b)). Once a state point x gets out of the orbit, it will be exposed to a new flow field $\dot{x}(x)$ which may be very different from the flow $\dot{x}(x_o)$ on the orbit. Suppose for instance (Fig. 5(c)) that a state point has come to a certain position outside the orbit where it may experience a stronger flow than on the orbit. Then the state point will get an excess velocity which may be in some cases large enough to cancel the deceleration effect due to diffusion. In this way a nonuniform phase disturbance can grow even if all the diffusion constants are positive.

Unfortunately such an instability has not yet been detected experimentally so far. Still a theoretical reaction model which may give rise to this kind of instability certainly exists. That is the Brusselator. As was stated before, a general expression for the phase diffusion constant ν may be obtained according to our nonlinear phase diffusion theory sketched in section 2. The Brusselator contains essentially three independent parameters A, B and D_y/D_x in usual notations. Here we use instead of B a bifurcation parameter ε defined by $B=B_c(1+\varepsilon)$ where $B_c=1+A^2$. Thus ν is a function of A, ε and D_y/D_x. The condition $\nu=0$ defines therefore a critical surface in the 3-dimensional space spanned by the above parameters. Figure 6 shows some of

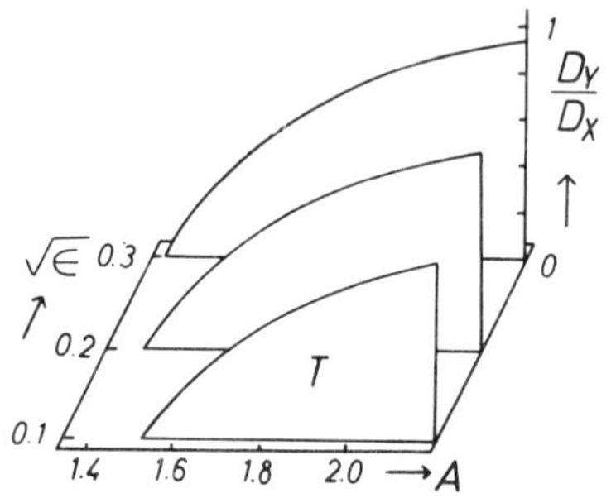

Fig. 6

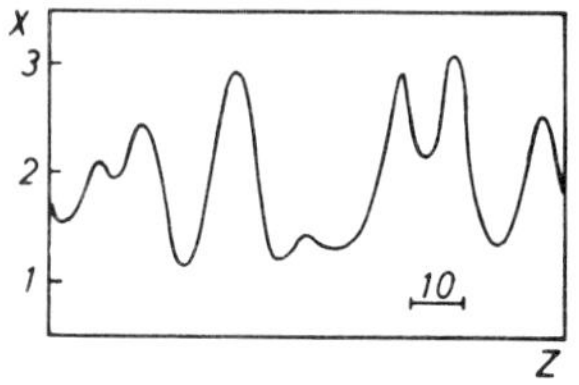

Fig. 7

such critical curves; T is the region where ν is negative. We confirmed by computer calculation for a 1-dimensional system with periodic boundary condition that the concentrations behave chaotically in space and time in the region T. Figure 7 shows an instanteneous pattern of the concentration x for the parameter values A=2.o, ε=o.1, D_x=1.o and D_y=o.o. Initially a certain nonuniform disturbance with a typical amplitude ~o.1 composed of various wavenumber modes was assumed. After a sufficiently long time the disturbance attains a much greater amplitude and persists to behave irregularly in space and time.

TDGL equation (3) will be much easier to analyse than the Brusselator for studying the chaotic behavior. Putting $w=P\exp[i(c_0-c_2)t+\phi_1]$, one may derive the nonlinear phase diffusion equation (11) with $\nu=1+c_1c_2$ and $\mu=c_2-c_1$. Thus, if $1+c_1c_2<0$, the solution of (3) is expected to behave irregularly. This was confirmed by computer calculation |13|.

Our nonlinear phase diffusion equation breaks down if $\nu<0$ because that equation reduces to a simple diffusion equation for q with a negative diffusion constant by the transformation $\phi_1=\nu\mu^{-1}\ln q$. One may ask whether it is possible to derive such an equation for ϕ_1 that does not break down if $\nu<0$. We have not yet succeeded in deriving this kind of equation from a general reaction-diffusion equation. This is possible, however, if we take as the starting equation TDGL and if $|\nu|<<1$. The result is |14|

$$\dot{\phi}_1=\nu\nabla^2\phi_1-\lambda\nabla^2\nabla^2\phi_1+\mu(\nabla\phi_1)^2 \qquad (24)$$

where $\lambda=(1+c_1{}^2)/2$. The only difference of the above equation from (11) is the appearance of the $\nabla^2\nabla^2\phi_1$ term. The fact that the solution from (24) behaves irregularly in space and time was again confirmed by computer calculation |14|. Owing to the extremely simple form of (24), that equation will also serve as a suitable model for the purpose of studying statistical mechanics of homogeneous and isotropic chemical turbulence |14||15|.

References

1. D.H.Sattinger: Topics in Stability and Bifurcation Theory (Springer, New York, 1973)
2. See, for example: J.H. Ferziger, H.G.Kaper: Mathematical Theory of Transport Processes in Gases (North-Holland Publ. Co., Amsterdam London, 1972)
3. Y.Kuramoto, T.Tsuzuki: Prog. Theor. Phys. 54,687(1975)
4. A.Wunderlin, H.Haken: Z. Physik B21,393(1975)
5. Y.Kuramoto: to be submitted to Prog. Theor. Phys.
6. Y.Kuramoto, T.Yamada: Prog. Theor. Phys. 56,724(1976)
7. A.N.Zaikin, A.M.Zhabotinsky: Nature 225,535(1970)
8. A.M.Zhabotinsky: Doklady Acad. Nauk SSR 157,392(1964); Biofizika 9,306(1964)
9. P.Ortoleva, J.Ross: J. Chem. Phys. 58,5673(1973)
10. N.Kopell, L.N.Howard: Studies App. Math. 52,291(1973); 56, 95(1977)
11. A.T.Winfree: Science 175, 643(1972); 181,937(1973) Sci. Am. 230,83(1974)
12. T.Yamada, Y.Kuramoto: Prog. Theor. Phys. 55,2035(1976)
13. Y.Kuramoto, T. Yamada: Prog. Theor. Phys. 56,679(1976)
14. T.Yamada, Y.Kuramoto: Prog. Theor. Phys. 56,681(1976)
15. H.Fujisaka, T. Yamada: Prog. Theor. Phys. to be published

Chemical Turbulence A Synopsis

O. E. Rössler

With 8 Figures

1. Introduction

'Tumbling' is a ubiquitous behavioral possibility in natural systems. The two oldest physical examples are a rippling flow of water, on the one hand, and the space-time behavior of three gravitating masses, on the other hand. A proof that stricty nonperiodic behavior is possible in the last-mentioned case was given by POINCARE [1] (who detected a 'homoclinic point' in a cross-section through the trajectorial flow). Later LORENZ [2] devised his well-known reduced equation for turbulent NAVIER-STOKES flows, which also possesses nonperiodic trajectories.

2. Turbulence in Non-distributed Systems

In recent years it became evident - starting with examples from reaction kinetics - that nonperiodic behavior is very easy to obtain in more than 2-variable systems: almost any combination of two 2-variable oscillators (for example, by contracting them to a 3-variable composed system) is appropriate at certain parameter values (see [3,4,5] for examples).

The most striking evidence for this potentiality is, perhaps, provided by the irregularly dripping water faucet (Fig.1). The 2 intertwined suboscillators are (a) the filling-and-discharging process (that is, the relaxation oscillator of droplet formation) and (b) the damped oscillation of the retracting residue (r), once a droplet has fallen.

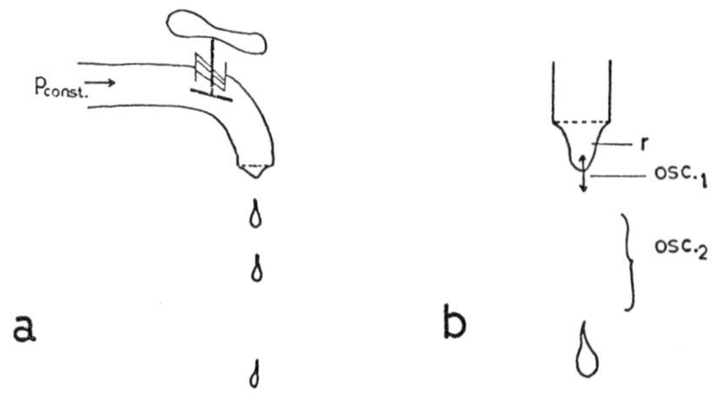

Fig.1 The irregularly dripping faucet (a) and detail (b). r = residue, osc.$_{1,2}$ = 2 oscillatory subprocesses.

Experimenting with faucets of all shapes suggests that there is always a flow-rate (somewhere in between regular dripping and the formation of a continuous thread of running water) where indeed the oscillations of the residue are strong enough and of long enough duration to affect the 'falling conditions' of the next droplet. The occurring 'entangling' between two oscillatory processes, such that they cannot cease coming into each other's ways, is quite easy to see.

Using this example as a conceptual model, it is not surprising that most nonlinear feed-back systems that are 'double-looped' and contain at least three variables, are capable of a complicated dynamics [3,6]; even certain apparently single-looped systems meet the same conditions [7].

Mathematically, two major classes of 3-variable turbulent (or 'chaotic' - meaning presence of nonperiodic plus infinitely many periodic trajectories [8]) systems have been distinguished [5]: the walking-stick map class (with the subclass of universal systems, cf. [3], to which the majority of the above-mentioned systems belong) and the sandwich map class (which contains the LORENZ equation plus several close analogues).

The basic classes of 4-variable chaos, on the other hand, have not been cataloguized. All that is known is that 'superchaos' [5] (with several subclasses) can be distinguished from ANOSOV flows (cf. [9]) and from the double-looped-torus map strange attractors of SMALE-RUELLE-TAKENS type (cf. [9,10]). For none of these 4-variable classes have concrete examples been indicated as yet.

3. Chemical Turbulence Sensu Stricto

If the term 'turbulence' is to be understood in a sense more closely related to its origin, it may be applicable as a general term to all kinds of nonperiodic phenomena that occur in spatially distributed dynamical systems. In this sense it would be synonymous to 'spatio-temporal chaos'.

The first and most straightforward possibility for "chemical turbulence" in this sense is, of course, provided by those distributed systems that consist, locally, of chaotic subsystems already.

An 'example' is depicted in Fig.2. It is the horizontal iron bar in the rain. The (basically) point-shaped faucet of Fig.1 is now replaced by a (basically) one-dimensional dripping system. The rail is assumed to be exactly horizontal, to have a completely smooth surface and to be of infinite extension. A fine misty rain is to constantly fall upon it from above, so that a connected film of water is formed around the bar, 'feeding' every point underneath at every moment.

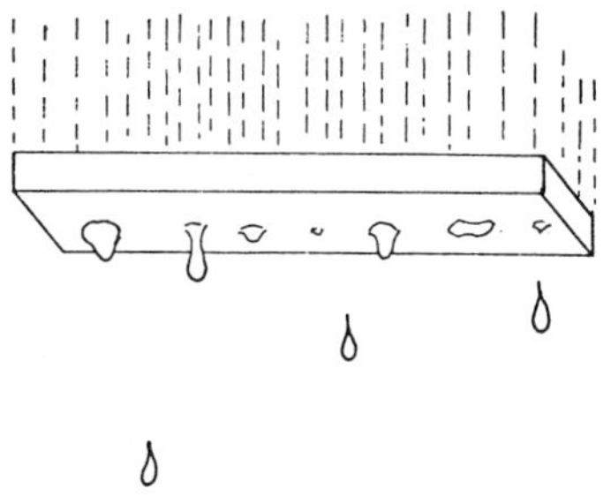

Fig.2 The irregularly dripping handrail (cf. text).

It is evident that such a system will produce nonperiodic behavior in both time and space. A first possible explanation is that the system consists of many chaotic local systems that are coupled in a cross-inhibitory manner.

However, there is a second, simpler, possibility: even if the local oscillators are ordinary limit cycle oscillators (corresponding each to a periodic faucet), still an irregular dripping in space and time can be expected. Reason: whenever a droplet forms due to sufficient local accumulation of water at some spot on the lower surface of the iron rail, it will grow by draining away water from neighboring spots, so that the budding of other droplets in its neighborhood is inhibited. Since this applies to every point, a very complicated (non-repetive) situation should apply again.

This observation of rainy days leads directly to the hypothesis that 2-variable limit cycle oscillators which are symmetrically coupled in a cross-inhibitory manner should in general be capable of spatio-temporal chaos.

4. Chemical Turbulence in 2-Variable Morphogenetic Systems (CI Turbulence)

Chaotic oscillations have been observed in a RASHEVSKY-TURING type morphogenetic system [11]. The simplest TURING system consists locally of a 2-variable potential chemical oscillator (the so-called TURING oscillator): Fig.3a. Here an autocatalytic substance A ("activator") leads to the formation of a second substance ("inhibitor") which catalyzes the outflux of A, that is, inhibits A. TURING [12] showed that the coupling of 2 or more such systems into a string by way of cross-inhibition (linear diffusion coupling between the inhibitors, Fig.3b) leads to standing patterns in space under appropriate parameter values ("morphogenesis"). TURING did not look for further dynamical implications of his model, however.

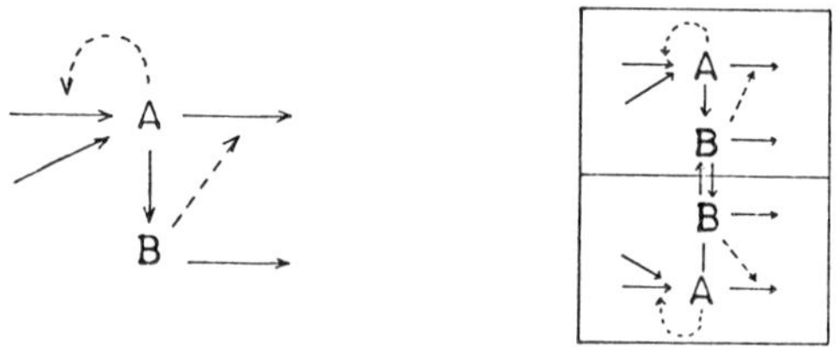

a b

Fig.3 The TURING oscillator (a), and the simplest RASHEVSKY-TURING morphogenetic system (b). The diffusion-type coupling between $\underline{A}$ (not depicted) has to be weaker than that between $\underline{B}$ (see [13,14] for details). Equations of the TURING oscillator [15]:

$$\dot{a} = f(a,b) = k_1 + k_2 a - k_3 ba/(a + K), \quad \dot{b} = g(a,b) = k_4 a - k_5 b.$$

As shown in Fig.4, a chaotic trajectorial flow is indeed also possible in such systems. For simplicity only the 2-cellular case is considered. More details, as well as an argument why this is typical 'spiral-type' chaos [4,5], are found in [11].

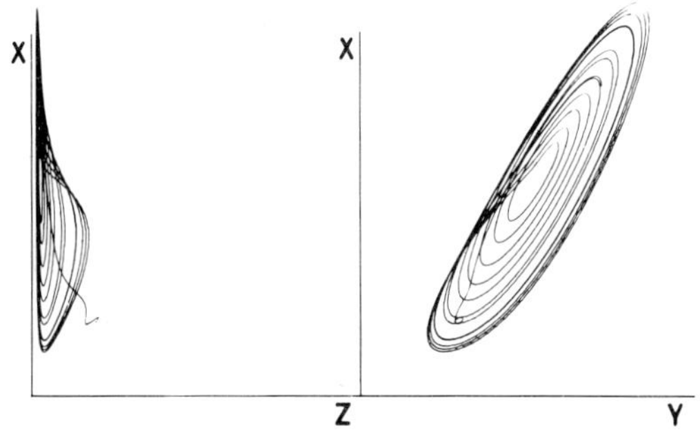

Fig.4 Chemical turbulence in the 2-cellular RASHEVSKY-TURING system of Fig.3b. Only two lateral projections of 4-dimensional state space are shown. Equation: $\dot{a} = f(a,b) + D_1(a' - a)$, $\dot{b} = g(a,b) + D_2(b' - b)$ for the first cell, and identically (but with primed and nonprimed variables interchanged) for the second cell. $\underline{f}$ and $\underline{g}$ as in Fig.3. Numerical simulation on a HP 9820A desk calculator with peripherals, using a standard RUNGE-KUTTA-MERSON integration routine (adapted by F. GÖBBER). Parameters: $k_1 = 1$, $k_2 = 8$, $k_3 = 6$, $K = 0.5$, $D_1 = 0.22$, $k_4 = 12$, $k_5 = 4$, $D_2 = 12$. Initial values: $a(0) = 0.61$, $a'(0) = 0.59$, $b(0) = b'(0) = 1.7$. $t_{end} = 30.9$. Axes: 0... 3.5 for a and a'; 0...7 for b.

Fig.5 shows the second prototypic morphogenetic system. Here the oscillation is not the result of (delayed) self-inhibition of $\underline{A}$ via perturbation of an inhibitor $\underline{I}$, but of (delayed) self-inhibition of $\underline{P}$ (an autocatalytic product) via depletion of its substrate $\underline{S}$. Such systems were also considered by TURING [12] already; the present example was studied by GIERER and MEINHARDT [16]; the PRIGOGINE group's oscillator [17] is equally appropriate. Again, morphogenesis is found under diffusion-coupling of cross-inhibition type (see [16] and [17], respectively).

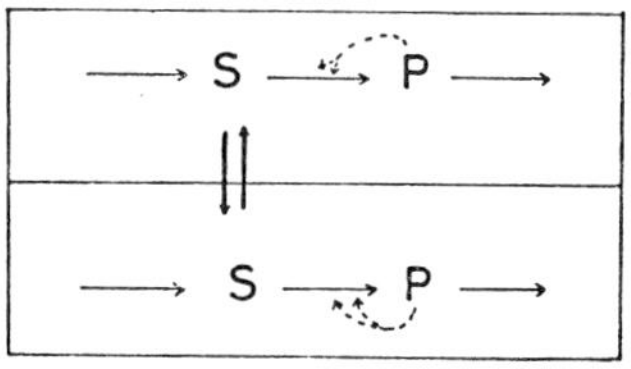

Fig.5 The simplest depletion-type morphogenetic system (see [16,15]). The larger diffusion constant applies to the 'inhibiting variable' (S) again.

Two-variable cross-inhibition type reaction diffusion media ('CI-media') in general allow for (symmetry-breaking) morphogenesis. The example of Fig.4 suggests that the same systems are in general capable of spatial turbulence also.

KURAMOTO and YAMADA [18] indeed found complicated spatio-temporal behavior in the diffusion coupled oscillator of PRIGOGINE and NICOLIS

[17] under numerical simulation of a corresponding partial differential equation. They termed their finding 'chemical turbulence', just as this was done independently in [11].

Thus, at least one type of chemical turbulence is rather well-established now. It occurs in reaction-diffusion media of cross-inhibition type.

5. Chemical Turbulence in 2-Variable Excitable Media (CA Turbulence)

Art Winfree observed an irregular 'meandering' of the spiral core in an excitable fluid [19], his modified BELOUSOV-ZHABOTINSKY reagent (Z-reagent) [20], and asked [21] whether he had observed chaos in the dish. His suggestion lead to the discovery of chaos first in abstract [3,4] and then in concrete well-stirred chemical reaction systems, including the ZHABOTINSKY reaction [22]. All of these well-stirred chaotic systems contain at least 3 variables. This allows the hypothesis that certain 3-variable cross-activating media are capable of 'turbulent' behavior.

Again, the question can be posed whether two variables are not sufficient in the CA case also. As shown in Fig.6, some preliminary evidence for 'meandering' has been obtained in computer simulation runs of a 2-variable excitable medium [23]. The system is in the 'relaxation type' regime ($\mu << 1$); the excitable local steady state is a stable focus. In the core region, successive positions of a concentration isohypsis of the first (activating) variable are shown. In Fig.6a,

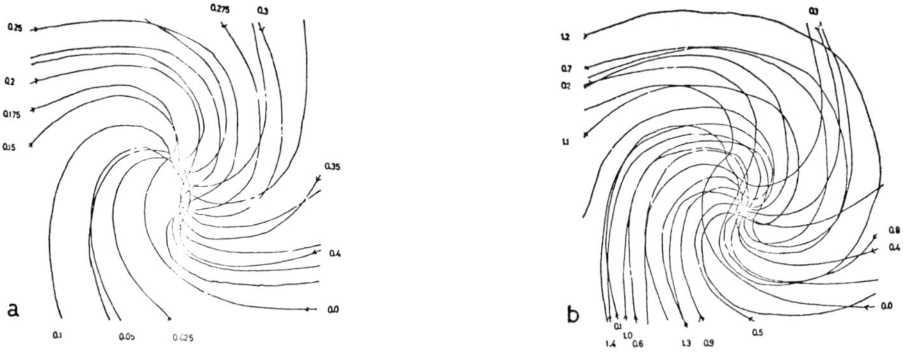

Fig.6 Irregular movements of the spiral core in a 2-variable excitable medium [23]. The simulated equation [15] is:

$\partial a/\partial t = f(a,b) - k_6 a^2 + D_1 \nabla^2 a$, $\partial b/\partial t = \mu g(a,b) + D_2 \nabla^2 b$, with $\underline{f}$ and $\underline{g}$ as in Fig.3. D_2 is put equal to zero for simplicity. Numerical simulation on the CDC 3300 of the Computing Center of the University of Tübingen, using an ADI CRANK-NICOLSON integration program [24]. The unit space is divided into 50 × 50 cells. Initial conditions as described in [24]. Parameters: $k_1 = 0.4$, $k_2 = 0.6$, $k_3 = 1$, $K = 1$, $k_6 = 0.04$, $D_1 = 0.11$, $k_4 = 1$, $k_5 = 2.285$, $\mu = 1/3000$.

roughly one turn of the spiral core is followed. The core hereby changes both position and form. Fig.6b covers roughly 3 1/2 rounds. The inserted numbers are in time units. Excitable media with similar parameter values have not been simulated before. The result requires further investigation.

6. 'Spreading Turbulence' in 2-Variable CI Media

Excitability in single-variable CA media leads to propagating fronts [25]. The local systems hereby switch toward another stable steady state, in the manner of a 'moving fold' [26]. 2-variable cross-activating media still show the same type of behavior, as far as triggering from a stable local steady (or quasi-steady) state is concerned. The regime toward which the switching occurs is hereby assumed to be another stable steady (or quasi-steady) state.

If the last-mentioned restriction is dropped, more complicated triggering events become possible. For example, a stable local system that is triggerable toward a stable limit cycle will, under CA coupling, determine propagation of a permanent oscillatory state into formerly 'quiescent' regions. Interestingly, CI coupling produces the same effect now.

With CI media there is another interesting possibility: even if the local systems are not triggerable, the whole system (consisting in the simplest case of 2 cells) may be so. This is a 'symmetry-breaking' phenomenon of 'higher' type. Such 'spreading morphogenesis' has not yet been described in the literature.

A candidate equation is obtained if, in the equation of the simplest RASHEVSKY-TURING system (see the caption of Fig.3), f(a,b) is replaced by $\bar{f}(c,b) \equiv f(a,b) - k_6 a/(a + K' + k_7 a^2)$. Depending on the parameters, the local system now is either a 4-steady state system (with a 'swallow tail' degenericity in parameter space), or a double-limit cycle system (triggerable from quiescence toward oscillation as described), or a 'trivial' system (with a single globally stable steady state in the non-negative quadrant of state space) just like the original TURING equation. In the last case, however, there is a small range of parameters now in which TURING's symmetrical (symmetry-breaking) saddle point is split in 2, with a stable homogeneous steady state in between (so that the whole system becomes tristable rather than bistable).

Since the whole trick (of replacing f by $\bar{f}$) was aimed at a change of qualitative behavior in the neighborhood of the symmetrical steady state only, there is a chance that the behavior farther away from this region stays essentially unaffected. Thus, both triggerable morphogenesis and triggerable turbulence are possible. Morphogenesis and turbulence thus remain closely linked phenomena in CI systems even when 'spreading' is concerned.

Excitable chaotic systems in general come in 2 varieties: one that remains chaotic forever after the triggering event, and one in which the chaotic regime is 'metastable' only, so that the system eventually relaxes toward either a stable steady state or a limit cycle. Such 'chaotic monoflops' [5] differ from ordinary monoflops in the unpredictability (for all practical purposes) of the timing of the switching-back event. YORKE [28] recently looked at a corresponding set of parameter values in the LORENZ equation, proposing the term 'preturbulence' for the phenomenon. COWAN [29] pictured epilepsy as an excitable (and spreading) case of turbulence - or preturbulence? - occurring in a population of neurons (which are known to be described by chemistry-analogous bulk equations; cf. [26]).

7. Chaos in 2-Variable CA-Systems with Fixed Boundaries

Fixed boundaries (and exogeneously maintained gradients, respectively) induce differing properties in the coupled subsystems. The 2-cellular case again is illustrative. If the 'left' cell has a higher spontaneous frequency than the 'right' cell (or if the left cell spontaneously oscillates while the right cell is merely triggerable as a monoflop, respectively), complicated 'beat' phenomena can occur. When the right-hand monoflop is triggerable anew only after a minimum time interval that is longer than the period of the left-hand oscillator, later and later parts of the excited state of the oscillator will trigger the monoflop on subsequent occasions, until eventually one beat of the oscillator is skipped by the monoflop. Then the process repeats from differing initial conditions, and so forth. Thus, an especially simple type of 'entangling' is encountered again.

A propos 'missing beats': missing heart beats are sometimes caused by the above-described effect; see [30] for a mathematical account. The fact that this is the simplest case of 'biased' CA turbulence and as such a general property of 2-variable excitable media, has not been noted before.

An excitable medium need not be of 'standard' type in order to produce turbulence under an exogeneous bias. For example, there is a special class of excitable media where the triggering variable at the same time provides the substrate for the next 'switch'. In that case two types of propagated waves are possible in the medium: the ordinary trigger (or 'domino') wave of more or less constant velocity (which occurs if one switch after another triggers the next) and an - in the limit infinitely fast - 'break-down' wave. For example, when the first element of the chain is switched off, this implies cutting-off the supply to all subsequent elements of the chain.

A system of this type is provided by the following single-variable equation: $a_t = j - ca/(a + K + k'a^2) + a_{xx}$ (see [15] for a corresponding reaction scheme). j may be zero if one of the boundaries is kept at a constant value a_c. In this case the border provides the sole influx to the chain.

If now a second variable is added in such a way that the local threshold for a's switching down decreases with the time elapsed since the last triggering (see [15] for an equation), again a chain of monoflops (or oscillators) coupled in the CA way and subjected to an exogeneously maintained gradient, are obtained - although with somewhat unusual properties.

Let us assume that the 2-variable system has a fixed (high a) boundary condition on the left. Then every local threshold-lowering is a function of both time and the position within the chain. If some element switches down first, all elements to its right are caught by a breakdown wave of the 'fast' type, while those to its left are subjected to downward triggering of ordinary type. The local systems in the switched-down areas subsequently 'recover' in the sense of becoming upwards-triggerable again. After a while, a 'setting wave' (of non-constant velocity) runs through the system. Thereafter, the whole process repeats, this time with some other local element taking the lead, and so forth.

This second principle for chemical turbulence in 'biased' CA-systems at the same time provides a 2-variable model for a qualitative understanding of the experimental results of GRAVES et al. [31], who observed spatio-temporal chaos in a fixed-boundary system. Even though the experimental set-up suggests the presence of 3 variables [31], it can be shown that 2 of these form a potentially bistable subsystem ('switch') that possesses the very properties of the above-described single-variable equation. The third variable then affects the local threshold as postulated.

At this place, the simplest example of a fixed-boundary turbulent system (with periodic source on one end) may finally be mentioned: seashore chaos. As shown in Fig.7, there are some shallow shores where arriving waves (after breaking) run up the shore for considerable distances. While the remnants of the respective last wave flow down the shore, the next wave arrives, and so forth. The preconditions for 'entangling' are fulfilled again, if the incoming waves have the same height, velocity and distance. By the way, sandpipers busy catching worms at the wet-dry boundary are fairly good at computing the extension of the next-arriving wave.

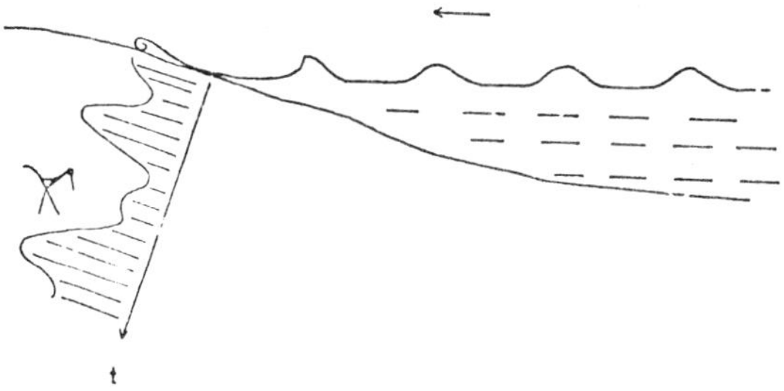

Fig.7 Seashore chaos.

8. Chemical Turbulence in Non-symmetrically Coupled Systems (IA Turbulence)

Besides symmetrical coupling (cross-activating and cross-inhibiting in the simplest cases), more complicated types of coupling also have to be taken into account. The simplest asymmetrical type is, apparently, the case of inhibition in one direction and activation in the other - as in a chain of autocatalytic reactions that 'feed' upon each other (Fig.8).

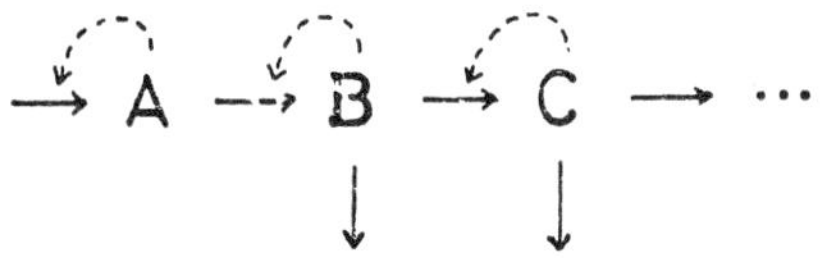

Fig.8 The simplest inhibition-activation system (one-dimensional case).

While cross-inhibition was realizable under linear diffusion coupling (by choosing TURING's trick of intercalating a second 'mediating' variable), a similar trick does not seem to be possible in the IA case. Therefore, only a highly non-canonic partial differential equation that in the limit corresponds to an infinite chain (or network) of IA elements can be derived.

There exists an extensive literature on VOLTERRA type equations (see the comprehensive review by Goel et al. [32]), although not under the specific point of view adopted here. KERNER [33] considered an ensemble of very many 'species' and successfully applied a statistical mechanical description. Hereby the fact that his equations admit a Hamiltonian was crucial. While this treatment in a sense presupposes chaos, the 'intermediary' problem whether few-variable systems also show turbulent behavior already, has not yet been addressed. A 3-variable candidate system has been mentioned in [3].

9. Conclusions

Deterministic spatio-temporal chemical turbulence is possible on the basis of cross-coupled chaotic subsystems of at least 3 variables. Concerning 2-variable chemical turbulence, CI, CA, and IA turbulence can be distinguished. CI turbulence is a characteristic potentiality of RASHEVSKY-TURING morphogenetic systems. CA turbulence occurs in excitable media without and with fixed boundaries. IA turbulence is a more complicated case.

The next problem will be the closer analysis of the individual prototypes. Local approximation techniques like those used in hydrodynamics (cf. [34]) will hereby prove useful. One aim will be to find those locally applying chaotic ordinary differential equations in 3 and 4 variables that represent the most 'natural' types of chaos both in chemical and hydrodynamics turbulence. In view of the many types of chaos and superchaos abstractly possible [5], an empirical 'selection rule' is needed.

References

1. H. Poincaré, Les Méthodes Nouvelles de la Méchanique Céleste, Vols. 1-3. Paris 1899. Reprint Dover, New York, 1957.

2. E.N. Lorenz, Deterministic Nonperiodic Flow. J. Atmos. Sci. 20, 130 (1963).

3. O.E. Rössler, Chaotic Behavior in Simple Reaction Systems. Z. Naturforsch. 31 a, 259 (1976).

4. O.E. Rössler, Chaos in Abstract Kinetics: Two Prototypes. Bull. Math. Biol. 39, 275 (1977).

5. O.E. Rössler, Different Types of Chaos in Two Simple Differential Equations. Z. Naturforsch. 31 a, 1664 (1976).

6. O.E. Rössler, Toroidal Oscillations in a 3-Variable Abstract Reaction System. Z. Naturforsch. a (in press).

7. O.E. Rössler, Chaos in a Modified Danziger-Elmergreen Equation (in preparation).

8. T.Y. li and J.A. Yorke, Period Three Implies Chaos. Amer. Math. Monthly 82, 985 (1975).

9. S. Smale, Differentiable Dynamical Systems. Bull. Amer. Math. Soc. 73, 747 (1967).

10. D. Ruelle and F. Takens, On the Nature of Turbulence. Commun. Math. Phys. 20, 167 (1971).

11. O.E. Rössler, Chemical Turbulence: Chaos in a Simple Reaction-Diffusion System. Z. Naturforsch. 31 a, 1168 (1976).

12. A.M. Turing, The Chemical Basis of Morphogenesis. Phil. Trans. Roy. Soc. London B 237, 37 (1952).

13. O.E. Rössler and F.F. Seelig, A Rashevsky-Turing system as a Two-cellular Flip-flop. Z. Naturforsch. 27 b, 1444 (1972).

14. O.E. Rössler, A Synthetic Approach to Exotic Kinetics, With Examples. Lecture Notes in Biomathematics 4, 546 (1974).

15. O.E. Rössler, Basic Circuits of Fluid Automata and Relaxation Systems (in German). Z. Naturforsch. 27 b, 333 (1972).

16. A. Gierer and H. Meinhardt, A Theory of Biological Pattern Formation. Kybernetik 12, 30 (1972).

17. I. Prigogine and G. Nicolis, On Symmetry-breaking Instabilities in Dissipative Systems. J. Chem. Phys. 46, 3542 (1967).

18. Y. Kuramoto and T. Yamada, Turbulent State in Chemical Reactions. Progr. Theor. Phys. 55, 679 (1976).

19. A.T. Winfree, Spatial and Temporal Organization in the Zhabotinsky Reaction. In: Advances in Biological and Medical Physics (to appear).

20. A.T. Winfree, Scroll-shaped Waves of Chemical Activity in Three Dimensions. Science (Wash.) 181, 937 (1973).

21. A.T. Winfree, Personal Communication 1975.

22. O.E. Rössler and K. Wegmann, Chaos in the Zhabotinsky Reaction (in preparation).

23. O.E. Rössler and C. Kahlert, The Winfree Effect: Meandering in a 2-Variable 2-Dimensional Excitable Medium (in preparation).

24. H.R. Karfunkel and F.F. Seelig, Excitable Chemical Reaction Systems. I. Definition of Excitability and Simulation of Model Systems. J. Math. Biol. 2, 123 (1975).

25. P. Ortoleva and J. Ross, Theory of Propagation of Discontinuities in Kinetic Systems with Multiple Time Scales: Fronts, Front Multiplicities, and Pulses. J. Chem. Phys. 63, 3398 (1975).

26. O.E. Rössler, Chemical Automata in Homogeneous and Reaction-Diffusion Kinetics. Lecture Notes in Biomathematics 4, 399 (1974).

27. O.E. Rössler, Two-Variable Excitable Morphogenesis (in preparation).

28. J.L. Kaplan and J.A. Yorke, Preturbulence in a Regime Observed in a Fluid Model of Lorenz. Preprint 1977.

29. J. Cowan, Personal Communication 1977.

30. H.D. Landahl, A Mathematical Model for First Degree Blocks and the Wenckebach Phenomenon. Bull. Biophysics 33, 27 (1971).

31. D.J. Graves, N. Yang and R.A. Tipton, Unusual Modification of Immobilized Enzyme Kinetics Caused by Diffusional Resistance. Preprint 1976.

32. N.S. Goel, S.C. Maitra and E.W. Montroll, Nonlinear Models of Interacting Populations. Academic Press, New York, 1971.

33. E.H. Kerner, A Statistical Mechanics of Interacting Biological Species. Bull. Math. Biophysics 19, 121 (1957).

34. H. Haken, Analogy Between Higher Instabilities in Fluids and Lasers. Phys. Lett. 53 A, 77 (1975).

Continuous Chaos

O. E. Rössler

With 14 Figures

While it is known that 3-variable continuous dynamical systems are 'infinitely' richer in their behavioral capabilities than two-variable ones (see, e.g., (1)), a systematic attempt at finding the simplest prototypes has yet to be made. Hereby a machinery for 'composing' higher-dimensional systems out of lower-dimensional subsystems such that the overall behavior remains predictable will be helpful. Liênard's 'decomposition' method (into slow submanifolds and fast foliations, see (2)) can be 'turned around' (3) for this purpose.

In two-dimensional systems, 'complicated' dynamical behavior is possible with reference to an exogeneous disturbance (which in a sense corresponds to a third variable): If 2 basins in state space are tightly wrapped around an unstable limit cycle which separates them from a third basin, any perturbation that is just sufficient to kick the system out of the third basin will lead to an 'unpredictable' response (4,5).

In 3 dimensions, the simplest possibility to obtain complicated trajectorial behavior on the basis of the 'building-block principle' mentioned above has been illustrated in Fig.1. Here a bistable slow manifold (as it occurs for example in Thom's cusp (4)) is combined with a two-dimensional unstable focus. A system of this sort was first considered by Khaikin (6,7), who dubbed it a 'universal circuit'. In spite of the telltale name, the system's typically three-dimensional capabilities were apparently overlooked at first. Until recently, the system was known only for its capability to exhibit either 'nearly linear' or 'relaxation type' oscillations on the turn of a single parameter - just as some 2-variable systems do (8). The decisive new feature, indicated

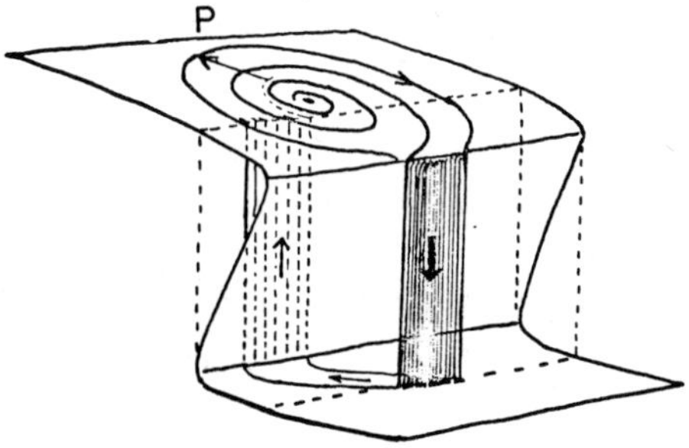

Fig.1 Spiral-type chaos in a universal circuit.
P = Poincaré radius.

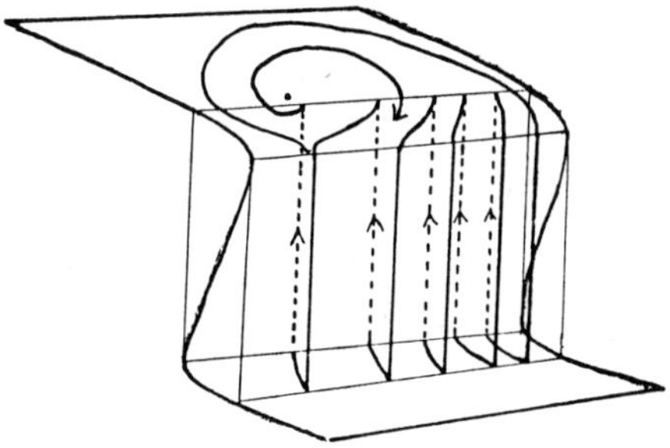

Fig.2 Screw-type chaos in a universal circuit.

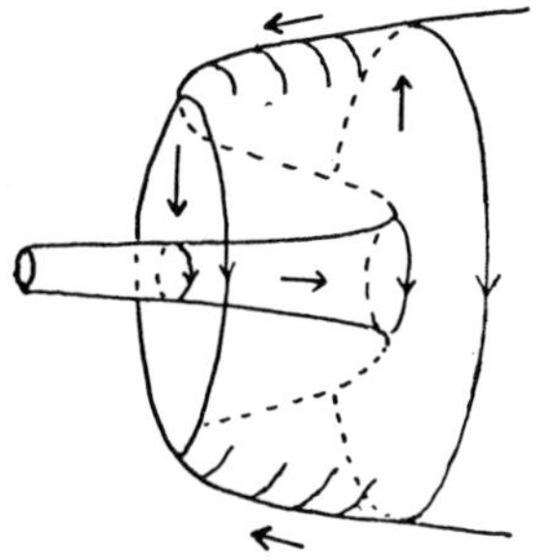

Fig.3 Toroidal oscillation in a generalized Liénard-van der Pol oscillator (see text).

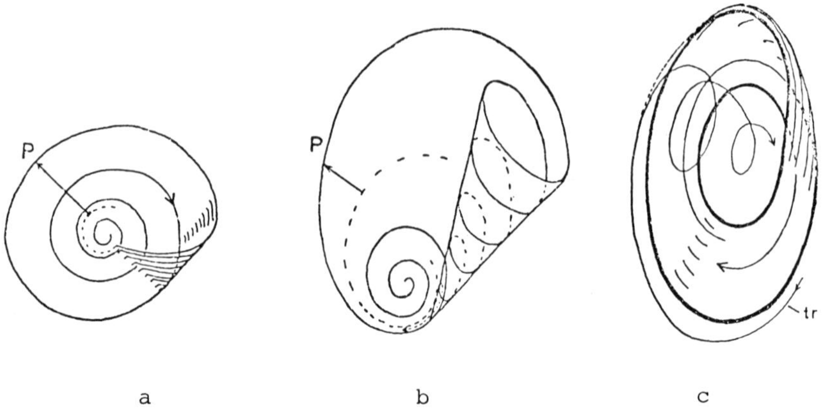

Fig.4 Paper models of less idealized flows: a) spiral-, b) screw-, c) torus-type chaos. tr = transient.

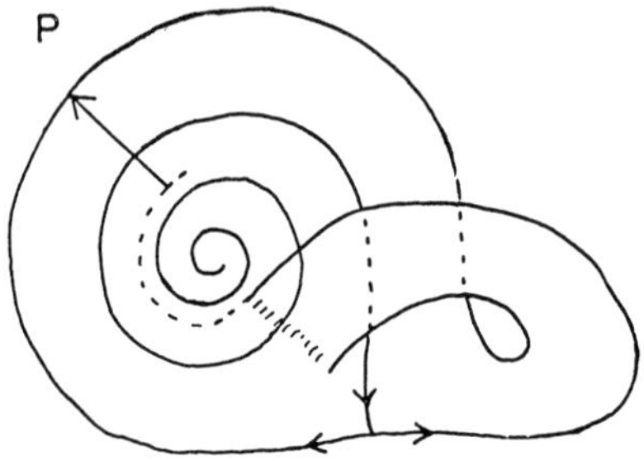

Fig.5 'Inverted' spiral type chaos

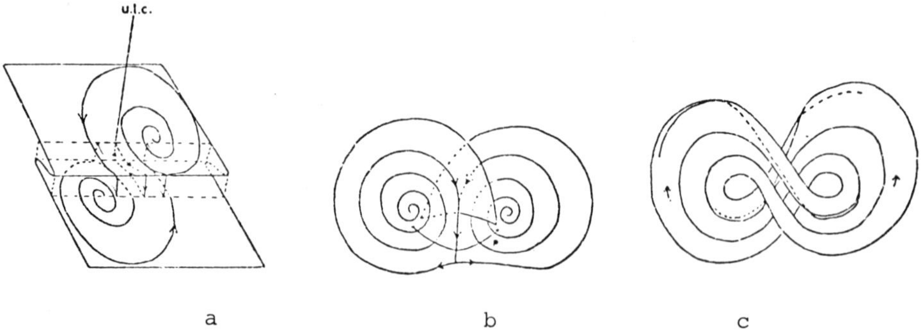

Fig.6 'Composite chaos': a) double-screw type chaos, b) Lorenz type chaos, c) interlinked double-spiral type chaos. u.l.c. = unstable limit cycle. P = Poincaré map.

in Fig.1, is the 're-injection principle' that is made possible by the roundabout way around the hysteresis loop (9). If the two (2-variable) regimes on the two stable branches of the slow manifold are not identical, which is the generic case, the re-injection can, for example, have the properties displayed in Fig.1. The result is a complicated 'entangling' of trajectories which in the limit can be described by a single-variable discrete dynamical system, the Poincaré map P (see Fig.1). Analogous discrete systems were recently considered by Li and Yorke (1o), following earlier work of Julia (11), Ulam (12), Lorenz (13), and Sharkovskii (14). Li and Yorke (10), who proved that such discrete systems possess both nonperiodic and an infinite number of periodic trajectories (almost all of them unstable), coined the term 'chaos' in order to characterize the behavior of these discrete systems. Continuous systems that possess similar maps as cross-sections can, accordingly, be said to produce 'continuous chaos'.

Universal circuits also allow for another type of entangling: Fig.2. This type of continuous chaos is not of 'spiral type' but of 'screw type' (15). Again, a (this time more complicated) Li-Yorke map is embedded as a cross-section.

A third possibility to create chaos on the basis of a 2-dimensional slow manifold and a fast foliation is sketched in Fig.3. Here we have a Liénard-type hysteresis oscillator between two (quasi-) stable regimes which, however, are limit cycles rather than limit points. If one of the two thresholds is not strictly rotation-symmetrical, the former toroidal oscillation gives way to an entangled (chaotic) flow again (15, 16).

The branched 2-dimensional flows considered so far correspond to semi-dynamical systems only (since the flows are non-invertible). When the condition "Liénard parameter $\mu \to 0$" is dropped, ordinary dynamical systems are obtained. These can, again, be approximated by 'paper models', if hereby the fact that the paper now has 'finite width' is kept in mind. This leads to the flows of Fig.4. By drawing these flows on a sheet of paper, cutting them out and gluing them together as prescribed, small 'analogue computers' are obtained which are quite helpful as an aid to understanding what really happens in these simple machines.

There is one more paper model of comparable simplicity: Fig.5. This system is analogous to the first of Fig.4, but the orientation of the re-injected part of the flow has been reversed. Note that this flow now contains an intrinsic saddle point.

Assuming that the set of 'simplest' possibilities is already exhausted with the preceding examples (which is hard to prove, however), the next-simplest examples will be of 'composite' type: Fig.6. The first possibility shown is an implication of the universal circuit principle again: two screw-type regimes have 'merged' end-to-end. The second possibility (Fig.6b) consists in the combination of two 'inverted' spiral-type systems. Interestingly, the Poincaré map through this system (as indicated in the Figure) is identical with that applying to Fig.5. The third possibility (Fig.6c) shows two intertwined chaotic flows of ordinary spiral type. They are running through each other like two links of a chain. Many more possibilities are open. All kinds of 'chains', 'rings' and 'bunches' of coupled chaotic subsystems of one or several types can be conceived of.

If the above examples may be taken as the beginning of a systematics of 3-variable chaotic flows, it is apparent that their behavior falls into two main categories: one where the Poincaré map is (singly or multiply) 'folded', and one where it is 'cut'. Thus, the cross-sections through the flows of the majority of the pictures shown above are of the folded type, as illustrated in Fig.7. All of these maps belong to the class of Li-Yorke mappings. Only the 2 flows of Fig.5 and Fig.6b have the cross-sections seen in Fig.8a,b. The corresponding piecewise linear maps (Fig.7d and Fig.8c) have been considered by Ulam (12) already. Especially that of Fig.8c is well-known from applications: the so-called modulo-counters used in digital computers for the generation of 'pseudo-random numbers' are based on this map. The fact that the numbers so generated are pseudo-random, that is, repeat after a finite time interval, is solely a consequence of the finite number of digits carried by digital computers (17). Fig.9 shows the corresponding 'non-idealized' maps, whereby the fact that the underlying flows possess 'finite width' (that is, are invertible) has been taken into account. The map of Fig.9d, for which the name 'sandwich map' has been proposed (18), is not a diffeomorphism, since the middle line is contracted to a point between one iteration and the next (and then cut in two). Recently, Guckenheimer (19) and Williams (20) proved that such maps determine a strange attractor in the sense of Ruelle and Takens (21), that is, produce a (in a certain sense, (22)) structurally stable attractor that is neither a point nor a closed trajectory. For the diffeomorphism of Fig.9a termed walking-stick map (18), a similarly strong result has not been

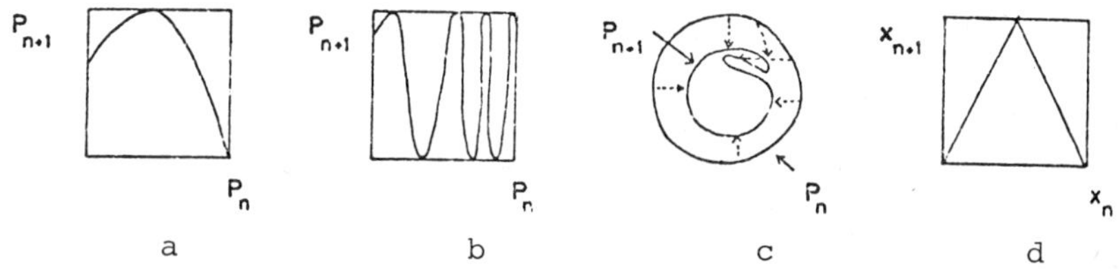

Fig.7 Idealized (one-dimensional) maps describing, as Poincaré cross-sections, a) spiral-type chaos, b) screw-type, and c) torus-type chaos. d) A corresponding piecewise linear map.

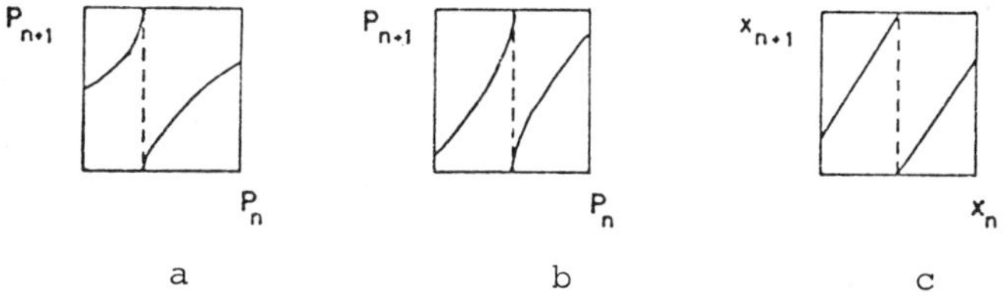

Fig.8 One-dimensional maps as in Fig.7, describing a) inverted-screw type chaos and b) Lorenz type chaos. c) A piece-wise linear analogue.

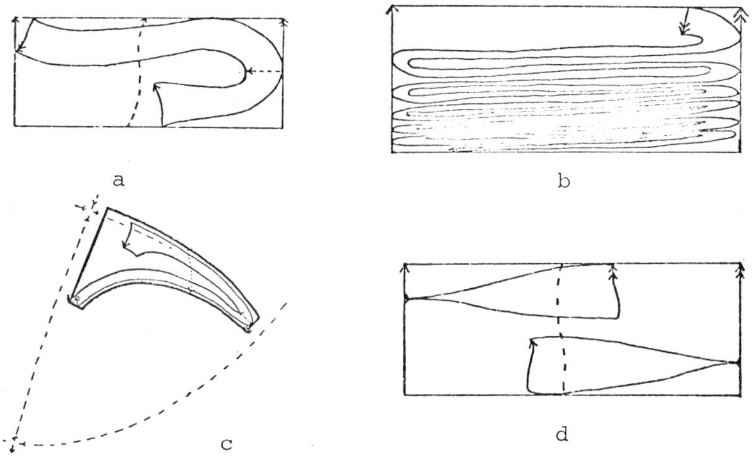

Fig.9 Nonidealized (2-dimensional) maps describing a) spiral type chaos, b) screw type, c) torus type chaos (compare Fig.7a-c), and d) inverted spiral (and Lorenz) type chaos (compare Fig.8a,b). Map a = 'walking-stick map', map d = 'sandwich map'. The arrows are to facilitate identification.

obtained. It could be shown, however, that for a sufficient 'overlap' of the walking-stick's handle, a double 'nonlinear horseshoe map' is formed in the second iterate of the map: Fig.10 (23). This map contains, just as Smale's original (in part linear) horseshoe map (24) does, a homoclinic point (H in Fig.10, (23)). Such an intersection point between the stable and the unstable manifold of a saddle-like fixed point in a diffeomorphism was first described by Poincaré (25) as occurring in the reduced 3-body problem. A single such intersection implies an infinite number of further intersections on subsequent iterations, and hence a trajectory of infinite period. In its neighborhood, always an infinite number of periodic trajectories exists (26) as well as an uncountable number (Cantor set) of nonperiodic trajectories (24) with continuous spectrum. It could not be shown, however, that generically none of the adjacent periodic trajectories is attracting. The majority of 3-variable chaotic systems (namely, all those whose asymptotic regime involves a folding in state space) can, therefore, only be said to possess a 'strange quasi-attractor' (18): They are 'stochastic monoflops', triggerable toward nonperiodic transients of unpredictable lengths, but finally relaxing toward a periodic attractor of pseudorandom character.

So much about 3-variable continuous chaos. The 4-variable case is much less understood presently (except for the Hamiltonian case (27) which is related to the 3-variable toroidal case). Just two basic facts are known: a) there can be structurally stable strange attractors even if the (now 3-dimensional) cross-section is a diffeomorphism; b) new types of chaos, standing in an analogous relation to three-variable chaos as 3-variable chaos does to 2-variable oscillation, are possible ('super chaos' (18)). Fig.11 shows three nontrivial 3-dimensional Poincaré maps. The first is a torus (T^2) mapped 'linearly' onto itself in such a way that a saddle-like (single) fixed point is formed. Then all points generically become intersection points of the saddle's two stable manifolds, that is, homoclinic points. For more details on such 'Anosov flows', see (24). The second map (Fig.11, middle) consists of a solid torus that is once folded before being put back into itself. It determines a structurally stable strange attractor (21, 24). The third map corresponds to the most straightforward possibility of super chaos. Its implications have yet to be derived.

Concerning the existence of examples, differential equations realizing typical 4-variable chaos have yet to be found. There exist, however, simple equations illustrating the above <u>three-variable</u> types of chaos. Fig.12 shows the simplest known examples for spiral and screw and torus

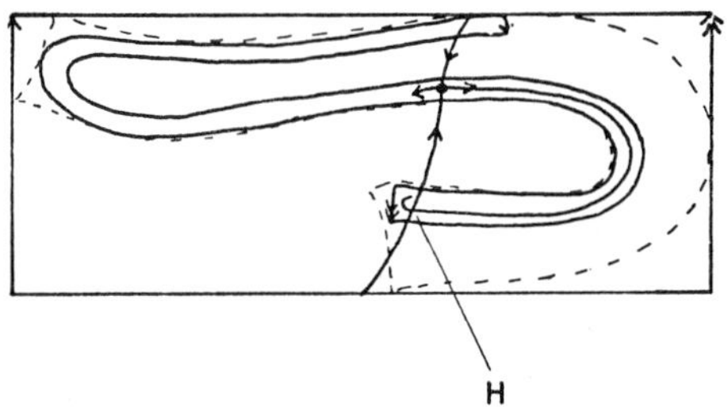

Fig.10 'Nonlinear horseshoe maps' formed in the second iterate of the walking-stick map (23). H = homoclinic point.

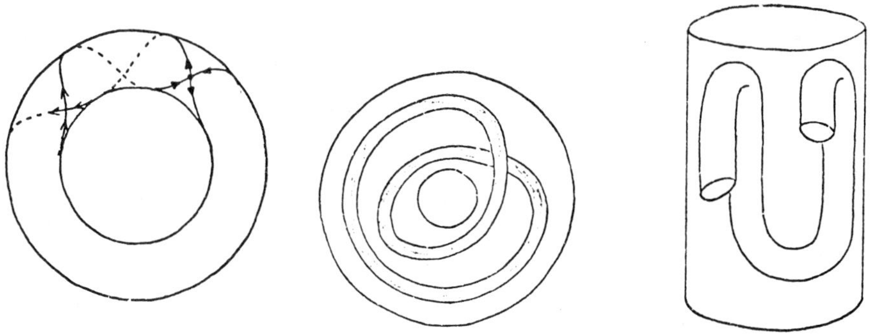

Fig.11 Examples of 3-dimensional chaos-generating diffeomorphisms.

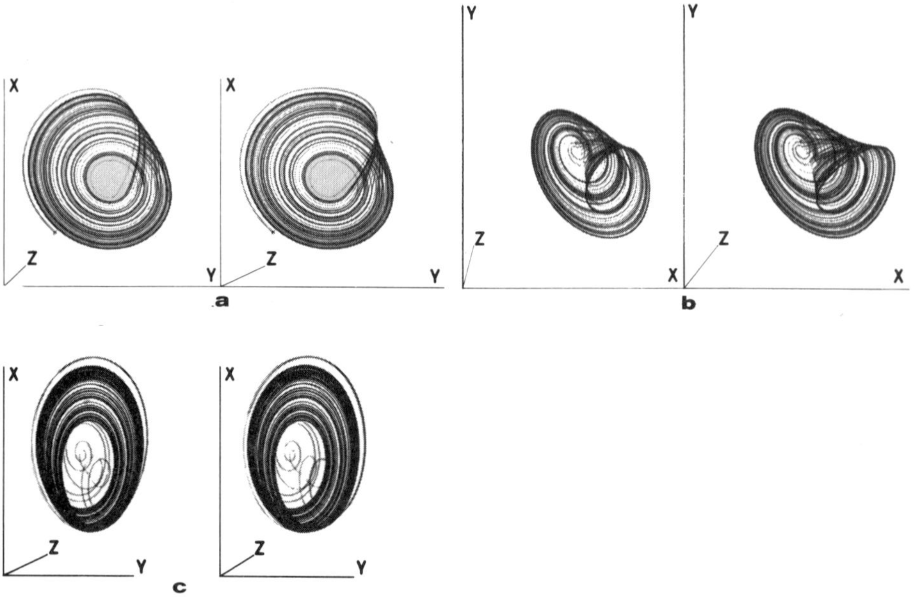

Fig.12 Stereoplots of the trajectories of 2 simple 3-variable ordinary differential equations. Numerical simulation on a HP 9820A desk calculator with peripherals, using a standard Runge-Kutta-Merson integration routine. Parallel projections. These 'wire models' are best viewed with 'crossed eyes': Try to fixate a pencil about 10 cm before your eyes such that, of the four blurred pictures behind it, the two innermost merge (although remaining blurred); then just wait for them to get into focus.
Equation for picture a):

$$\dot{x} = -y - z, \quad \dot{y} = x + ay, \quad \dot{z} = b + z(x-c)$$

(see (28)), where $a = b = 0.2$, $c = 5.7$, $x(0) = y(0) = -0.7$, $z(0) = 0.1$, $t(\mathrm{end}) = 325$, axes: $-14 \ldots 14$ for x and y, $0 \ldots 28$ for z.
Equation for picture b):
the same as for a), but with $a = 0.55$, $b = 2.2$, $c = 4$, $x(0) = 1.9$, $y(0) = -0.6$, $z(0) = 1$, $t(\mathrm{end}) = 447$, axes: $-10 \ldots 10$ for x and y, $0 \ldots 10$ for z.
Equation for picture c):

$$\dot{x} = -y - z, \quad \dot{y} = x, \quad \dot{z} = 0.275(1-x^2) - 0.2z$$

(see (34) for a more complicated analogue), where $x(0) = 1$, $y(0) = -1.4$, $z(0) = -0.4$, $t(\mathrm{end}) = 408$, axes: $-20 \ldots 20$ for x and y, $-10 \ldots 10$ for z.
These 3 trajectorial flows 'realize' spiral, screw, and torus type chaos (compare Fig.4a-c).

type chaos. Fig.13 gives two examples of 'inverted' spiral type chaos, while Fig.14 illustrates the three simplest 'composite' types of chaos.

The preceding equations suggest that composite chaos, as well as sandwich map chaos, require at least two second-order nonlinearities (or one of third order) while walking-stick map chaos does with one of second order. One of the examples listed above is known since a rather long time: the Lorenz flow (13) of Fig.14b. Flows with Lorenzian behavior were also described by Cook and Roberts (30), Malkus and Robbins (31,32) and in (18). The last-mentioned equation (underlying Figs.13a and 14c) has the asset of being 'decomposable' (18). Moore and Spiegel's equation (33) incidentally also yields Lorenz behavior (for $R = 20$, $T = 6.3$); there are two 'unnecessary' twists involved, however. The example of torus type chaos (Fig.12c) has not been described before; a typically toroidal initial transient can be seen in the Figure. The flow of Fig. 14a also forms a new type. A sketch of a flow drawn by Smale (35) for an electric circuit (incidentally of universal circuit type) is possibly related, but has 2 stable foci and is not chaotic. Double screw chaos will deserve a special analysis because of the fact that its Poincaré map involves an infinite number of foldings in the first iterate already.

There are two reasons why the consideration (and classification) of simple chaotic differential systems is of interest. One is the fact that such behavior apparently is ubiquitous among more-than-two variable nonlinear dynamical systems as they occur in all branches of science (15). Poincaré, who first saw the possibility of complicated dynamical behavior, already stressed that homoclinic points (or, as we saw, folded cross-sections) should be the rule rather than the exception in nonlinear systems. Therefore, chaos will have to be excluded rather than proved in all future modelling of natural systems by nonlinear ordinary differential equations of more than two variables (cf. (36)).

The second reason is a more indirect one: chaos is ubiquitous also among distributed systems. The simplest distributed systems consist of more or less identical subsystems. Partial differential equations constitute the simplest class of such 'synergetic' systems (37). The best-known chaotic example in this class is a turbulent flow of water (38). Recently, lasers (39,40) and 2-variable reaction-diffusion systems (giving rise to 'chemical turbulence' (41-44)) have been added to the list. One of the best ways toward understanding the problem of turbulence is, apparently, through applying local approximation techniques which lead to ordinary differential equations of 3 or 4 or more variables (see 13,

Fig. 13 a and b

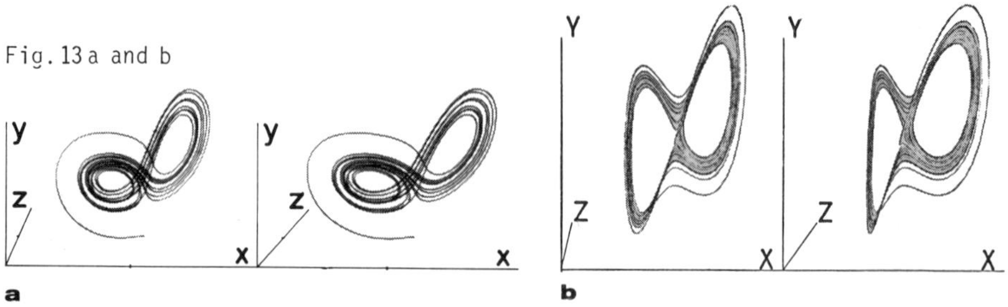

Fig. 14 a-c

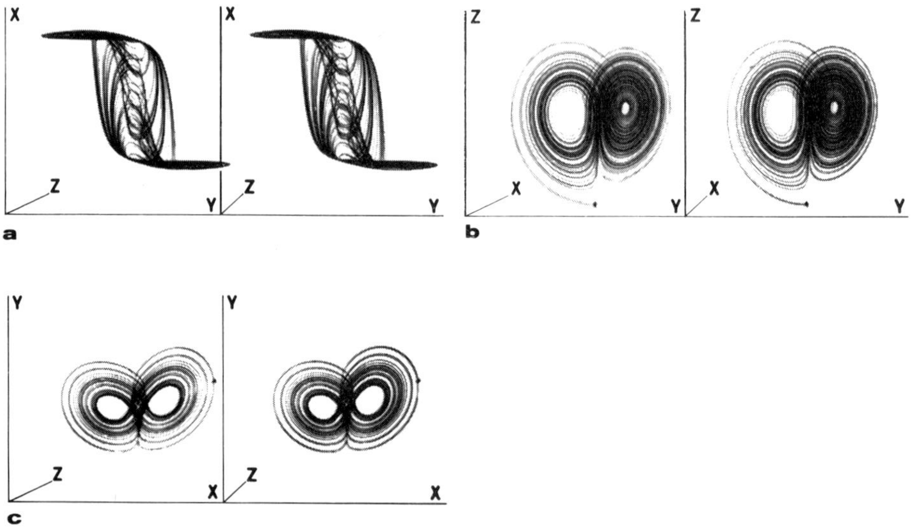

Fig.13 Two further examples. Stereoplots as in Fig.12.
Equation for picture a):

$$\dot{x} = x - xy - z, \quad \dot{y} = x^2 - ay, \quad \dot{z} = bx - cz + d$$

(see (18)), where a = 0.1, b = 0.08, c = 0.38, d = 0.0015, $x(0) = y(0) = z(0) = 10^{-6}$, t(end) = 330, axes: -1.6...1.6 for x, 0...1.6 for y, -0.08...0.08 for z.
Equation for picture b):

$$\dot{x} = \sigma(y - x) + a, \quad \dot{y} = x(r - z) - y, \quad \dot{z} = xy - bz$$

(Lorenz equation (13), modified by putting $a \neq 0$), where a = 2.2, σ = 4, r = 80, b = 8/3, x(0) = 5.1, y(0) = -13.72, z(0) = 52, t(end) = 197, axes: -50...50 for x, -70...70 for y, 0...140 for z.
These two trajectorial flows 'realize' inverted spiral type chaos (compare Fig.5).

37,38). Thus, equations of all of the described types are eligible still for the role of being 'most natural' (that is, occurring most frequently in simplified models of turbulent systems).

To sum up, Table 1 gives an attempted systematization of the material presented above. (Some topics and references not covered by the preceding text have been added in order to give not too incomplete a list.)

Table 1 Classification of presently known types of deterministic chaos

Types of Chaos

I. Continuous Chaos

- A. Chaos in 3-variable differential equations
  - a) Walking-stick map chaos
    1. spiral type
    2. screw type
    3. torus type
    4. double screw type
  - b) Sandwich map chaos
    1. inverted spiral type
    2. Lorenz type
  - c) Combined (folded and cut) map chaos
- B. Chaos in 4-variable differential systems
  - a) Anosov flows
  - b) Folded-torus map flows
  - c) Super chaos

Fig.14 Three examples realizing composite chaos of double-screw, inverted double spiral (Lorenz), and interlaced spirals type, respectively (compare Fig.6). Stereoplots as in Fig.12.
Equation for picture a):

$$\dot{x} = -ax - y(1-x^2),\quad \dot{y} = \mu(y+0.3x-2z),\quad \dot{z} = \mu(x+2y-0.5z)$$

where $a = 0.03$, $\mu = 10$, $x(0) = -1$, $y(0) = 0.55$, $z(0) = 0.12$, $t(\text{end}) = 6040$, axes: $-15\ldots15$ for x, y, z.
Equation for picture b):
Lorenz equation (see Fig.13b) if $a = 0$, $\sigma = 10$, $r = 28$, $b = 8/3$ (as in (13)), with $x(0) = z(0) = 0$, $y(0) = -0.01$, $t(\text{end}) = 92$, axes: $-40\ldots40$ for x and y, $0\ldots60$ for z. (Remark: The same behavior occurs in equation (Fig.13a) if $d = 0$.)
Equation for picture c):
as in Fig.13a (18,28), with $a = 0.04$, $b = 0.06$, $c = 0.326$, $d = 0$, $x(0) = +(-)\ 0.007$, $y(0) = 0.8$, $z(0) = +(-)\ 0.01$ (the minus sign is for the 'left' flow), $t(\text{end}) = 178$, axes: $-1\ldots1$ for x and z, $0\ldots2$ for y. (Remark: The same behavior occurs in the Lorenz equation (29).)

C. Chaos in 2-variable partial differential equations
   a) Cross-inhibition media ('CI') turbulence (43)
   b) Cross-activation media ('CA') turbulence
   c) Inhibition-activation media ('IA') turbulence
   d) Hydrodynamic (and laser) turbulence
D. Chaos in single-variable functional differential equations (45)

II. Discrete Chaos

A. Chaos in single-variable difference equations (46)
B. Chaos in multiple-variable difference equations
   a) based on endomorphisms (11,47)
   b) based on diffeomorphisms (48)
   c) based on conservative endo- and diffeomorphisms (26,27,49)

III. Mixed Chaos

A. Chaos in implicit differential equations (9,50)

References

1. Andronov, A.A., Leontovich, E.A., Gordon, I.I., and Maier, A.G., Theory of Bifurcations of Dynamic Systems on a Plane, Wiley: New York 1972. First Russian ed. 1967. See p. XXIII.
2. Zeeman, E.C., In: Towards a Theoretical Biology, Vol. 4, pp. 8-67 (C.H. Waddington, ed.), University Press: Edinburgh 1972.
3. Rössler, O.E., Lecture Notes in Biomathematics 4 (Springer-Verlag), 546 (1974).
4. Thom, R., In: Towards a Theoretical Biology (C.H. Waddington, ed.), Vol. 3, pp. 89-116, Aldine Publ. Comp.: Chicago 1970.
5. Rössler, O.E. and Denzel, B., Two-variable Stochastic Flip-flop in Abstract Kinetics (in preparation).
6. Khaikin, S.E., Zh. Prikl. Fiz. 7 (6), 21 (1930).
7. Andronov, A.A., Khaikin, S.E., and Vitt, A.A., Theory of Oscillators, p. 725, Pergamon: New York 1966. (First Russian ed. 1937.)
8. Rössler, O.E. and Hoffmann, D., 4th Int. Biophysics Congress, Moscow 1972 (Abstracts Vol. 4, p. 49).
9. Rössler, O.E., Z. Naturforsch. 31 a, 259 (1976).
10. Li, T.Y. and Yorke, J.A., Amer. Math. Monthly 82, 985 (1975).
11. Julia, G., J. Math. Pur. Appl., Série 8, 1 (1918).
12. Ulam, S.M., In: Mathematical Models in Physical Sciences (S. Drobot, ed.), pp. 85-95, Prentice Hall: Englewood Cliffs 1963.
13. Lorenz, E.N., J. Atmos. Sci. 20, 130 (1963).
14. Sharkovskii, A.N., Ukranian Math. Jour. 20, 136 (1968).
15. Rössler, O.E., Bull. Math. Biol. 39, 275 (1977).
16. Rössler, O.E., Biophys. J. 17, 281a (1977). (Abstract.)
17. Hull, T.E. and Dobell, A.R., Soc. Ind. Appl. Math. 4, 230 (1962).
18. Rössler, O.E., Z. Naturforsch. 31 a, 1664 (1976).

19. Guckenheimer, J., In: The Hopf Bifurcation (J.E. Marsden and M. McCracken, eds.), pp. 368-381, Springer-Verlag: New York 1976.

20. Williams, R.F., The Structure of Lorenz Attractors (preprint).

21. Ruelle, D. and Takens, F., Commun. Math. Phys. 20, 167 (1971).

22. Guckenheimer, J., Structural Stability of Lorenz Attractors. (Preprint.)

23. Rössler, O.E., Syncope Implies Chaos in Walking-stick Maps, Z. Naturforsch. a (in press).

24. Smale, S., Bull. Amer. Math. Soc. 73, 747 (1967).

25. Poincaré, H., Les Méthodes Nouvelles de la Mécanique Céleste, Vols. 1-3, Paris 1899; reprint Dover: New York 1957.

26. Birkhoff, G.D., Acta Math. 50, 359 (1927).

27. Gumowski, I., R.A.I.R.O Automatique 10, 7 (1976).

28. Rössler, O.E., Phys. Lett. 57 A, 397 (1976)

29. Rössler, O.E., Phys. Lett. 60 A, 392 (1977).

30. Cook, A.F. and Roberts, P.H., Proc. Camb. Phil. Soc. 68, 547 (1970).

31. Malkus, W.V.R., EOS. Trans. Am. Geophys. Union 53, 617 (1972).

32. Robbins, K.A., A New Approach to Subcritical Instability and Turbulent Transition in a Simple Dynamo, Proc. Camb. Phil. Soc. (in press).

33. Moore, D.W. and Spiegel, E.A., Astrophys. J. 143, 871 (1966).

34. Rössler, O.E., Z. Naturforsch. 32 a, 299 (1977).

35. Smale, S., J. Differential Geometry 7, 193 (1972).

36. Rössler, O.E. and Wegmann, K., Chaos in the Zhabotinskii Reaction (preprint).

37. Haken, H., Synergetics, Springer-Verlag: New York 1977.

38. Joseph, D.D., Stability of Fluid Motions, Vols. I and II, Springer-Verlag: New York 1976.

39. Haken, H., Phys. Lett. 53 A, 77 (1975).

40. Graham, R., Phys. Lett. 58 A, 440 (1976).

41. Kuramoto, Y. and Yamada, T., Progr. Theor. Phys. 56, 679 (1976).

42. Rössler, O.E., Z. Naturforsch. 31 a, 1168 (1976).

43. Rössler, O.E., Chemical Turbulence - A Synopsis, In: Synergetics (H. Haken, ed.), Lecture Notes in Physics (in press).

44. Kuramoto, Y., Chemical Turbulence and Chemical Waves, In: Synergetics (H. Haken, ed.), Lecture Notes in Physics (in press).

45. Mackey, M., Biophys. J. 17, 281a (1977). (Abstract.)

46. May, R.M., Nature 261, 459 (1976).

47. Guckenheimer, J., Oster, G., and Ipatchki, A., The Dynamics of Density Dependent Population Models. Theoret. Popul. Biol. (in press).

48. Newhouse, S. and Palis, J., Astérisque 31, 44 (1976).

49. Moser, J., Stable and Random Motions in Dynamical Systems with Special Emphasis on Celestrial Mechanics, Princeton University Press: Princeton 1973.

50. Takens, F., Lecture Notes in Mathematics 535, 237 (1976). Springer-Verlag.

Morphogenesis

Morphogenesis Persistence of Organized Structures and Breakthrough of New Structures

H. Kuhn

With 5 Figures

General Considerations

The basic difference between non living and living structures is that living structures act as having the aim to survive (they act purpose oriented) and they have a knowhow to survive in their environment, which has a complicated structure (they are adapted to their environment). This new quality of living systems is based on the Darwinian mechanism of multiplication of forms, occurence of mutants, survival of fit mutants. This mechanism is realized in all living systems by the same genetic apparatus.

In answering the question of how this apparatus could have originated, the problem is essential how molecules can build up functional systems of increasing complexity. It is well known [1] that the simulation of primeval conditions greatly favours the formation of essential building bricks of living systems and thus the main problem is finding a mechanism by which a genetic apparatus could have evolved from such building bricks.

It is of interest to study in a quite general sense possibilities how molecules can cooperate and how organizational systems of ever greater complexity can evolve. One should find driving forces of selforganization A subsequent problem is to find possibilities of a chemical realization of proposed theoretical models. Unfamilar ways are necessary in attempting to understand the evolution of the genetic apparatus: long causal chains of physico-chemical steps must be invented which cannot be compared with known facts before the last step. Drastic simplifications are necessary to grasp the development of complicated mechanisms. It would be quite wrong to seek the answer in some profound theoretical concept: Seen in this light the main problem is to find a sensible starting point and to invent a reasonable pathway, in which the development of ever more complex forms can be shown to be plausible as a physico-chemical process.

It has been attempted to compare the initiating process in the evolution of life with the process of the formation of dissipative structures such as the Benard structures, which are obtained in a liquid in a flat container when heated from below and when a sufficiently large heat gradient is reached.

This type of selforganization does not lead to structures which evolve to functional systems of stepwise increasing complexity. This aim is reached by systems that make use of Darwin's principle: multiplication of forms, sporadic occurence of mutants, selection of certain mutants. This process requires a reproduction mechanism for some species, whereby this species must be of sufficient complexity to ensure that some mutants, like the original form, are able to reproduce themselves. A mutant may have a smaller or larger survival probability than the original form. In the first case the mutant will disappear again after some time, and therefore will play no role. In the second case the mutant will be multiplied and will gradually displace the original form.

A simple possibility to realize such a mechanism can be imagined by considering molecular strands consisting of several kinds of monomers, and energy rich monomers, which associate with the strands under favourable conditions and polymerize at that matrix to form complementary strands. The double strands are assumed to separate into single strands which assume convoluted structures under appropriate conditions, depending on the monomer sequence in the strand. Certain convoluted forms will be able to survive better than others, and by repeated multiplication, mutation and selection an accumulation of strands with certain sequences should be reached (Fig.1 above).

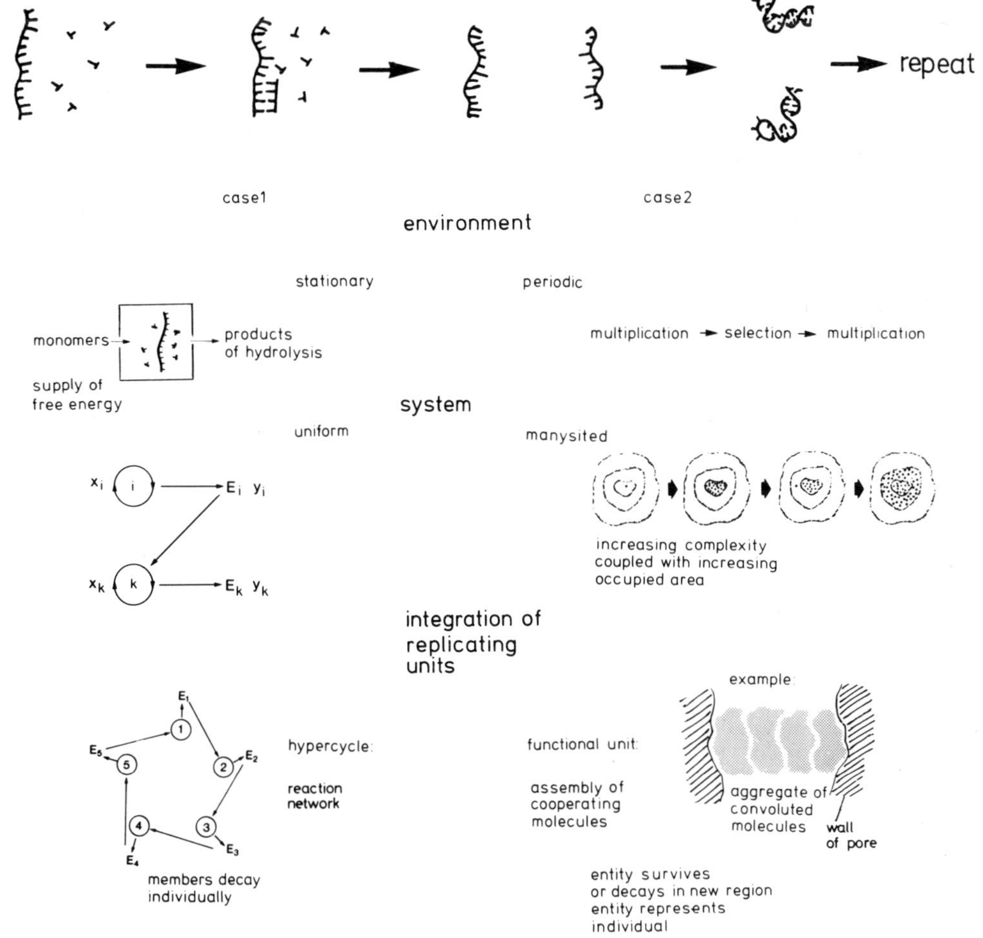

Fig.1 Drive to morphogenesis inherent (case 1); by structure of environment in space and time (case 2).

We may proceed starting from two fundamentally different model assumptions (case 1 and 2):

Case 1: A homogeneous system, or a heterogeneous system with uniformly distributed reacting components, is considered which is stationary supplied by free energy. The system may be a solution containing molecular strands able to replicate. The free energy is supplied in the form of energy-rich monomers. The strands multiply by replicating as described above, and by the hydrolysis of the strands a selection mechanism is given. The products of hydrolysis flow off. This case was considered by Eigen [2] who investigated the detailed population development of strands of different sequences.

After a certain evolution has taken place these nucleic acid like strands are assumed to be the matrices for the formation of specific polypeptides. The ring i in Fig.1 symbolizes a replication cycle, and E_i symbolizes a polypeptide with enzymatic activity. The concentration y_i of E_i depends on the concentration x_i of the corresponding nucleic acid, e. g. $y_i = c_i x_i$.

E_i is assumed to catalyze the formation of a nucleic acid k, the formation rate of that acid then being given by

$$\frac{dx_k}{dt} = (a_k + b_k\, y_i)x_k - d_k x_k \qquad (1)$$

Eigen demonstrated that a cooperative effect integrating such systems is reached by a distinct mode of coupling: enzyme E_1 produced by replicating system 1 activates replicating system 2. Enzyme E_2 activates replicating system 3, etc. and a last enzyme finally activates system 1 (in the case of Fig.1, enzyme 5). Such a hypercycle has all-or-non growth kinetics, i. e. a hypercycle that has succeeded is not in competition with others. Therefore, according to Eigen, Darwinian evolution does not take place during the hypercyclic phase.

A hypercycle, similar to the Benard structure, is a dissipative structure depending on the stationary flux of energy-rich monomers. It can be maintained under the condition given by the Prigogine-Glansdorff principle. This principle then appears as the necessary condition of evolution and the formation of a hypercycle as a necessary step. The basic question in evolution appears as a thermodynamic problem [2].

It is not easy to see how an evolution to ever greater complexity should be possible in this way. The molecules should have the tendency to replicate and select under constant environmental conditions. The replication is favoured by open single strands or structures that can be easily opened to accept monomers, while on the other hand the selection requires rigidity of the convoluted forms leading to properties depending specifically on the sequence. It seems easier to trigger selection by a mechanism which brings the molecules periodically into an environment that favours multiplication and an environment which forces them to assume specific convoluted forms with specific survival properties. For maintaining the process of multiplication, mutation and selection a sophisticated program of environmental changes seems necessary, which drives again and again a complicated concerted reaction.

A second difficulty in the case considered above is caused by the fact, that there is no outside stimulus to increase complexity. The homogeneous system (or the heterogeneous system with uniformly distributed reacting components) is subject to a constant flux of free energy. Thus the molecules are considered to develop by inherent reasons. It is not easy to see what the stimulus to increased complexity should be.

Case 2: Let us now assume [3] that a very specific structure in space and time initiates and stimulates selforganization, i. e. the mechanism basic to living systems - the adaption to a highly structured environment - is assumed to be the driving force of evolution even of the first replicating molecules. We may consider an environment which drives the molecules alternately between replication and selection phases.

By this periodicity and specificity of the environment a mechanism is set going by which the molecular systems adapt increasingly better to the environment, and a state will be reached in which the number of molecules originating in the multiplication phases will compensate for the number of molecules disappearing in the selection phases. The systems are now trapped at their evolutionary level.

The stimulus of an evolution to higher complexity is given by a manysited, heterogeneous environment, a spacial structure which shows dispersion in form and shape. Certain areas in this landscape, in which a development of the replicating molecules thus obtained is not possible, can be populated by more complex forms. This mechanism of shielding and thus protecting more complex forms by the presence of different areas is considered to be the driving force to a directed evolution of systems of increasing complexity: The first systems are restricted to a very particular area with a specific periodic environmental structure. A neighbouring area with slightly different environmental structure can be populated only by systems that can multiply also in this neighbour area. They must have a slightly increased complexity, otherwise the neighbour area would have triggered the formation of first simple forms. In the neighbour area the more complex systems can easily develop, since they have no competitors. In this manner the populated region is gradually increased and the complexity is increased accordingly, since presently unpopulated areas can be populated quite generally only by more complex forms.[1)]

The driving force in this approach, periodicity in time and manysitedness in space, is excluded by choice of boundary conditions if a homogeneous system (or a heterogeneous system with uniformly distributed reacting components) in a stationary state is considered as the basic model.

[1)] Obviously, a system of increased complexity is only necessary for populating a new area if the presently existing systems do not fit. A region may be discovered at a later stage, that could have been populated by earlier forms. Such a particular region could have been populated much earlier if it would have been reached. It does not contribute to the gradual increase of complexity of the evolving systems by gradually increasing the populated areas, but the existence of such areas does not contradict the basic concept.

Many new ecological niches for simple systems are not present before later stages of evolution, e. g. ecological niches for bacteriophage and virus like systems are restricted to an environment containing ribosomes etc. for translating their genetic information into the protein language.

In the present model the outside stimulus is the forming power of the main organizational structures evolving with time. Each individual process in this development is thermodynamically unproblematic. The basic problem in evolution of life is not a problem in thermodynamics. The main question is finding the principle breaks through of organizational structures of molecular systems under appropriate physical conditions. Obviously, this problem has thermodynamic aspects.

The evolving organizational systems of cooperating molecules survive or die away as entities (functional units). The entity represents the individual. In a certain area associates of interlocking convoluted forms of simple molecules may be the evolving species, and free molecules, which do not match into such an associate diffuse away, thus escaping the particular area. Such an associate then survives as an entity: formation rate and decay rate of this entity are the determining quantities. The survival probability of a functional unit is a new quality depending on the system of cooperating molecules and the particularities of the environment. In the case of a hypercycle, on the other hand, the replicating molecules forming the members in a reaction cycle, survive or die away as individuals. This is expressed by the system of equations (1). The growth rate of a hypercycle is a function of the growth and decay parameters of each member and the pair coupling parameters determining the influence of members on others. The hypercycle is a reaction network, while the functional unit is a system of spacially assembled molecules, this entity acting as an individual (Fig.1).

The evolution of biological systems is preliminated by the evolution of the planets, i. e. by the development of rotating bodies with manysited surface structure. It is of particular interest that these properties (periodicity and manysitedness), according to the present model, stimulate the initiation of the biological evolution if particular chemical conditions are fulfilled (presence of liquid water, temperature range likely for making a transition between open and convoluted structures of appropriate molecules easily possible).

The subsequent evolution of macromolecules can be described as a stepwise increase of the populated area by increasing complexity. Each breakthrough into a new area is initiated by the accidental formation of a new organizational structure, which enables survival in this area. The new area is populated (convergent phase). Later, the new organizational structure is only slightly changed by increasing adaption to the environment (divergent phase) until a new fundamental change initiates the sudden transition to a new convergent phase.

The formation of aggregates of interlocking molecules discussed above is an illustrative example of a breakthrough of a fundamentally new organizational structure: Molecules with advantageous convoluted structures cannot be selected, since the probability of error in the reproductive process is too great, unless some highly effective selection mechanism is in operation for such molecules, which removes each defective speciment immediately after the replication phase. Thus the systems seem to be trapped at their evolutionary level.

Such a highly effective selection mechanism is given with the simultaneous appearance of interlocking convoluted forms. If a faulty chain constituent is built into the new strand on replication, the molecule will have a changed convoluted form and no longer fit into its place within the associate. Its place is taken by a piece that fits correctly. This trick ensures that the aggregate reproduces with fair constancy although reproductive flaws may arise quite frequently in the single molecules. The aggregate can multiply. Thus the barrier at this evolutionary stage is passed by some fundamental change in the organizational structure: With this transformation the fund of genetically transmitted information is greatly enhanced. The monomer sequence of all molecules which agglomerate into the associated form remains unchanged in all positions relevant for the convoluted structures: Thus, the information implied in the aggregate is genetically fixed.

The problem to be considered now is to check the basic concepts in some illustrative examples. We have given elsewhere [4] a detailed model pathway demonstrating the possibility of the development of a genetic apparatus. This pathway consists of a causal chain of simple and transparent physico-chemical model steps which are shown to be reasonable by simple quantitative approximations. In the following we confine ourselves to a qualitative discussion of the very first steps in the evolution of macromolecules.

Initial Steps of Selforganization

Let it be supposed that energy rich monomers for instance mono-nucleotides, have accumulated in some very particular place. The partial condensation of monomers on drying up, results in the formation of short polymer strands. This assumed initial situation is not unrealistic. It is easy to imagine that such a solution could form repeatedly in particular places, for instance in the vicinity of a meteorite of the class of carbonaceous chondrites, where nucleic bases have in the past been proved present alongside aminoacids [1]. It is easily possible that energy rich monomers are produced in a particular place and are accumulated specifically in an other particular place again and again by previous drying,redissolving, adsorption and desorption processes. This starting situation can be assumed at many different places.

Most of these spontaneously formed polymer strands would be useless as a matrix for replication. Let it be assumed that strands which could form a double helix with a complementary strand, would be capable of replication under suitable environmental conditions. Energy rich monomers would then be bound on such a strand condensing together due to the cooperativity in the resulting double helix. Such strands, which can serve as matrices, could have by chance d-ribose built into each monomer, and the monomers would have to be joint with each other appropriately.

Let us consider such a spontaneously formed strand and look at some given position. The probability that it has the proper building block correctly connected with the neighbours is about $\alpha = 10^{-2}$ and in the case of N = 10 building blocks the probability to obtain a strand able to replicate with all elements being correctly matching is $p = \alpha^N = 10^{-20}$. A litre of solution can be assumed to contain a total of about 10^{20} spontaneously formed chains, and hence about one molecule capable of replication.

This probability decreases exponentially with length of strands and is so minute where longer strands are concerned, that only short strands could be involved in the initial process of evolution.

The required replication will only occur under very special conditions. In order to have as concrete a mental picture as possible to adhere to, two sorts of complementary building block pairs will be postulated, as in real biological systems, where we have the pairs GC and AU, with G and C bridged by three hydrogen bonds, A and U less firmly by only two hydrogen bonds.

A nucleation centre for the new strand should now form at the end of the chain, from which polymerisation could proceed along the strand forming the matrix, but no nucleation centre must arise anywhere along the rest of the strand.

A suitable candidate for the initial process could therefore be a strand which would carry monomers forming strong hydrogen bonds at the end only.

In order that these strands could start to replicate and continue to reproduce, periodic temperature changes of a very particular kind will have to be assumed.

First it will have to be high enough for single strands to exist. Then it will have to fall sufficiently to allow the nucleation centre to form at the end of the chain, falling again until the double helix can gradually form along the matrix, starting at the nucleation centre. If this process is to proceed as faultlessly as possible, building in as few false monomers as infrequently as possible, it must occur not far from equilibrium, just below the melting point of the double helix, where double strands start to turn into single strands. Just as with the growing of perfect single crystals, the necessary temperature has to be strictly maintained. After the double helix has formed, the temperature must rise again, so that single strands can form and the process start afresh. Prebiotic molecules, in particular polypeptides, can play an important role as catalysers in the replication process.

Experimental efforts to realize these processes would be of interest. Methods to synthesize the proposed strands of nucleic acid are available, and by experiments with such strands a periodic program for initiating replication might be found. This program could then be applied to a mixture of polymers obtained by spontaneous condensation of monomers. It should lead to the selection of replicating strands. The solution should become optically active. The sign of the optical rotation, being determined by the initiating event, would depend on chance.

This periodic change in the state of the environment has to follow a very precise program, if replication with a minimum of error is to be achieved. This would be realisable on a prebiotic earth, if the region concerned were situated now in sunshine, now in the shade of one or other neighbouring rock. Temperatures would then fluctuate according to a definite, periodic program, depending on the fortuitous arrangement of surrounding objects. Quite a different program would prevail even in closely adjoining regions (where surrounding objects produced a different interplay of light and shade). An immense variety of programs would be realized within quite a narrow region. One or other of these many different programs would lead to the requisite, highly specific periodic reaction sequence, for statistical reasons.

Let the region be assumed to be situated in a very finely porous rock, awash with energy rich monomers. The monomers can diffuse into the pores of this substrate, but polymer strands will tend to be retained, given a sufficiently fine degree of porosity. They will therefore accumulate gradually in this substrate as the replication process proceeds, until a stationary state is reached, with as many strands leaving the region in any given period, as are newly formed by replication.

Supposing now, that the pores in an adjoining region are a little larger, so that the strands diffuse away too rapidly to reproduce there. If two strands happen to condense to form a longer strand, the molecule will be more voluminous. It no longer diffuses as rapidly out of the new region, so that more and more molecules gradually accumulate in this region due to replication, until a stationary state has again been reached, with as many molecules newly formed as escape in the same period by means of diffusion.

The systems could thus extend their range simply by lengthening the chain. They are able to reproduce in the largerpored substrate, where shorter chains could not maintain themselves.

Knowledge - a Measure of the Value of a Genetic Information

With the proceeding evolution the populated area is extended and the systems are increasingly better adapted to the environment (Fig.2): the evolving systems gradually increase their know-how to survive in the environment of given spacial and temporal structure. Any system that develops by multiplication, mutation and selection by some fittnesstest has a new quality not present in the preceeding steps of the planetary evolution: To act as an entity in a purpose-oriented manner, aiming an increase of the know-how to survive. This new quality appears with the formation of the first replicating molecules and retains basic in each succeeding evolutionary step. This quality is what distinguishes living systems from non living. Living systems are usually defined by giving a list of properties typical for biological systems. We prefer to define a living system as an object with this new quality.

It must be possible to measure a new quality by a new quantity.

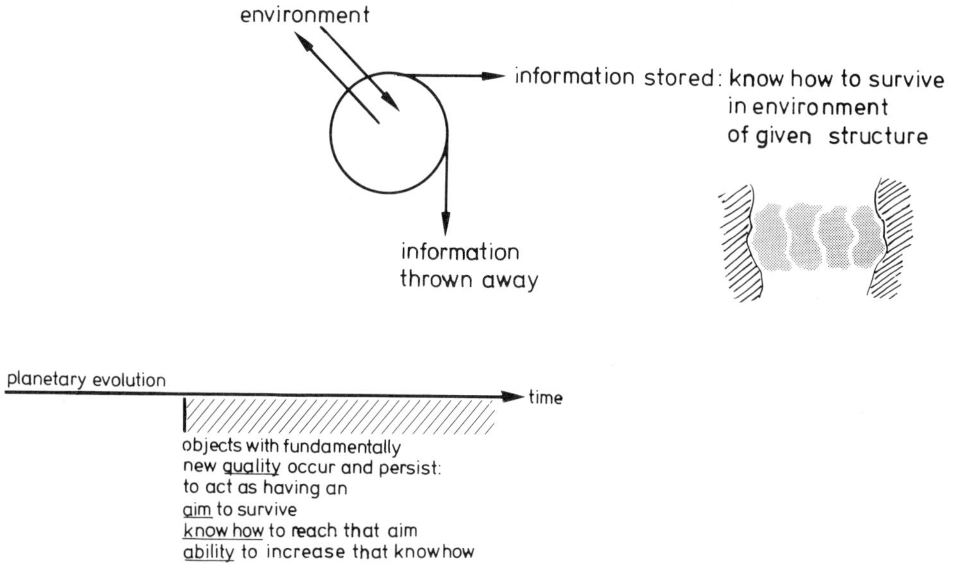

Fig.2 Darwinian mechanism (multiplication, sporadic mutation, selection of fit mutant) introduces new quality, measured by quantity K (knowledge).

This new quantity shall be called knowledge. The knowledge is a measure of the pragmatic or semantic value of the genetic information in a given evolutionary state. This value is not measured by the Shannon information. The number of bits used to express the genetic information does not measure how useful this information is for surviving in the given environment. Knowledge also is not measured by complexity. With growing evolution the complexity of the systems increase, and thus complexity has often been considered as a measure of the evolutionary level. However, complexity is increased quite generally with the number of possibilities and therefore with the number of particles. It is no measure for the adaptness of the system to the environment. A system can be complex, but useless in the sense discussed here.

Knowledge however is knowledge about the environment and can be considered only in relation to a given spacial and temporal environmental structure. Knowledge must be defined in such a way, that it measures the know-how to survive in an environment of given structure. It should measure the state of adaption, reached by a system that has evolved by multiplication, mutation and selection.

Knowledge characterizes a given evolutionary state. The evolutionary state shall be defined in analogy to the thermodynamic state which we consider first.

A body is repeatedly heated from absolute zero to a certain temperature T (Fig.3a). Each time its final microstate and the evolutionary route to reach that state are different, but the body is considered to be in the same thermodynamic state. The trick in thermodynamics, leading to a great simplification in the description of a complex system, is reached by considering an immense number of distinct microstates as realization possibilities of the same macrostate. A body in its internal equilibrium has a single realization possibility at zero temperature ($P = 1$ at 0° K and $P = \exp(S(T)/k)$ realization possibilities at temperature T). We consider the "evolution" of microstates when heating the body from 0° K to T in steps. In the i-th step (temperature T_i) the heat Q_i flows from the temperature bath to the body. The entropy S at T is given by

$$S \geqslant \sum_{o}^{T} \frac{Q_i}{T_i} \tag{2a}$$

Now let us consider in a Gedankenexperiment the evolution of macromolecules in an environment of given spacial-temporal structure. We wait for a given number of periods or generations. This number Z measures the degree of evolution (Fig.3b). When repeating this experiment each time another thermodynamic state is reached, but the evolving system will have essentially the same functions in its environment. For instance in a certain number of periods in an environment of a certain structure we may reach the stage where associates of convoluted chain molecules are present. Each time a completely different monomer sequence will be obtained, but the successful systems will always carry the same function: to form aggregates of interlocking molecules which are safe against diffusion and therefore survive in the scares porous regions. By definition, these different possibilities are realizations of the same evolutionary state. These thermodynamic states are the microstates of the same evolutionary state. The evolutionary state is clearly not a state in the thermodynamic sense, but in defining this state we make use of the trick used in thermodynamics to grasp an essential feature of a complex system by an averaging description.

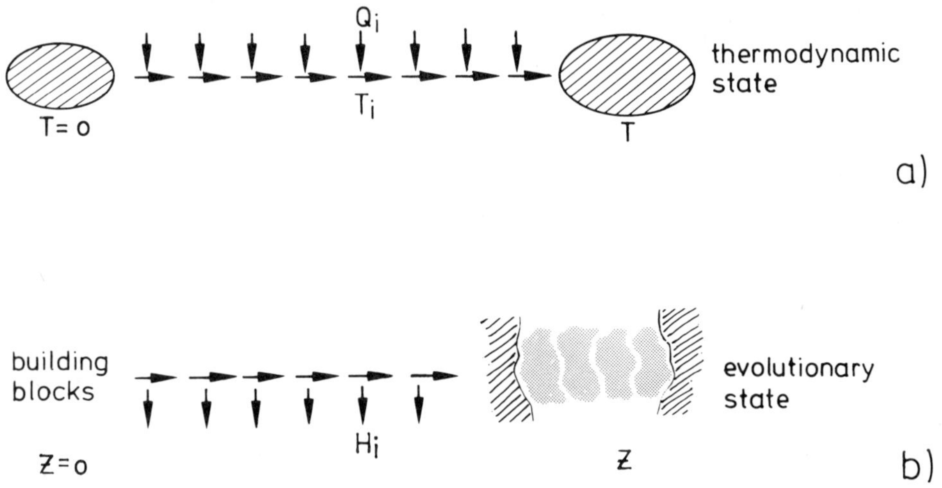

Fig.3 Evolution of thermodynamic microstate (realization possibility of thermodynamic state) (a)
and evolution of functional unit (realization possibility of evolutionary state (b).

A measure of the knowledge in a given evolutionary state should have the course in Fig.4 (K is given in the ordinate, the degree of evolution Z in the abszissa. The knowledge is practically constant as the systems are trapped at their evolutionary level, and it increases abruptly, as soon as a mutant is formed, which enters a new area to be populated. The knowledge increases suddenly with the beginning of each new convergent phase. This corresponds to the sudden increase of the entropy at each phase transition when heating a body. In each evolutionary state the populated area in the average has reached a certain size A. Correspondingly, a body at constant pressure has a certain volume V in a given thermodynamic state. The populated area is given by the geometrical size in the example discussed here, but a new ecological niche within the populated volume may as well represent the freshly discovered area.

For finding a measure for the knowledge in a given evolutionary state we consider the simple example of the selection of molecular strands with appropriate convoluted forms. For increasing knowledge, information has to be thrown away: The number of bits for describing the building plans of each individual discarded in the selection process. In the case of a strand i this information H_i is given by its sequence. The name information is misleading in this context since there is no receiver. H_i is the information that would be obtained by an observer that would take strand i and determine its sequence.

A hypothetical path can now be imagined, consisting of selective steps, leading from the simplest possible replicative molecule to the kind of system which has been considered, perhaps the aggregate in Fig.3b, or to some other form with a different micro-structure, representing the same evolutionary state. The sum ΣH_i can then be calculated, representing the statistical average of the information thrown away by degrading those strands which are gradually eliminated in the evolution process. For all pathways of this kind, leading to a certain evolutionary state given by a certain populated area A, the information discarded along one of these pathways has the minimum value K:

$$K \leqslant \Sigma \bar{H}_i \tag{2b}$$

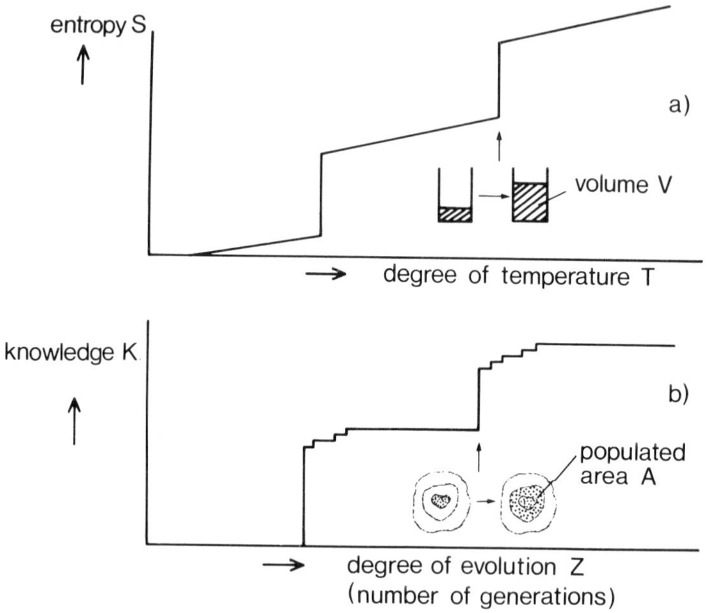

Fig.4 Increase of entropy S with increasing temperature T (a); increase of knowledge K with increasing number of generations Z (b).

K denotes the knowledge of the form.

It should be noted that $\Sigma \bar{H}_i$ is solely the information for the blue prints thrown away. If the entropy production in the selective phases would be counted as well, we would not obtain a measure for the knowledge increment. In these phases, all individuals are subject to a suitability test, which makes different demands on free enthalpy, depending on prevailing conditions. In some evolutionary steps, the systems which spend energy economically have the greatest survival chances, in others, the system which waste energy have the best chance to pass this suitability test.

If a form is trapped at its evolutionary level by remaining in the same environment and not changing its function, the magnitude K will remain unchanged within statistical limits: K has been defined as the minimum of information which has to be discarded in the average, in order that a particular function may be reached. As long as the function is unchanged, K remains constant. The breakthrough of a new structure is connected with a sudden increase of K, since in the average we must throw away many molecules during the divergent phase until the new organization structure appears by chance and is able to succeed (Fig.4).

The first jump from K = O to a distinct value K_1 reflects the appearence of the first replicating molecule among spontaneously forming strands, which occurs in the favourable region and is able to succeed. K_1 is the information to be thrown away in the average, until this chance event happens. A replicating molecule may appear and its offsprings may die out again in a few generations. This event is of no value for the evolution process. In calculating K it must be taken into account that only a fraction of favourable new strands actually succeed. A computer simulation to calculate K in simple cases has been given elsewhere [6].

By the averaging procedure in defining the evolutionary state, K can reflect only very general lines in evolution, but ΔK may be a useful quantity to compare distinct evolutionary steps.

These concepts can be applied to evolutionary processes in a quite general sense, such as the evolution of concepts. This evolution follows the same pattern as the evolution of molecules, but proceeds with highly increased speed: We construct logic pictures and select those which lead to the best fit with a sum of facts. In constant repetition we obtain conceptual structures which describe a multitude of facts with particular aptitude.

From such considerations new views concerning the general problem of the physical limits in the measuring process can be obtained. A limit is given by Heisenberg's uncertainty principle, and according to Szilard the entropy production in the measuring process (ΔS) cannot be smaller than the entropy decrease due to the measurement (gain of information H):

$$H \leqslant \Delta S \tag{3a}$$

Szilard [5] has considered a simple automata as model for an intelligent being. A better model for an observer is a system which has a device to observe, a device to store the observational data and a machinery for ordering these data by using an evolutionary process (Fig.5):

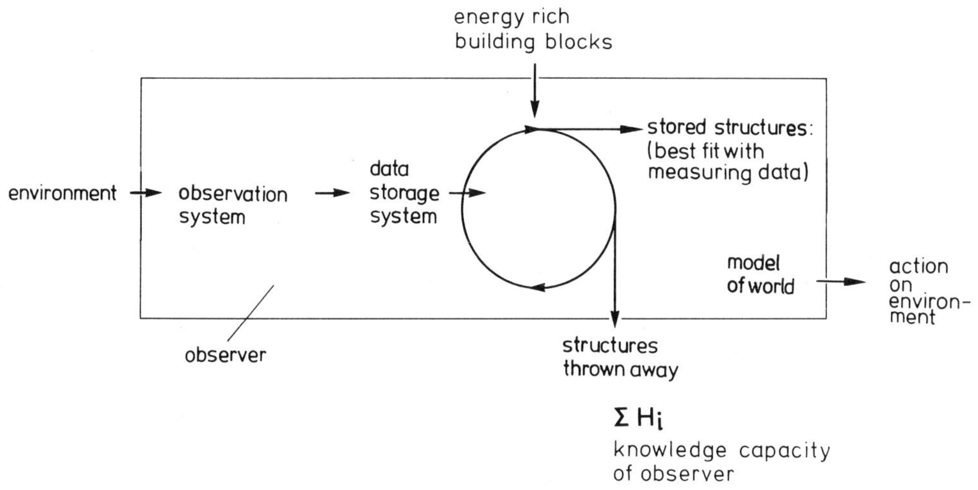

Fig.5 Knowledge capacity of observer. Maximum gain of knowledge $\Delta K \leqslant \Sigma H_i$.

This assumed machinery produces structures and selects those which fit best to the stored experimental data. By multiplication of structures, mutations due to reproduction errors by chance, and selection, structures will be obtained that will be more and more adapted to the stored observational data. These structures constitute the observers internal model of the world. The apparatus has an appropriate device to act on the environment according to its model of the world: it acts intelligent, different to Maxwell's demon in Szilard's case. It should be remembered that the axiom systems in mathematics and the general formulations in physics are free inventions. This process of inventing and selecting structures by trial and error is simulated by the model observer.

There must be a fundamental limit in the knowledge ΔK achieved by the apparatus. This limit is given by the construction and lifetime. A great number of structures are produced in all selection cycles during the lifetime of the apparatus and are thrown away. Their information content ΣH_i shall be called knowledge capacity. The gain of knowledge by the apparatus, ΔK, cannot be larger than its knowledge capacity

$$\Delta K \leqslant \Sigma H_i \tag{3b}$$

Eq. (3b) gives the maximum knowledge to be obtained by the observer.

The present approach is based on the assumption, that the laws of physical chemistry are sufficient for understanding the evolution of living systems, and that therefore the selforganization of matter is not a fundamental physical problem. This statement given elsewhere [6] has been critizised by Manfred Eigen [7]. The statement is concerned with the conceptual framework of physics and means that no fundamental problems occur within this framework, since the basic Darwinian mechanism appears as an obvious molecular process in the proposed model systems. We do not call in question that fundamental problems appear at higher levels in the hierarchic order of structure formation, and in fact focuss our effort to the search in this direction.

In the present view, the typical property of living systems, to adapt to an existing environment by a Darwinian mechanism, appears with the first replicating molecules and is basic to each subsequent evolutionary step. On the other hand, according to Eigen, the Darwinian evolution of molecules is succeeded by an integrating hypercyclic phase. In this phase a hypercycle takes over the regime by a non-Darwinian all-or non-selection. This phase is followed by a differentiating phase in which a compartimentation and again a Darwinian evolution occurs.

Eigen has mentioned in connection with the above criticism that one must carefully distinguish between theoretically derived logic necessities and models, and Eigen considered the non-Darwinian hypercyclic phase as a derivable, logic necessary, integrating intermediate step.

The necessity of a non-Darwinian phase is restricted to the particular case of spacial uniformity, thus excluding a Darwinian competing situation, which would demand regions of different concentrations of reactants. If evolution is an increasingly better adaption of the evolving systems to a given environment, each evolutionary theory must depend on a model describing the environmental structure. Logic requirements refer to models and it cannot be distinguished between theoretically derived necessities and models. What we have considered in model 2 as the driving force in evolution, is omitted in model 1 by the particular model assumptions.

We find that the Darwinian evolution of molecules can change over to a Darwinian evolution of functional units (assemblies of molecules acting as individuals, which multiply and die away), and therefore that a non-Darwinian hypercyclic phase is not a logic necessary integrating intermediate step in the selforganization of matter. We do not call in question the importance of hypercyclic reaction schemes in discussing many processes in biological structures and in ecosystems. Both modes of mutual support (hypercyclic or predator-preycyclic interaction where components decay as individuals, and formation of functional units where the entities decay) are important in evolutionary integration processes.

We have considered possibilities how molecules combine to produce more and more complex functional systems. The driving force for the selforganization of matter is seen in a specific environmental structure. The search for system properties important for selforganization may be useful as a guide line for future experiments to realize molecular functional systems.

1 K. Dose und H. Rauchfuss, Chemische Evolution und der Ursprung lebender Systeme, Wissenschaftl. Verlagsges., Stuttgart 1975; M. Calvin, Chemical Evolution, Clarendon Press, Oxford 1969.

2 M. Eigen, Naturwissenschaften 58, 465 (1971).

3 H. Kuhn in: Synergetics, Cooperative Phenomena in Multi-Component Systems, ed. H. Haken, Teubner-Verlag, Stuttgart 1973; Angew. Chem., Internat. Ed., 11, 798 (1972); Naturwissenschaften 63, 68 (1976).

4 H. Kuhn in: Biophysik - Ein Lehrbuch, S. 662, ed. Walter Hoppe, Wolfgang Lohmann, Hubert Markl, Hubert Ziegler, Springer-Verlag, Heidelberg 1977.

5 L. Szilard, Z. Physik 53, 840 (1929).

6 H. Kuhn, Ber. Bunsengesellschaft 80, 1209 (1976).

7 M. Eigen, Ber. Bunsengesellschaft 80, 1059 (1976) see p. 1080.

The Spatial Control of Cell Differentiation by Autocatalysis and Lateral Inhibition

H. Meinhardt

With 9 Figures

The organization of development is a major unsolved problem in biology. Beginning with a more or less homogeneous egg, more and more complex patterns of differentiated cells are reproducibly formed. How this development has to proceed must be genetically determined. However, the genetic material in each cell of an organism can be assumed to be the same. We have a degree of understanding of how DNA determines the amino acid sequence of proteins and how it is involved in the regulation of the transcription, but we have less understanding of how these processes in turn determine development.

Pattern formation out of more or less homogeneous initial conditions is by no means a peculiarity of biological systems. Stars, spiral nebula, clouds, lightning, creeks and rivers, sand dunes, hailstones and crystals may serve as examples. Common to all these pattern formation processes is that small deviations from homogeneous conditions have a strong positive feed-back such that they reinforce their own growth. Proverbial is the avalanche. To get a stable pattern, the self-amplification must be complemented by an antagonistic process which ensures that the surrounding of such an autocatalytic center is prevented from being "infected". A longer ranging inhibitory effect has to spread out from an activated center and may result from a depletion of material necessary for the autocatalysis or by an inhibition which is produced as a by-product of the autocatalysis.

On the basis of autocatalysis and lateral inhibition we have proposed a theory (GIERER and MEINHARDT [1-3]) which can account for the formation of stable patterned distributions of substances. Local concentration maxima can be the signal to initiate, for instance, local cell differentiation or the evagination of a cell sheet, or can act as an area of attraction for chemotactic sensitive cells. In the present article the basic principle of the theory will be briefly summarized and the main properties of the proposed reaction schemes will be illustrated with computer simulations. Similarities to biological pattern formation will be shown.

Let us assume two substances, the autocatalytic activator $a(x,t)$ and its antagonist, the inhibitor, $h(x,t)$. To obtain the criterion for determining which interactions will lead to a patterned distribution, one can proceed as follows [2]:

The production of a may depend on the concentration of both substances:

$$\frac{\partial a}{\partial t} = f\,(a,h) \tag{1}$$

To avoid oscillations, the inhibitor should equilibrate quickly and, to provide the lateral inhibition, it should have a higher diffusion rate. The concentration of h is then determined by the average a concentration ($\bar{a}$) in the area from which h is derived. The change of a can be approximated by

$$\frac{\partial a}{\partial t} = f\,(a,h(\bar{a})) \tag{2}$$

A homogeneous equilibrium of both substances would exist at the concentrations a_o, h_o if

$$\left(\frac{\partial a}{\partial t}\right)_{a_o} = f\left(a_o, h(a_o)\right) = 0 \tag{3}$$

This equilibrium would be stable if a deviation is regulated back:

$$\left(\frac{\partial f}{\partial a}\right)_{a_o} + \left(\frac{\partial f}{\partial h}\frac{\partial h}{\partial a}\right)_{a_o} < 0 \tag{4}$$

But a local increase of a will increase further if

$$\left(\frac{\partial f}{\partial a}\right)_{a_o} > 0 \tag{5}$$

while the average a concentration will be maintained by the relation (4). In other words, a patterned distribution of both substances is formed since a local increase of activator cannot be compensated for by the increasing inhibitor concentration, because the inhibitor diffuses more rapidly into the surroundings regulating down the inhibitor there, while, at the activated center, the activator concentration increases further until, for instance, the loss by diffusion is equal to the net activator production. The criteria (4) and (5) allow us to invent reactions which will lead to pattern formation. Let us assume a very general reaction scheme:

$$\frac{\partial a}{\partial t} = c\,\frac{a^r}{h^s} - \mu a + D_a \frac{\partial^2 a}{\partial x^2}\,; \qquad \frac{\partial h}{\partial t} = c\,\frac{a^t}{h^u} - \nu h + D_h \frac{\partial^2 h}{\partial x^2} \tag{6}$$

where c is a constant, μ the decay rate of the activator and D the diffusion constant. The relation (5) demands for the local instability

$$r > 1 \tag{7}$$

That means the activation must be non-linear. This is intuitively reasonable, since the autocatalysis must be stronger than the decay $-\mu a$. To obtain the global stability, the relation (4) requires

$$\frac{st}{u+1} > r - 1 \tag{8}$$

The following equations fulfill this requirement:

$$\frac{\partial a}{\partial t} = c\,\frac{a^2}{h} - \mu a + D_a \frac{\partial^2 a}{\partial x^2} + \rho_o\,; \qquad \frac{\partial h}{\partial t} = c a^2 - \nu h + D_h \frac{\partial^2 h}{\partial x^2} \tag{9}$$

Or, if the inhibitor has a quadratic influence on the activator production, the inhibitor production itself can be linear:

$$\frac{\partial a}{\partial t} = c\,\frac{a^2}{h^2} - \mu a + D_a \frac{\partial^2 a}{\partial x^2} + \rho_o\,; \qquad \frac{\partial h}{\partial t} = c a - \nu h + D_h \frac{\partial^2 h}{\partial x^2} \tag{10}$$

The small activator-independent "basic" activator production ρ_o is added in (9) and (10) to initiate the system in the absence of any activator. As we have shown, this type of reaction leads to pattern formation which has much similarity to that observed in biology. Its properties will be demonstrated here with some computer simulations of (9): The pattern can be monotonic (Fig.1,3), symmetric (Fig.2) or periodig (Fig.4, 5). In two dimensions the substances can be graded in only one dimension (Fig.3). Furthermore it has been shown [4,5] that if the response of the cells to this prepattern is included, experiments concerning the segment determination during early insect development (see Fig.6) and the formation of netlike structure, such as blood vessels or leaf veins (Fig.8,9), can be described.

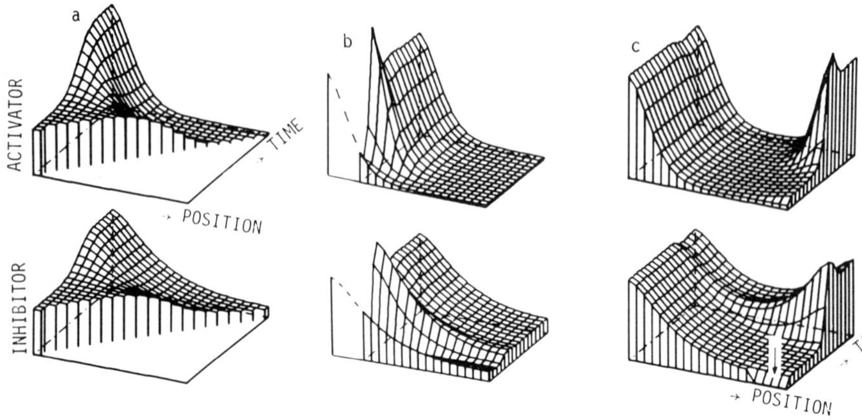

Fig.1 *Formation of an activated region, its regeneration and its unspecific induction.* We have assumed a linear array of cells and an interaction of the autocatalytic activator and its faster diffusing antagonist - the inhibitor - according to (9); the concentrations are plotted as function of space and time. a) In a growing field the homogeneous distributions will become unstable if the spatial extension exceeds the range of the activator: A small local increase of the activator concentration, caused for instance by a random fluctuation, will increase further, since the additionally produced inhibitor diffuses rapidly into the surroundings regulating down the activator production there. However, at the activated region, the activator concentration will increase further until some natural limit is reached - for instance, if the loss by diffusion is equal the net production. The high activator concentration will appear at one boundary, since, due to the lateral inhibition mechanism, the activated and non-activated parts will appear at maximum distance from each other. Which boundary will become activated is arbitrary as long as virtually no asymmetry is imposed (which is rarely the case in biology). The marginal activation is also maintained if further growth occurs. Therefore, cells can get different signals if a critical size is surpassed and the monotonic gradient formed may be the basis for polarity - a frequently observed tissue property. b) After removal of an activated area, the inhibitor is no longer produced and, after its decay, a new activation is triggered by the remnant or basic activator concentration; the activator profile regenerates. If, for instance, in a Hydra, the high activator concentration is the signal for head formation, this mechanism explains the reformation of that signal after head removal. (Moreover, a set of transplantation experiments with Hydra tissue [6] can be described correctly [1,2]). c) If the inhibitor outside of an activated area is reduced (arrow), the basic activator production ρ_0 can trigger a second activation. Such an activation may be triggered by very unspecific manipulations, such as cell poisoning or by leakage of the inhibitor through an injury. Unspecific activation has been a puzzling and disappointing observation in the investigation of organizing substances [7], but according to our theory it is a straightforward consequence of the mechanism which enables the formation of spatial differences.

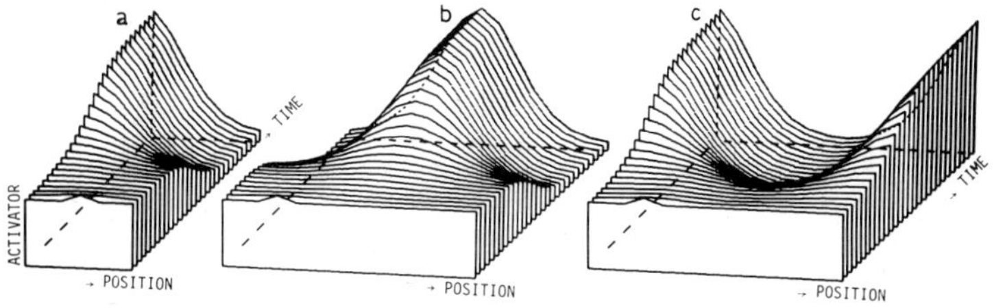

Fig.2 *Formation of monotonically graded and symmetric pattern.* a) In a small area, an activator increase in the center will disappear since the inhibitor has not enough space to diffuse in (the boundaries are assumed to be tight) and thus will smother the activation. Random fluctuation - 1 % from cell to cell in the constant c, Eq.(9) - can trigger a maximum at one boundary. b,c) In a larger field, symmetrical patterns are favored, either with a maximum at the center (b) or with a maximum at each end (c). Such patterns can - as is shown - also emerge if the deviation from the equilibrium was placed in an asymmetric position. In some developing systems, a symmetric pattern can be evoked if the pattern formation process is disturbed quite late, after a certain size has already been attained. For instance, YAJIMA [8] found symmetrical embryos with either two heads or two abdomens after centrifugation of certain insect eggs (see also Fig.6).

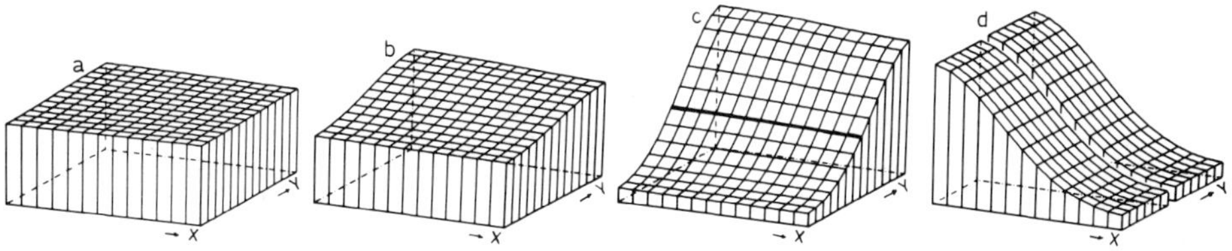

Fig.3 *Stages in the formation of a one-dimensional gradient in a two-dimensional field.* In a two-dimensional field of the size of the activator range, a graded activator distribution in one dimension but with a nearly constant concentration profile in the other will be formed (a-c). Such a system could provide positional information in one dimension. The second dimension can be specified in the following way: If, for instance, the high and low concentrations specify different structures and a diffusion barrier is placed in between (indicated by a heavy line in Fig.c), the pattern will reorient itself in the dimension perpendicular to its original orientation (Fig.d; see also [9]), since now the longest extension of the field is oriented perpendicular to the first gradient. In agreement are experiments of SANDER [10] with the eggs of a leafhopper: The dorso-ventral axis is specified only *after* the antero-posterior axis is organized.

As discussed in more detail in Fig.4 and 5, a sharp activator peak may be used to initiate local events like formation of hairs or bristles on insect epidermis, leaves or stomatas in plants, buds or tentacles in Hydras, or may act as a signal for local tissue evagination in the lung or the salivary gland [15].

In contrast, the monotonic activator or inhibitor distribution as shown in Fig.1 or 3 would provide a unique concentration of a substance (or a set of substances for more dimensions) to a particular cell such that the cell can detect its location within the tissue and behave appropriately. - The cell would have "positional information"[16].

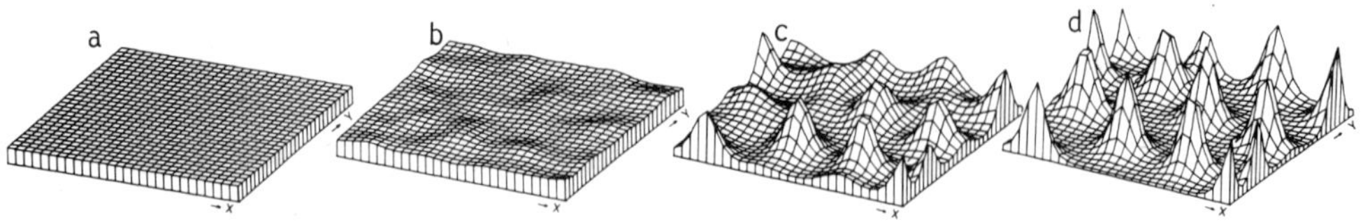

Fig.4 *Stages in the formation of periodic activator maxima in a two-dimensional field.* If the total size is much greater than the activator and inhibitor range, several activator peaks are formed. If the pattern formation does not start until a certain size is obtained, the pattern is somewhat irregular: To begin with, the mutual inhibition of the incipient activator peaks is small, peaks may appear too close together and some of them in the end may disappear. A shift of an activator peak after its full development to a more optimal location is nearly impossible, since every peak is surrounded by a cloud of inhibition. Such somewhat irregular structures in which nevertheless a maximal and minimal distance is maintained, can be observed in the patterns of stomatas [11] - the openings in the leaf epidermis - or in the pattern of hairs of an insect [12]. In agreement with the model in both these cases, the pattern formation starts only after a certain growth period is completed.

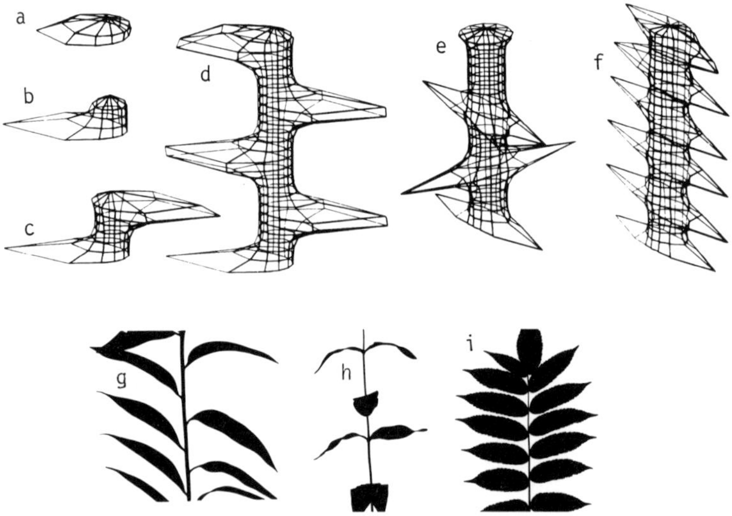

Fig.5 *Regularly spaced peaks - a simulation of phylotaxis.* Activator maxima with a regular spacing appear if the pattern-forming process operates during growth. To simulate a growing shoot of a plant, a cylindrical arrangement of cells which grows by the division of the uppermost ring of cells is assumed. High activator concentration should initiate a leaf. Random fluctuation may trigger the first maximum (a), after further growth (b), the next maximum is triggered opposite (c), since there the inhibition is at its lowest level. The final result is an alternate (distichous) pattern. If the diameter of the shoot is higher or the inhibitor range is lower, two maxima can appear simultaneously, especially if an inhibitory effect arises from the growing tip. The next pair of peaks appears perpendicular to the first, and so on. The result is an opposite (decussate) pattern (e). The pattern of budding in Hydra may be governed by a similar mechanism [3]. Parallelly arranged maxima (f) will appear, if growth is so fast that the cells have some memory that their ancestors were originally activated or if activator diffusion is facilitated in axial direction. Biological examples corresponding to the patterns d, e, f are shown in g, h, i (Fig. according to [13,14]).

An example where the formation of several different structures may be controlled by the graded distribution of one substance could be found in early insect development. Since the egg is filled with nutritional substances, normal development can be disturbed severely, for instance by constriction or UV irradiation, and still a development to individually recognizable segments is possible. Some of these very detailed experiments (for review see [17-19]) offer an excellent opportunity to test a model. We have shown [4] that many such experiments can be quantitatively described under the following assumption: (i) A high activation is located at the posterior pole. (ii) The inhibitor which spreads out from this activation is the controlling agent - the morphogen. Its absolute concentration determines which segment will be formed. (iii) The cell responds to the morphogen concentration by a stepwise and irreversible change in the determination state from more anterior to more posterior structure, until the determination obtained corresponds to the morphogen concentration.

The power of the gradient model will be demonstrated here only with one example. After the irradiation of a small portion of the anterior pole of a *Smittia* egg, a completely symmetrical embryo is formed, with a second abdomen in the place of the head [20]. But also the segment in the center of the egg is altered, even if this part has not been irradiated. According to the model, abdominal structures are formed at high inhibitor concentrations (Fig.6). The formation of a second abdomen would indicate a second activation. An unspecific activation is possible by lowering of the inhibitor concentration (see Fig.1c). And this may be - directly or indirectly - the action of the UV irradiation.

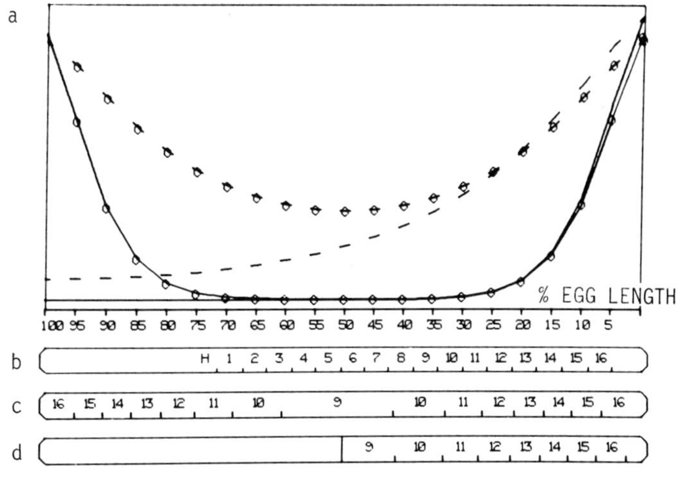

Fig.6 *Explanation [4] of the double abdomen formation [20] in the insect Smittia.* From experiments with earlier and later ligations the approximate diffusion ranges of the morphogens can be estimated and the resulting activator (—) and inhibitor (--) distribution is shown in Fig.a. The approximate location from which cells are derived for a particular segment is indicated in Fig.b (H = head, 7-16: abdomen). If, by UV irradiation, a second maximum is triggered at the anterior pole, a completely symmetrical distribution of both substances is formed (marked in Fig.a with circles). But since in this case the inhibitor is produced at both poles, the inhibitor concentration in the center increases and, instead of the segment 5, the segment 9 is formed here, which is in essential agreement with the experimental results [20]. In addition, if by a ligation the egg is separated into two halves, segment 9 is also formed at the posterior site of the ligation (d) [17]. The fact that the same segment is formed at the center after the different manipulation is a natural consequence if the segment pattern is controlled by the graded distribution of a diffusible substance: The ligation introduces a diffusion barrier, the inhibition accumulates to the same level as it would if a second activation had been formed: in both cases the net flow of the inhibitor through the center is zero. This may be the best evidence available that a long-range control by a diffusible substance exists.

The last application which should be mentioned here is the formation of filaments and their organization into netlike structures. Such structures are present in almost every higher organism; blood vessels or leaf veins are examples. The filaments can consist of linearly arranged differentiated cells, such as in leaves, or of long extended extrusions of individual cells, such as in nerves. The mechanism we have proposed [5] for the formation of netlike structures will be explained for the special case where undifferentiated cells undergo an ordered differentiation along a line.

We have seen that a sharp local activator maximum can be formed by interaction of an activator $a(x,y,t)$ and an inhibitor $h(x,y,t)$

$$\frac{\partial a}{\partial t} = c\,\frac{a^2}{h}\,s - \mu a + D_a \Delta a + \rho_o y \qquad (11)$$

$$\frac{\partial h}{\partial t} = c a^2 s - \nu h + D_h \Delta h \qquad (12)$$

(the new terms s and $\rho_o y$ will be explained below). The local high activator concentration should be the signal for a cell to differentiate. The differentiation can be described by

$$\frac{\partial y}{\partial t} = da - ey + y^2/(1+fy^2) \qquad (13)$$

In the absence of activator, y can only be at two stable states, at zero, or due to the saturating autocatalysis, at high concentration. In the presence of sufficient activator the low y concentration will become unstable and the cell will irreversibly "differentiate" into a high y-state. Let us now assume that the purpose of the net with in the organism is to remove a substance (which may be auxin in the case of plants [21] or Nerve Growth Factor [22] in the case of adrenergic nerves [23]).

The substance s is assumed to be produced everywhere with the constant rate c_o, decays with the rate γ and disappears in the differentiated cells with the rate $\varepsilon s y$ and diffuses:

$$\frac{\partial s}{\partial t} = c_o - \gamma s - \varepsilon s y + D_s \Delta s \qquad (14)$$

A depression of s will be formed around a differentiated cell and if - as written in (11) - the activator production depends on s, the activator peak will escape the s-depression and be shifted to that neighbour cell which contains the highest s concentration and so on. Filaments of indefinite length can be formed and branches can be initiated if the differentiated cells have a small basic activator production ($\rho_o y$ in (11)). This process is illustrated with a numerical solution of (11-14) in Fig.7 and 8. For more details, such as the modes of reconnection, limitation of the maximum net density or the homing of a growing filament into a particular target area, see [5].

The mechanism outlined above is not the simplest possible one. Two different substances have been employed for inhibitory action on the activator production: The inhibitor, to keep the activator maximum localized; and the depleted substance s to shift the activator peak away from the differentiated cells. Both inhibitory actions can be taken over by the substance s alone. But this has the consequence that branching can only be accomplished by a split of one activator peak into two, causing a bifurcation of a growing filament. Sprouting out from an existing filament is no longer possible, since the differentiated cells must be so inhibitory. This simpler pattern has a counterpart in the evolutionary early dichotomous branching pattern (Fig.9) which can still be found in the ginkgo and some ferns. If this model is correct, the inventic of a separate inhibitor during evolution would have made the leaf mechanically more stable since it allows reconnections, and would have enabled intercalary growth since the growing intersticies can be provided with new branches.

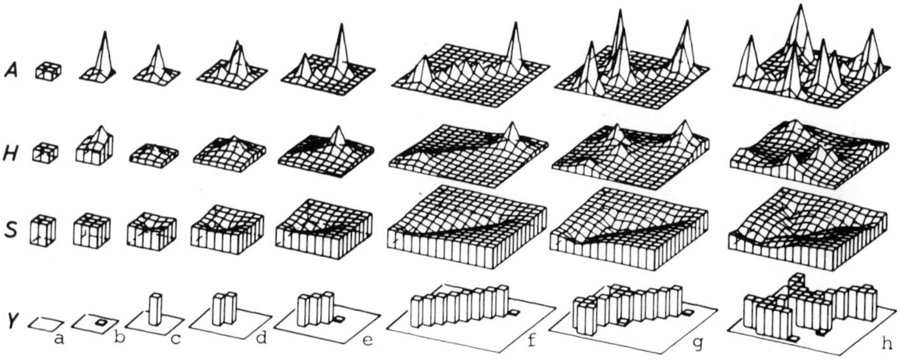

Fig.7 *Formation of a netlike structure.* The interaction of the activator (A, upper row) with the inhibitor (H) leads to the formation of a sharp maximum (a,b), the corresponding cell is differentiated (switching Y, bottom row, from low to high concentration). The differentiated cells remove the substance (S). Since the activator production depends on S, the activator peak is shifted and a neighbouring cell becomes activated and finally differentiated (c,d). Long filaments of differentiated cells are formed (e,f). If the distance between two growing tips is high enough, the inhibition along a filament may become so low that the basic activator production can trigger a new peak which escapes the S valley as fast as possible and a new branch is initiated. The branch will sprout out at 90° as long as no other constraints such as proximity to the margins or other filaments are imposed.

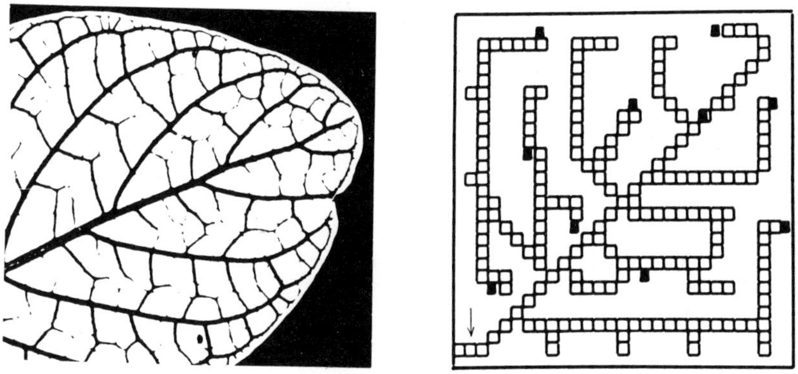

Fig.8 *Natural and simulated leaf pattern.* The very bright appearance of the major veins of the plant *Fittonia verschaffelti* (which are shown here in a negative) allow a comparison with a pattern generated by the numerical solution of (11-14). One differentiated cell (↓) was assumed to be initially present. The line first extends in a diagonal direction in order to obtain more distance from the margins. Branches try to grow out at 90° but are deflected by the margins, since no substance s can be obtained from beyond. Successive branches are deflected by the earlier branches and grow out, therefore, also at 45°. Reconnections are formed mainly between higher order branches.

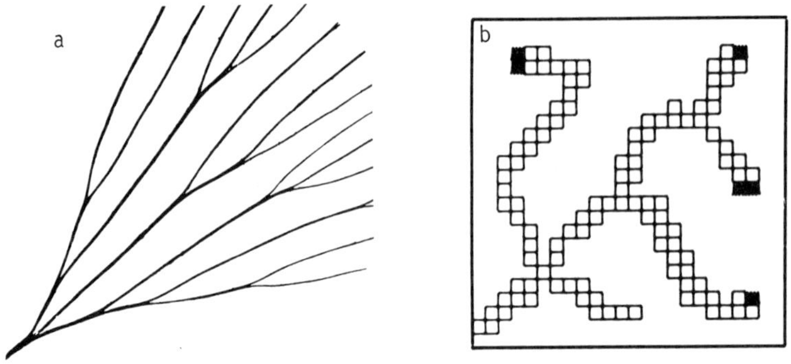

Fig.9 *The dichotomous branching pattern* is an evolutionary early form and can still be found in some ferns and in the *Ginkgo*. Fig.a shows the veination of *Kingdonia* [24]. Characteristic is that branching occurs solely in the form of bifurcations and not by lateral branching. This pattern can be simulated (Fig.b) with only two differentiation-controlling substances [5]. The first fork - in the leaf as well as in the simulation - is a triple fork, since at small sizes, the relative growth by division of marginal cells is high and two splittings of the activator maximum follow quickly upon one another.

In conclusion, striking and very different patterns can be reproducibly formed out of an initially almost homogeneous distribution. The proposed mechanism includes only biochemical processes, such as mutual activation and inhibition of substances and of course diffusion, which are known to exist in biochemical systems. Pattern formation appears to be a process of self-organization; it is brought about by the intrinsic properties of the substances and their mutual interaction, and these properties can easily be coded by the genes. The pattern formation observed in the computer simulations shares many similarities with that observed in biological systems and it is to be hoped that the theoretical framework will facilitate the biochemical identification of the pattern-forming processes.

References

1. Gierer, A., Meinhardt, H.: Kybernetik 12, 30, 1972
2. Gierer, A., Meinhardt, H.: Lectures on Mathematics in the Life Science 7, 163. Providence, Rhode Island: Amer.Math.Soc. 1974
3. Meinhardt, H., Gierer, A.: J.Cell Sci. 15, 321, 1974
4. Meinhardt, H.: J.Cell Sci. 23, 117, 1977
5. Meinhardt, H.: Differentiation 6, 117, 1974
6. Wolpert, L., Hicklin, J., Hornbuch, A.: Symp.Soc.exp.Biol. 25, 391, 1971
7. Spemann, H., Mangold, H.: Arch.f.mikr.Anat.u.Entw.Mech. 100, 599, 1924
8. Yajima, H.: J.Embryol.exp.Morph. 8, 198, 1960
9. Gierer, A.: Curr.Top.devl.Biol. (in press)
10. Sander, Wilhelm Roux Arch. Entw.Mech.Org. 167, 336, 1971
11. Bünning, E., Sagromsky, H.: Z.Naturf. 3b, 203, 1948
12. Lawrence, P.: Adv.Insect Physiol. 7, 197, 1970
13. Meinhardt, H.: Ber. dtsch.bot.Ges. 87, 101, 1974
14. Meinhardt, H.: Rev.Physiol.Biochem.exp.Pharm. (in press)
15. Spooner, B.S.: Bioscience, 25, 440, 1975
16. Wolpert, L.: J.theor.Biol. 25, 1, 1969
17. Sander, K.: In: Cell Patterning, Ciba Symp. No. 29, pp.241-263. Amsterdam: Association of Scientific Publishers. 1975
18. Sander, K.: Adv.Insect Physiol. 12, 125, 1976
19. Counce, S.J., Waddington, C.H. (eds): Developmental Systems: Insects. London, Academic Press 1972

20. Kalthoff, K., Sander, K.: Wilhelm Roux Arch.Entw.Mech.Org. 161, 129, 1968
21. Jost, L.: Z.Bot. 38, 161, 1942
22. Levi-Montalcini, R.: Science, 143, 105, 1964
23. Hendry, I.A., Stöckel, K., Thoenen, H., Iversen, L.C.: Brain Res. 68, 103, 1974
24. Foster, A.: Amer.J.Bot. 47, 684, 1960

Regelmäßig periodische Verteilung im Raum

H. J. Maresquelle

Biologische Probleme : 1. Auf einer Oberfläche : Haare auf dem Schädel, Federn auf dem Vogelkörper, Flecke auf dem Felle des Leoparden ... ; 2. Längs einer Achse : Sreifen auf dem Felle des Tigers oder des Zebras ; aber auch, bei allen Wirbeltieren, embryologische Erscheinung der späteren Metamerie auf dem anfänglich kontinuierlichen Embryo, mit sehr wichtigen Folgen ebenso für Nervensystem und Blutgefässsystem, wie für Skelett : Rippen, Wirbeln.

Schon untersuchte Probleme : 1. Spaltöffnungen (Stomata) auf Blättern (Erwin BÜNNING, Tübingen) ; 2. Hier werden nur die zwei gründlicheruntersuchten Erscheinungen behandelt : Agregationszentren bei Dictyostelium discoīdeum ; Konvektionssaülen im Bénard-Experiment (Vergleich dazwischen bei KELLER und SEGEL, J. Theor. Biol. 26, 399, 1970).

Erste Frage : wie soll man sich die Kooperativität bei Dictyostelium und im Bénard-Experiment vorstellen ? Darf ich annehmen, dass Kooperativität darin besteht, dass raümlich abgetrennte Teilsysteme zur Bildung derselben gemeinsamen raümlichen Struktur mitwirken ? Dass also jene "raümlich abgetrennte Teilsysteme" entweder Moleküle sein können (bei chemischer Kooperativität), oder bewegliche Lebewesen (bei Dictyostelium), oder noch grössere makroskopische Wassermassen (beim Bénard-Experiment, da ja Konvektionsbewegungen nur makroskopisch entstehen können und kein molekularer Ursprung möglich ist) ?

Zweite Frage : wie kann man, ausgehend von beiden untersuchten Erscheinungen, zu einem gemeinsamen allgemeinen mathematischen Modell gelangen ?

Wir nehmen folgendes an : eine Erscheinung E (entweder Konvektionssaüle, oder Amöbenagregation) kann überall vorkommen ; doch, wenn einmal irgendwo E vorhanden ist, wird E in der nächsten Nähe unmöglich. Jedes E-Bezirk wird von einer "non E"-Hemmungszone umgeben. So ist es im Bénard-Experiment, wo jede steigende Konvektionssaüle von einer Zone nach unten fliessender Wasserfaden umgeben wird ; ebenso bei Dictyostelium, wo jede Agregationszone von einer leeren Zone (S. KELLER und SEGEL) umgeben wird. Wenn nun sich mehrere solche runde Systeme (E umgeben von "Non E") zusammenpressen, gibt es natürlich ein hexagonales Netz, welches als hochwahrcheinlich erscheint und keinen Auspruch auf Struktur erheben darf. Die Dignität einer Struktur gehört dagegen dem Elementarsystem E - non E.
Nun, wie könnte man solche Gedanken mathematisch ausdrücken ?
Zwar gibt es hier Beziehungen zu den Fragestellungen nach H. MEINHARDT. Doch besteht folgender Unterschied : MEINHARDT sucht nach einer chemischen Formulierung (Aktivator, Inhibitor, Autokatalysis), was ja hochwahrscheinlich für viele biologische Differenzierungen gültig sein wird. Hier wird im Gegenteil eine ganz abstrakte Lösung gewünscht.

Warum ein solcher Unterschied ? Weil gerade bei unseren Fällen die chemische Deutung unmöglich ist : selbstvertändlich im Bénard-Experiment,aber auch bei Dictyostelium, wo die chemische Verursachung eine ganz andere ist. In beiden Fällen haben wir nämlich "Hemmung ohne Hemmstoff" : bei Bénard weil das absteigende Wasser dem aufsteigenden den Weg absperrt ; bei Dictyostelium, weil eine Amöbenanhaüfung in einem Punkte die nächste Umgebung so schlimm entleert, dass weitere Anhaüfungen daneben unmöglich sind. Wir müssen also den Begriff "Hemmung" in abstrakter Weise behandeln, ohne ihm die chemische Darstellung aufzuprägen, welche bei der chemischen Weltanschauung der jetzigen Wissenschaft so üblich ist. Ich hoffe dass die Mathematik, mit ihrer so scharf abstrakten Betrachtungsweise, uns behilflich sein wird !

Note of the editor

Though Prof. Maresquelle did not participate at the Workshop I included his comment because of its general interest. I hope that the ideas I presented in my introductory talk might serve as a first step to find an answer to Maresquelle's question about an abstract formulation of pattern formation.

Biological Systems

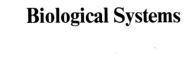

Neurosynergetics

J. D. Cowan*

With 10 Figures

1. Introduction

Herman Haken's introduction of the term "synergetics" [1] for the study of, among other things, that cooperative interaction of many subsystems which generates ordered structures or activity, will strike a reminiscent note with most neurobiologists: The term "synergy" was introduced by SHERRINGTON [2] nearly 100 years ago, in exactly the above sense, in his analysis of muscle systems and their spinal control mechanisms. In this paper I discuss, not synergetic motor systems, but sensory systems, in particular the visual system, from a synergetic point of view.

2. Neuronal Cytoarchitectonics

A very simplified representation of the sensory-motor system of a higher vertebrate is shown in Fig. 1. It consists of four cellular aggregates

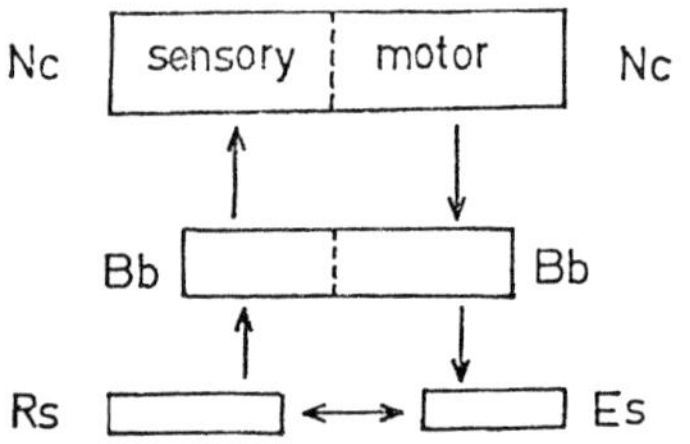

Fig. 1 The vertebrate sensory-motor system. Rs = receptor surfaces, Es = effector surfaces, Bb = "between" brain or diencephalon, Nc = neocortex

arranged in an ascending and descending hierarchy. The lowest level comprises the receptor surfaces, receptors and peripheral neurons, and the corresponding effector surfaces, "motor" neurons and muscles. The second level comprises the so-called "between" brain or *diencephalon* , a complex of neuronal nets that serves as a relay for signals passing between the peripheral receptor and effector surfaces, and the third level neuronal nets of the neocortex. In general, as one ascends the hierarchy from periphery to center, the neuronal nets become more complicated. For example, in the primate visual system there are of the order of 10^6 neurons in the retina, perhaps 10^7 neurons in the lateral geniculate nucleus and pulvinar, and about 10^8 neurons in the visual neocortex. Given that there are two retinas, this implies a multiplication factor of not more than

* Permanent address: Department of Biophysics and Theoretical Biology, The University of Chicago, Chicago IL 60637, USA.

one order of magnitude in going from the periphery to the cortex, in the number of neurons representing a given visual direction. I shall return to this point later. For the present it suffices to note that the visual cortex, as a typical piece of cortex, is some 2 - 3 mm thick, is close-packed with neurons and sattelite cells, the neuronal packing density being some 5.10^4 neurons per mm^3, and it has a total surface area of about 800 mm^2. Of course the cortex is very convoluted, since the primate brain has to fit into a smallish skull. Fig. 2 for example shows a coronal section of the occipital cortex, the posterior pole of which consists of the visual cortex.

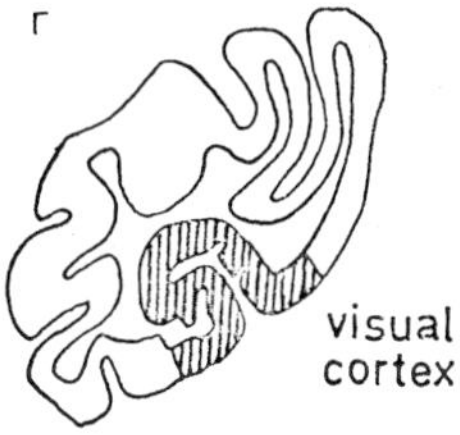

Fig. 2 Coronal section of occipital neocortex, showing the visual cortex

However, the most striking feature of the visual and other parts of the neocortex, is not so much the high packing density, nor the large number of cortical afferents (incoming nerve fibers), $1.25 \cdot 10^3$ per mm^2, nor the correspondingly large number of cortical efferents (outgoing nerve fibers), but the very high density of intracortical connexions. The most recent estimates of this density are of the order of upto 4.10^4 synaptic terminals (i.e. terminals made by fibers from other, mostly cortical, cells) per cell. It is such a coupling density that distinguishes the primate cortex from even the most advanced and complex microcomputer yet fabricated, and that leads to the inference that any overall understanding of the functioning of such a system will require a nontrival understanding of the complex kinetics of intracortical activities generated in neuronal nets of high cellular and coupling densities. In addition, within such nets there is a very high degree of spatial order, both local and long-range. There are perhaps 50 differing neuronal types in visual neocortex, arranged in highly ordered spatial patterns [3]. Fig. 3 shows an example of this complexity. All afferent fibers are assumed to be excitatory in their mode of action on the cortex. However within the neocortex there may be neurons, the action of which is inhibitory rather than excitatory. Two such cell types are shown in Fig. 3, an inhibitory "basket" cell in layer IVCß, and a more localized inhibitor in layer IIIA. So far as excitatory cells are concerned, there are two main types, "spiny" stellates which act mainly to filter and relay incoming signals from the *diencephalon*, and "pyramidals," that serve to transmit outgoing signals to other parts of the system. Figure 3 shows a stellate cell in layer IVCß, relaying to pyramidal cells mainly in layer IIIB, but also stimulating pyramids in layers VA and VI. The pyramid in layer VA is of interest in that it is an interneuron, like the inhibitory cells of layers IIIA and IVCß, relaying its signal along its output progress or axon, only as far as the cells of layers II, IIIA and IIIB within the same local patch of cortical tissue. It is evident that all forms of local interactions (++, +-, -+, --) are possible within such an organization. It seems to be the case that longer ranged intercortical interactions are mutually excitatory.

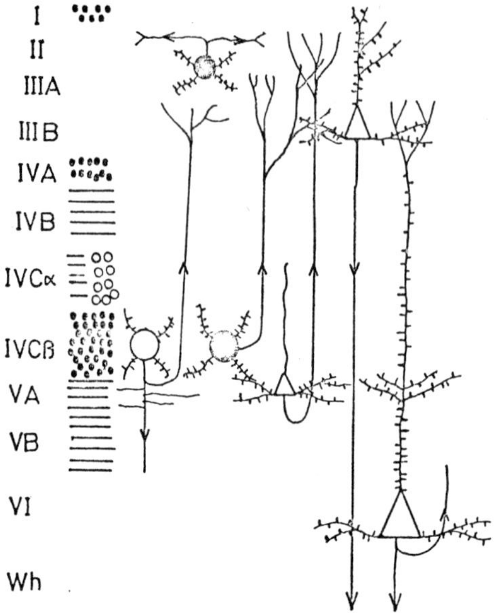

Fig. 3 Diagram showing one or two of the more important neurocortical cell types, and their interaction patterns. Incoming fibers of various kinds are shown on the left, the numbers of the cortical layers on the left. Excitatory relay cells are shown as open cycles, inhibitory interneurons as filled circles, and excitatory pyramidal cells, the principal output cells of the neocortex, as open triangles. Cell dendrites, the site of "synapses" with other cells, are shown with spines on them.

3. Receptive Fields and Cortical Order Parameters

So much for cytoarchitectonics! I now want to outline some of the remarkable physiological discoveries which have been made in the last 20 years or so, concerning the functional organization of cat and primate visual systems. The starting point is the concept of a receptive field: that part of the receptive surface which, when stimulated, will elicit a response from a given cell, as measured via a microelectrode. Such a concept is of course an abstraction of a net property rather than a single cell property, and needs to be used with considerable care, particularly for cortical properties. Nevertheless it has proved to be an extremely useful idea. The most important discoveries are summarized in Fig. 4.

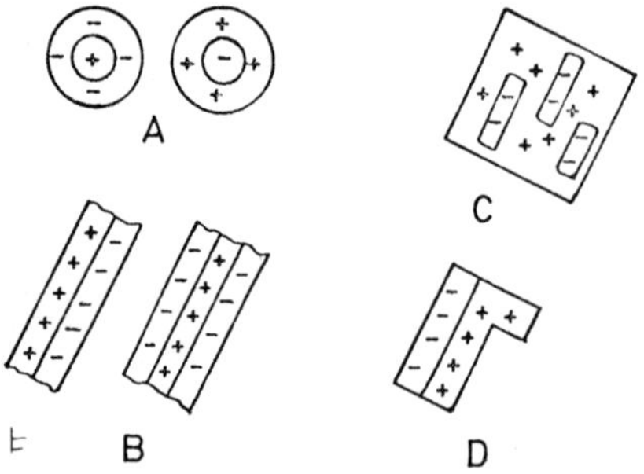

Fig. 4 Various types of receptive fields found in the cat and primate visual systems. A. Radially symmetric *on* and *off* center fields. B. *Simple* oriented *edge* and *bar* fields. C. *Complex* field. D. *Hypercomplex* field.

The *on* and *off* center fields depicted in Fig. 4A are typically found in the retina and lateral geniculate nucleus. Thus these nets respond to either light spots on dark backgrounds, or conversely, to dark spots on light backgrounds[4]. Fig. 4B shows a receptive field that is termed *simple*. It was HUBEL and WIESEL'S discovery of the existence of such oriented fields in the cat visual cortex [5], that triggered the modern era of physiological research into such properties. *Simple* receptive fields have been associated with the spiny stellate cells of layer IVCß. Such cells therefore respond optimally to a light edge or bar on a dark background, or conversely, provided such stimuli are correctly oriented, and cross the receptive field at a fixed velocity and direction of movement, or if stationary, are "flashed" on and off at an appropriate rate. HUBEL and WIESEL also discovered the *complex* and *hypercomplex* fields depicted in Figs. 4C and 4D. The *complex* field differs from the *simple* field in that it comprises several regions within which an edge or bar, appropriately presented, will stimulate the associated cortical element. Such fields have been associated with the pyramidal cells of layers II, III, V and VI. *Hypercomplex* fields contain excitatory patches of more complicated shapes than either edges or bars, for example, corners; otherwise they are similar to *simple* fields in their properties, and they have been associated with various pyramidal cells of the visual cortex. It should also be noted that there are also numerous non-oriented fields in the cortex, with both *on* and *off* center properties. The basket cells of layer IVCß have such fields, for example. It seems clear that the various receptive fields encountered are determined by the local details of net interactions, either by afferent topology, or by the local topology of intracortical interactions, or by both factors. It is still an open question as to exactly how such fields are constructed.

Superimposed upon all the local order implied by the existence of such net properties, there is also a very large amount of long range order. Thus each retina is "mapped" onto the lateral geniculate nucleus in a continous, retinotopic fashion, so that order relations on the retinas are preserved. The two retinal maps lie, one on the top of the other, in the lateral geniculate. The geniculate then maps continously onto the visual cortex via layer IV, in such a way that the two retinal maps are interleaved as shown in Fig. 5. Thus, although each retina is

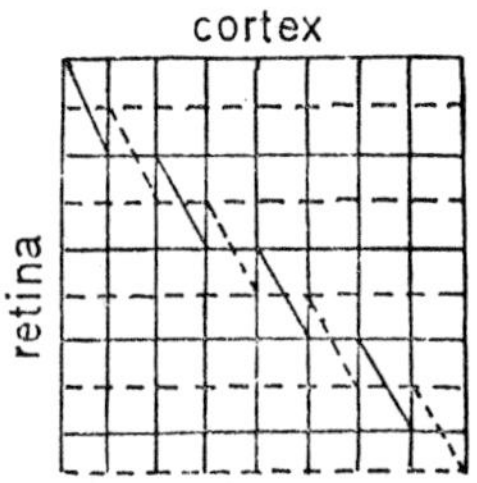

Fig. 5 Structure of the retino-geniculo-cortical map. —— = right retina, --- = left retina.

continuously represented, so far as visual field coordinates are concerned, at the cortical level the representations are discontinuous, thus permitting the interleaving of the two maps. So far as the nature of each individual retinal map is concerned, although it is continuous, the metric is non-euclidean, i.e., there is a systematic distortion of the retinal map, such that retinal (polar) coordinates and cortical (retangular) coordinates are connected by the equation

$$x = \alpha^{-1} \ln(1/2(\alpha r+(\alpha^2+\beta^2 r^2)^{1/2})) \qquad y = r(\alpha^2+\beta^2 r^2)^{-1/2}\theta \qquad (1)$$

where α and β are constants. Such a transformation derives from the fact that although the packing density of neurons in the visual cortex is approximately translation invariant, the packing density of the output cells of the retina, the axons of which project to the lateral geneculate nucleus, is proportional to $(\beta^2+\alpha^2r^2)^{-1}$. Two limiting cases are of interest. At the center of the visual field, corresponding to $r = 0$, (1) reduces to $x = r/\beta$, $y = r\theta/\beta$. Thus the action of the dilatation group, the representation of which is $U = r\frac{\partial}{\partial r}$, reduces to $x\frac{\partial}{\partial x} + y\frac{\partial}{\partial y}$, the dilatation group in rectangular coordinates. Thus close to the center of the visual field, the identity transformation operates, and the map is essentially undistorted. However, sufficiently far from the origin, when $\beta \ll \alpha r$, (1) reduces to $x = \alpha^{-1} \ln(\alpha r)$, $y = \theta/\alpha$, so that $r\frac{\partial}{\partial r} = \alpha^{-1}\frac{\partial}{\partial x}$. Thus the dilatation group transforms into the group of translations parallel to the y axis, and the retino-cortical transformation is the complex logarithm, as first suggested by FISCHER [6] and elaborated recently by SCHWARTZ [7].

Of course such maps, strictly speaking, reflect only the mean packing densities of the relevant cellular populations. There is a great deal of local scatter as well. However even the scatter is ordered [5]. Fig.6, for example, shows the width of the excitatory regions of *simple* cortical receptive fields, measured along a vector perpendicular to the long axis of the field, plotted as a function of retinal coordinates. It will

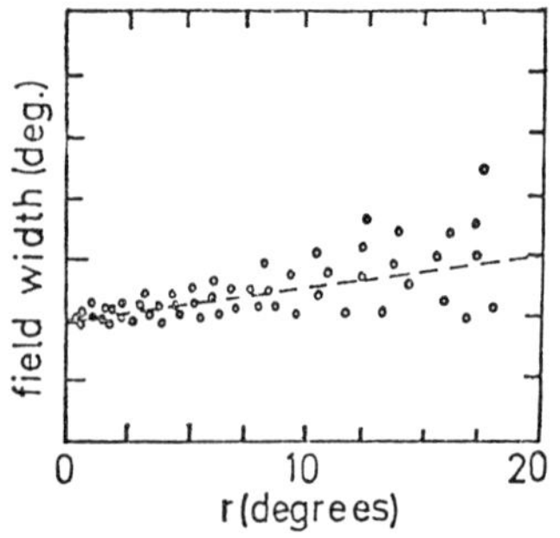

Fig. 6 Receptive field width of *simple* cortical cells, as a function of retinal eccentricity

be seen that the scatter of field widths increases linearly with r, and in fact so does the scatter of field positions at which receptive fields of a given constant width w are found. That is, for sufficiently large r, both $\Delta w/r$ and $\Delta r/r$ are constant. All these details can be derived from the following simple growth model. Assume that the retina grows in a radially directed fashion, from the center outward, and that there is an annular growth region at the retinal margin where cells proliferate. The assumption that a constant number of cells are made at each region, plus or minus a constant scatter, leads to a mean retinal packing density of $(\beta^2+\alpha^2r^2)^{-1}$, and to the observed scatter properties [8].

There is yet another order present in the visual cortex, the details of which are of great interest. Fig. 7 shows for example, the results of a typical microelectrode penetration of the primate visual cortex [5]. It will be seen that the orientational selectivity of cells encountered in such a penetration changes in a piecewise continuous fashion., i.e. there are sequences in which the orientational selectivity changes continuously from cell to cell, by some 20°, punctuated by breaks in which there are large changes in preferred orientation. There seems to be no interaction between the orientation map so revealed, and the ocularity maps schematized in Fig. 5, but the investigations (by

HUBEL, WIESEL ET. AL) are still in an early stage. In any event it will be clear that the retino-cortical system is very highly ordered indeed, and that any theory which seeks to explain the functioning of such a system must take account of such order properties.

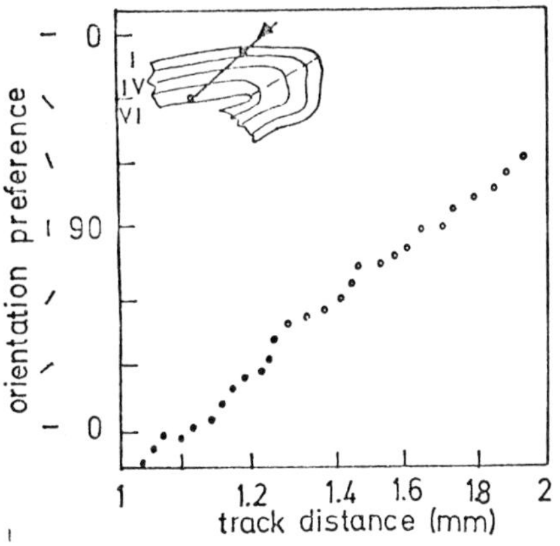

Fig. 7 Orientation selectivity of area 17 neurons, as a function of electrode penetration distance. Filled circles = left eye dominance, open circles = right eye dominance

4. Psychophysics and the Representation of Events in Neuronal Nets

It is clear that there is an enormous gap to be bridged in linking the microstructure so far discussed, with the overall functioning of the visual system, as revealed by psychological experiments. Certain perceptual tasks however, such as the detection of contrast and contour, are sufficiently simple that an integration may be possible of the facts of neuroanatomy and neurophysiology, with those revealed by psychophysical investigations. Fig. 8 shows, for example, the contrast sensitivity of the human visual system, as revealed by threshold measurements

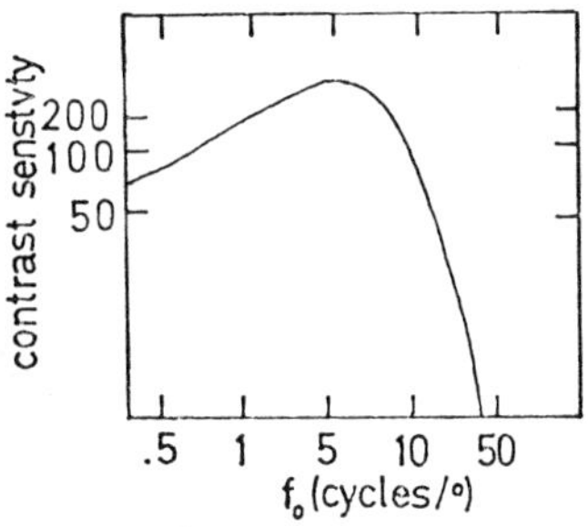

Fig. 8 Contrast sensitivity of the human visual system as measured by sinewave gratings

of the visibility of sinusoidally modulated contrast patterns of the form $o(r) = l_0(1+m\cos(2\pi f_0 r))$, CAMPBELL and ROBSON [9]. l_0 is the mean luminance of the pattern, $m = (l_{max}-l_{min})/l_{max}+l_{min})$ is the contrast, and f_0 is the spatial frequency of the stimulus, measured in cycles per degree of visual angle.

It will be seen that the system exhibits a band-pass frequency tuning characteristic. Thus there is both a high frequency cut-off and a low frequency one. The high-frequency attenuation is caused, ultimately, by the finite size of the excitatory receptive field centers depicted in Fig. 4, and the low-frequency attenuation by the action of the inhibitory surrounds of such receptive fields. Exactly how these properties emerge from the interactions of cells in the visual pathway is the problem to be solved.

Let K(f) be the transfer function depicted in Fig. 8. Then if the visual system were linear and spatially homogeneous, its response to the object $o(r) = l_0(1+m\cos(2\pi f_0 r))$ would be $l_0|K(0)| + l_0 m|K(f_0)| \cos((2\pi f_0 r)+\Omega(f_0))$ where $|K(f)|$ is the attenuation and $\Omega(f)$ the phase shift of the filter K(f). Of course simple spatial filtering is not contrast detection; there must also be a detector somewhere in the system. The simplest assumption is that there is a contrast detector, such that $m|K(f)| \geq \mu$ for detection. Thus a simple model of the overall functioning of the contrast detecting mechanism would be a band-pass filter, followed by a one parameter contrast detector. Any suprathreshold stimulus is presumed to trigger a response in the detector. Observations by CAMPBELL and MAFFEI [10] suggest an electrophysiological basis for such a response. Fig. 9 shows the voltage recorded from the occipital surface of the human scalp (the site of the visual cortex) to the sinewave grating o(r), averaged over 10^3 trials. It will be seen that the

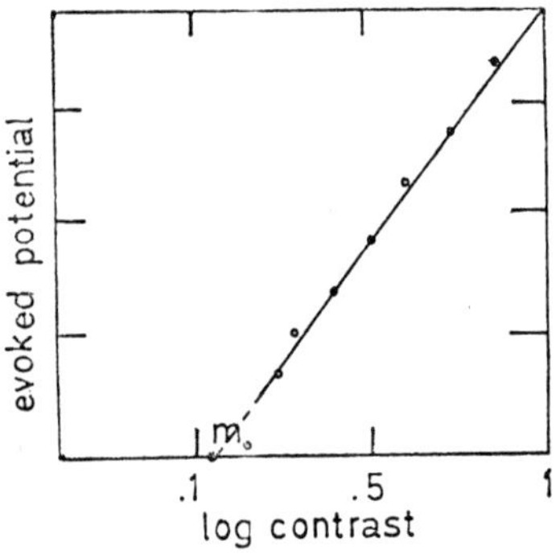

Fig. 9 Averaged evoked potential as a function of log contrast

response is a linear function of ln m, suggesting that there is also a logarithmic nonlinearity in the visual system, prior to the spatial filter K(f). However the most striking observation is seen if the graph is extrapolated to zero voltage; the corresponding contrast m_0 equals the psychophysically determined contrast threshold m_T. Thus in this experiment, there is a 1:1 correspondence between a perceptual event, the detection of a contrast pattern, and a physiological one, the occurence of an electrical discharge. Detection can therefore be taken as the escape of neuronal activity in the cortex from a ground state to an excited state, there being a threshold defined by the parameter μ, for such a transition.

The implication of this for any theory of representative activity in neuronal nets, is that there must be switching between states, occuring in such a way as to correspond to the facts of psychophysics. McCULLOCH and PITTS [11] were the first theorists to confront this problem, but their formulation dealt more with the logic of neuronal switching, and the logical representation of events in a Boolean algebra, and not with the dynamics of spatio-temporal activity in such nets. It is in the dynamics that a connexion with psychophysics lies. Such a theory was developed fairly recently, WILSON and COWAN [12,13] in terms of nonlinear integral equations of Hammerstein type. The result is a nonlinear field theory relating a source distribution $f(x,t)$ of activated neurons, to a temporally filtered field of incoming pulses of excitation $\psi(x,t)$. There are various ways to express such a field theory. One form, COWAN and ERMENTROUT [14] is

$$V(x,t) = \lambda\int_{-\infty}^{\infty}\int_{-\infty}^{t} \exp[-\alpha|x-x'|-\beta(t-t')]\sigma[V(x',t')]dx'dt'+P(x,t), \quad (2)$$

where the voltage $V(x,t)$ is related to the field quantity $\psi(x,t)$ by the equation

$$V(x,t) = \int_{-\infty}^{t} \exp[-\beta(t-t')]\psi(x,t')dt', \quad (3)$$

where σ is a sigmoidally shaped nonlinear function, and where $P(x,t)$ is an external stimulus. The solutions of these equations exhibit a variety of switching properties characteristic of nets of transistor-like elements, just as do the pattern formation equations of GIERER and MEINHARDT [15]. Both bifurcation theory and catastrophy theory have been used to study such equations [14]. The results may be summarized as follows. A net consisting of only excitatory cells acts as a non-local bistable "flip-flop". That is, the only stable states are the ground-state, zero or low-level background activity, or else a constant spatially uniform level of activity everywhere. Switching between these states can occur, however the time course of state transitions is non-exponential. Conversely, a net consisting of inhibitory cells only, can support only local monostable activity: although a spatially localised "excited" state can be induced in response to a suprathreshold stimulus, it is not stable, and the induced activity will decay in an exponential fashion. Evidently nets comprising two cell types, one excitatory, the other inhibitory, can support various combination of the above: both local and propagated switching and cycling are possible [12]. The properties of nets consisting of more than two cell types are even more complex.

In this paper I consider only the conditions necessary for the "escape" of neuronal activity from the ground-state. For all such events there is a threshold surface $\Delta P\Delta t \geq F(\lambda,\alpha,\beta,\sigma)$, where ΔP is the stimulus increment, expressed as a current, and Δt its duration, such that below this surface escape cannot occur, and conversely, above it escape can occur along an appropriate time-course. In simple cases the form of the threshold function F can be calculated explicitly [14]. For example in the case of a net of excitatory cells, the threshold function takes the form:

$$F(\varepsilon) = (\varepsilon/2-1)^{1/2} + 2(1/3)^{3/2}(\varepsilon/2-1)^{3/2}\Delta T, \quad (4)$$

where the coupling parameter ε is a known function of λ, α, β and σ. Thus for brief stimuli, the first term dominates, and for long-lasting ones, the second term dominates. In general, for stimuli that are not spatially constant, the spatial "filtering" properties of the net (in the above case contained in the kernel $\exp[-\alpha|x-x'|]$), and of the afferent map (contained in $P(x,t)$) will affect the firing threshold condition. Thus if $P(x,t)$ is derived from $o(x,t)$ by the transformation

$$P(x,t) = \int_{-\infty}^{\infty}\exp[-\gamma|x-x'|]\ o(x',t)dx' \quad (5)$$

then for a brief stimulus $\Delta o(x)$ of duration Δt, the firing condition is:

$$[\int_{-\infty}^{\infty}\exp[-\gamma|x-x'|]\Delta o(x')dx']\Delta t \geq (\tfrac{\varepsilon}{2}-1)^{1/2}. \quad (6)$$

Thus if $\Delta o(x) = l_0(1+m\cos(2\pi f_0 x))$, then (6) yields the condition:

$$[2l_0/\gamma+(2l_0\gamma/\gamma^2+4\pi^2 f_0^2)\, m\cos(2\pi f_0 x)]\Delta t \geq (\frac{\varepsilon}{2}-1)^{1/2}. \qquad (7)$$

Similar, but more complex conditions obtain for long-lasting stimuli, involving the net kernel $\exp[-\alpha|x-x'|]$ as well. In effect one has to solve (2) to obtain such conditions.

It should be evident that the relationship between threshold firing conditions in neuronal nets, as discussed above, and the processes underlying contrast detection, is not simple and direct. As a matter of fact, recent work, [9], indicates that the one parameter contrast detector model previously discussed, does not predict the outcome of most detection experiments. Instead a model has been proposed (see [8] for details) in which the responses of many filter-detector systems are statistically summed. The basic idea is as follows. Given that there are many filter-detector systems or "channels" as they are usually called, let p_i be the probability that any single channel will detect a given object $\Delta o(x)$. On the assumption that detection, overall, is an incoherent process, it follows that the overall detection probability P is bounded by the function $1-\prod_i(1-p_i)$, where the product is taken over all available channels. QUICK [16] introduced the function $p_i = 1-\mathrm{two}(-V_i^p)$ where V_i is the response of the i th channel. It follows that the overall detection probability $P = 1-\mathrm{two}(-\sum_i V_i^p)$. This may be written as $1-\mathrm{two}(-||\underline{V}||^p)$ where $||\underline{V}|| = (\sum_i V_i^p)^{1/p}$, the p th norm of the vector $\underline{V}$. The utility of such a choice of detection function is that if m is the stimulus contrast, so that $P = 1-\mathrm{two}(-||m\underline{V}||^p)$, and if the threshold is taken to correspond to $P = 1/2$, then the contrast threshold m_T is given by $m_T||\underline{V}|| = 1$. Thus the contrast sensitivity function m_T^{-1} is given directly by the norm of $\underline{V}$.

To connect these considerations with the anatomical and physiological facts previously discussed, the filter-detector systems of channels described above are taken to be local net properties of the cortex. Then instead of the discrete norm introduced above, the continuous norm $(\int dx|V(x)|^p)^{1/p}$ is introduced. Norms of this type, incorporating the scatter of channels at each point of the visual field, can be shown to give a good fit to the available data, and to provide a means to assess the overall functioning of the retino-cortical system, at least for contrast detection. One or two points are worth noting. Firstly the norming operation is carried out in cortical coordinates, not in retinal ones. Thus there is an implicit weighting of channels with respect to their packing density in the visual field, and in addition, the filter function associated with each channel $K(f,x)$ also changes its form as a function of position in the visual field. In such a fashion, the spatial inhomogeneity of the visual field is taken into account. Secondly, because of such a weighting function, embodied in the coordinate transformations given in (1), it follows that care must be taken in the choice of object functions to be used in psychophysical experiments. Thus to image a cosine wave on the cortex, the function $\Delta o(r) = l_0(1+m\cos(2\pi f_0\alpha^{-1}\ln(1/2(\alpha r+(\beta^2+\alpha^2 r^2)^{1/2}))))$ must be used, instead of the function $\Delta o(r) = l_0(1+m\cos(2\pi f_0 r))$. Finally, it should be noted that the norm introduced is a purely phenomenological description of the detection process. It remains to develope a true nonlinear net theory along the lines discussed above for such a process, in which the incoherent detection of afferent signals is taken to be a cellular property, perhaps of *complex* cells, as many recent experiments suggest.

5. Neuronal plasticity

I now turn to the final topic of this paper, the fact that in many situations the parameters of neuronal nets, and therefore receptive field properties, change as a function of local neuronal activity. For example, the properties of the visual system seem to change radically when high, rather than low contrast patterns are viewed, BLAKEMORE, MUNCEY, and RIDLEY [17], and the overall contrast sensitivity function K(f) of the system is much flatter than the threshold sensitivity function depicted in Fig. 8. Adaptation experiments based on the prolonged viewing of high contrast sine-wave gratings of the form $\Delta o(r) = l_0(1+m\cos(2\pi f_0 r))$, GILINSKY [18], BLAKEMORE and CAMPBELL [19] show that a "notch" appears in the threshold contrast sensitivity curve, as shown in Fig. 10.

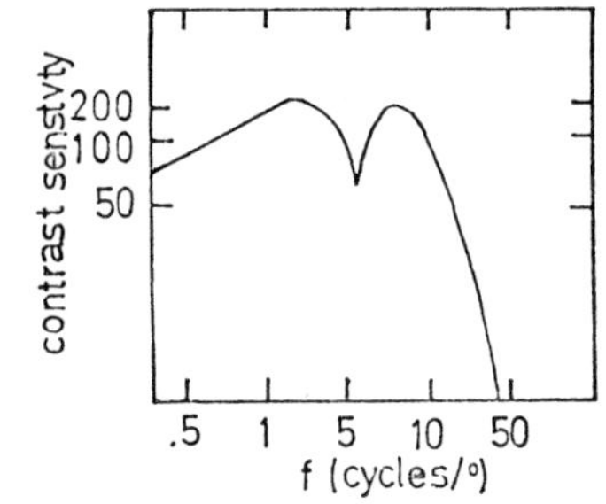

Fig. 10 Contrast sensitivity of the human visual system following prolonged viewing of a high contrast 5 cycles per degree grating

Various explanations have been proposed for this and related effects. The most interesting possibility is that the changes of threshold are the result of inter-channel inhibition, DEALY and TOLHURST [20]. The precise nature of such changes, however, remains unclear. It has been suggested that a modifiable HEBB synapse [21] is responsible for the changes, either in a feedforward configuration, WILSON [22], or in a feedback configuration, COWAN [23]. Such suggestions can be formalised in the following fashion. Let x and y be input and output vectors, respectively, associated with a neuronal net, and let E and I be associated excitatory and inhibitory coupling matrices, the coefficients of which are fixed. Let S be a matrix of modifiable coefficients. Then the following equations obtain:

$$\text{feedforward:} \quad y = E^t x - I^t x - \mu S^t x, \; \dot{S}^t = -\beta S^t + \alpha y x^t, \tag{8}$$

$$\text{feedback:} \quad y = E^t x - I^t y - \mu S^t y, \; \dot{S}^t = -\beta S^t + \alpha y y^t. \tag{9}$$

Similar equations obtain in a continuous formulation. Both equations can be shown to reproduce quite well, the notches found in threshold sensitivity functions, especially if the channel weighting function described early is utilised. However, the time course of the effects is quite different in the two cases. Consider the spatially uniform case, the time course of which will dominate the solutions. Then (8) and (9) reduce, respectively, to:

$$\dot{y} = -(\beta+\alpha\mu x^2)y + \beta|E-I|x, \tag{10}$$

and

$$\dot{y} = \beta y - \beta|1+I|(|E|x)^{-1}y^2 - \alpha\mu(|E|x)^{-1}y^4. \tag{11}$$

It will be seen that the first equation is linear in y, but that the second is a nonlinear Bernoulli equation. Thus the first system predicts an exponential decay in sensitivity, whereas the second predicts, approximately, a power law decay, of the form $(y/y_0) = (t/t_0)^{-1/3}$, KOHONEN [24]+. It is interesting to compare these predictions with those which would follow from the assumption of simple *fatigue* or *facilitation* instead of a modifiable HEBB synapse, both of which will produced appropriately notched threshold changes, with suitable channel weighting functions. A suitable *fatigue* model takes the following form:

$$y = \Omega_1(E^t x - I^t x - \Omega_2(\bar{y})), \quad \dot{\bar{y}} = -\beta\bar{y} + y, \tag{12}$$

where, in general, Ω_1 and Ω_2 are nonlinear functions. Combining these two equations gives:

$$\dot{\bar{y}} = -\beta\bar{y} + \Omega_1(E^t x - I^t x - \Omega_2(\bar{y})). \tag{13}$$

Evidently different time-courses can be obtained, simply by choosing appropriate forms for Ω_1 and Ω_2. Thus if Ω_1 and Ω_2 are linear, $\bar{y}$ will exhibit exponential decay, whereas if Ω_1 and Ω_2 are square-law detectors, $\bar{y}$ will show power law decay, with an exponent of -1/3, approximately, and y more complex behavior. Similarly, an appropriate *facilitation* model is:

$$\text{feedback:}\ y = E^t x - I^t y - \mu S^t y, \quad \dot{S}^t = -\beta S^t + \alpha y^t \tag{14}$$

the time course of which is approximated by the equation:

$$\dot{y} = \beta y - \beta|1+I|(|E|x)^{-1}y^2 - \alpha\mu(|E|x)^{-1}y^3 . \tag{15}$$

Such an equation again shows, approximately, power law decay, but of the form $(y/y_0) = (t/t_0)^{-1/2}$. It is an interesting possibility that the various possibilities may be eliminated by means of measurements of the time-course of the effects. In this connexion, it must be noted that measurements of the time-course of the GILINSKY effect are not sufficiently detailed as yet, to enable the various possibilities to be eliminated. However, there is a related after-effect, the well-known McCOLLOUGH effect [25], the time-course of which is known to follow very closely the minus one third power law, MACKAY and MACKAY [26]. The McCOLLOUGH effect is produced by presenting subjects with alternating horizontal and vertical gratings. The first grating consists of black and red stripes, the second of black and green stripes. After a few minutes of exposure, the subjects if presented with a horizontal, black and white grating, will report seeing black and green stripes, and if presented with a vertical black and white grating, will report seeing black and red stripes. The strength of the effect can be measured by color matching techniques, and its decay follows closely, the minus one third power law. Thus a fourth order nonlinearity is present somwhere in the system. It is an attractive possibility that such a nonlinearity is generated by the presence of a HEBB synapse, in the feedback configuration described, and that a similar mechanism is responsible for the GILINSKY effect, and for numerous other figural after-effects.

6. Discussion

It is clear that the (visual) brain is very highly ordered. Any comprehensive description of its functional organization will require a large number of order parameters, Integration of the enormous number of facts about the visual system provided by experiments, poses a formidable set

+ I am indebted to Dr. Kohonen for this observation.

of theoretical problems. The approach sketched in this paper is to try to find the simplest possible non-trivial representation of activity in neuronal nets. This can be achieved in terms of two sets of equations, a "fast" set representing the kinetics of neuronal firing patterns, and a "slow" set, the (local) kinetics of synaptic changes. In such a format (2) and (9) take the form

$$\dot{V} = -\beta V + \lambda S^t\sigma[V] + A^tP \tag{16}$$

$$\dot{S}^t = -\gamma S^t + \alpha\sigma[V]\sigma[V]^t \tag{17}$$

in which the time constant β^{-1} is much smaller than the time constant γ^{-1}. The solutions of (16) and (17), (S,V), provide information about the spatio-temporal evolution of the net activity V, and the underlying coupling matrix S. To use such a formalism it is necessary to make some assumptions about the global nature of the neuronal net, and about the afferent maps to it. To date, the experimental evidence strongly suggests that the (visual) cortex is spatially homogeneous, i.e., its properties are roughly translation invariant with respect to the surface coordinates of the cortex. HUBEL and WIESEL'S investigations of the functional anatomy of visual cortex [5], indicate that the cells in a unit of 1 mm^2 surface area, extending throughout the depth of the cortex, have a complete repertoire of filtered responses, suitable for the (local) analysis of visual stimuli. Thus (16) and (17) can be thought of as representing the interaction of such a cortical unit, called a hypercolumn [5], with the stimulus P(x,t). It follows that the visual cortex itself can be thought of as a very complex aperiodic crystal, with a unit structure comprising some 10^5 cells, probably with local inhibition within and between units, and (perhaps) long-range excitation between distal units.

To such a description must be added the appropriate coordinates. Given that the visual field is not spatially homogeneous it follows that the afferent maps must contain the inhomogeneity. As I have discussed, the receptive surface, in this case the retina, is the generator of the inhomogeneity, and the appropriate coordinates are approximately logarithmic, as given in (1). I have sketched some of the perceptual consequences of the existence of such coordinates. In any event (1), (16), and (17) supplemented with appropriate boundary conditions, and with sufficiently many degrees of freedom to take proper account of the scatter of filter properties, at each point (r,θ) of the visual field, may be said to constitute a master equation for cortical kinetics. So far as synergetic effects are concerned, it remains to be seen to what extent intracortical connectivity plays an important role in cortical action. The power laws found in the decay of perceptual after-effects such as the McCOLLOUGH effect, and many other observations, suggest that it cannot be negligible.

7. References

1 H. Haken: Synergetics, Springer-Verlag, Heidelberg (1977)
2 C.S. Sherrington: Man on his nature, Cambridge Univ. Press (1940)
3 J. Szentagothai: (In) Recent development of neurobiology in Hungary, (Ed) K. Lissak (1967)
4 S.W. Kuffler: Cold Spr. Harbor Symp. Quant. Biol. 17, 282-292 (1952)
5 D.H. Hubel and T.N. Wiesel: J. Physiol. 148, 574-591 (1959); J. Comp. Neurol. 158, 3, 295-308 (1974)
6 B. Fischer: Vision Res. 13, 2113-2120 (1973)
7 E. Schwartz: Biol. Cybernetics 25, 181-194 (1977)

8 J.D. Cowan: (In) The visual field (Eds) R. Held and E. Pöppel, MIT Press (1977)
9 F.W. Campbell and J.G. Robson: J. Physiol. 197, 551-556 (1968)
10 F.W. Campbell and L. Maffei: J. Physiol. 207, 635-652 (1970)
11 W.S. McCulloch and W. Pitts: Bull. Math. Biophys. 5, 115-133 (1943)
12 H.R. Wilson and J.D. Cowan: Biophysical J. 12, 1-24 (1972)
13 H.R. Wilson and J.D. Cowan: Kybernetik 13, 55-80 (1973)
14 J.D. Cowan and G.B. Ermentrout: (In) Mathematical problems in Biology (Ed) S. Levin, Math. Assoc. America (1977)
15 A. Gierer and H. Meinhardt: Kybernetik 12, 30-39 (1972)
16 R.F. Quick: Kybernetik 16, 65-67 (1974)
17 C. Blakemore, J.P.J. Muncey, and R.M. Ridley: Vision Res. 13, 1915-1931 (1973)
18 A. Gilinsky: J. Opt. Sci. Am. 12, 13-18 (1968)
19 C. Blakemore and F.W. Campbell: J. Physiol. 203, 237-260 (1969)
20 R.S. Dealy and D.J. Tolhurst: J. Physiol. 241, 261-270 (1974)
21 D.O. Hebb: The Organization of Behavior, Wiley, New York (1949)
22 H.R. Wilson: J. Theoret. Biol. 50, 327-352 (1975)
23 J.D. Cowan: (In) Neural mechanisms of Learning and Memory, (Eds) M.R. Rosenzweig and E.L. Bennet, MIT Press (1976)
24 T. Kohonen: Associative Memory, Springer-Verlag, Heidelberg (1977)
25 C. McCollough: Science 149, 1115-1116 (1965)
26 D.M. Mackay and V.M. Mackay: J. Physiol. 237, 38-39 (1973)

I thank the ALEXANDER-VON-HUMBOLDT STIFTUNG for support during the preparation of this paper.

Biological Control through Long Range Coherence

H. Fröhlich

1. Introduction

The analysis of the molecular structure of biological giant molecules has led to the result that systematic spacial order is absent in biological systems. It was often concluded then that no physical order exists in these materials. Modern physics, however, knows another type of order which may be termed dynamic order. In thermal equilibrium this type of order is dominant in superconductors and in superfluid helium. Thus X-ray analysis exhibits a spacial correlation between helium atoms similar to that in other fluids. Yet the dynamic order imposes conditions which lead to vanishing entropy as the temperature approaches zero in spite of this apparent disorder.

Order of this type is connected with certain phase correlations and may as well exist in non equilibrium systems. Examples are a sound wave or a laser system. The present article is based on the hypothesis that active biological systems exhibit this type of dynamic order [1-13]. Attempts have been made some time ago to specify this hypothesis so that it can be tested experimentally. A growing amount of experimental evidence seems to support the theoretical proposal. The difficulty of quantitative experiments on active biological systems is such, however, that caution must be exercised until the various experiments have been repeated by independent groups.

2. Biological Sensitivities and General Approach

From a certain point of view biological systems behave as ordinary physical systems of great complication. Thus all known biochemical transformations satisfy the usual rules concerning types of chemical events and changes of free energies. Already here, however, the high rate at which these events are catalysed by enzymes poses problems which have not been solved yet [14]. This should not imply that the basic laws of physics must be changed but rather that biological systems have developed extraordinary ways of making use of these laws. A most striking example is the eye which has been found sensitive to a single light quantum. The eye thus represents an image converter with ultimate sensitivity; but it is constructed of materials which no electronics engineer would dream of using for this purpose.

Other remarkable sensitivities are found in the brain of certain fish, sensitive to electrical signals of 10^{-8} volt/cm or homing birds using the earth's magnetic field [15]. In these cases the signals are so weak that extraordinary arrangements must exist to achieve sensitivity above the thermal noise.

Other remarkable features of the brain in this respect investigated at present refer to E.E.G (brain waves). It has been found that in certain periods many thousands of generators of E.E.G. act coherently [16]. Fourier analysis has shown that external signals tend to correlate the phases of various Fourier components [17,18]. Other remarkable properties of E.E.G. are discussed in [15].

Another remarkable feature is the control over cell division exerted on tissues which usually prevent occurrence of cancer. This clearly requires interactions between cells whose nature has not yet been established.

In attempting a novel general approach to biology from the side of physics clearly the first step should be the formulation of appropriate novel physical concepts. While one might then speculate on the use of such concepts for its interpretation of biological events that cannot be understood in a conventional manner I personally think that the main effort should first go into showing experimentally that the basic physical events required in these new concepts are actually present in biologically active systems.

3. The Physical Concepts and Their Consequences

Amongst material properties common to all biological systems the extraordinary dielectric properties should perhaps be given first rank [5]. Amongst these, again, most remarkable is the high electric field of the order of 10^7 volt/m maintained in most biological membranes. Biological membranes usually about 100 Å thick, are not homogeneous layers but contain a variety of proteins (and perhaps other molecules) dissolved in them. Under ordinary circumstances such a layer would break down electrically in fields of the strength mentioned above. At present the only established role of these fields is in nerve conduction. This, however, could be achieved with much weaker fields nor would this help to understand its maintenance in the membranes of other cells.

Biological molecules [19] when investigated outside their biological environment (e.g., in aqueous solution) have permanent dipole moments which are large (several hundred Debye units) compared with small molecules but not large considering the size of these molecules. Usually, however, they are highly polarisable in view of the attachment of counter ions to such molecules.

Arising from these dielectric properties the two basic concepts - assumed of great importance for biological activity - will now be formulated.

(i) There exists a metastable state with very high electric dipole moment.

(ii) Through metabolic energy supply coherent electric vibrations in the region of 10^{11} Hz can be excited.

The first principle follows under very general conditions when a material is highly polarisable electrically and deformable elastically - as discussed in [20-22].

The second principle arises when we consider mechanical oscillations of an electrically polarised material such as a membrane or a molecule in its state of high dipole moment. Possible vibrations will then be excited through supply of energy arising from metabolic processes. The above mentioned frequency arises by considering that linear dimensions of the order 10^{-6} cm are typical and that elastic properties can be expressed in terms of a velocity of sound of the order 10^4-10^5 cm/s. Clearly both smaller and larger dimensions are available so that the given frequency must be considered as a very rough order of magnitude.

The coherence of the excitation may be supported by model calculations [2,4,7] which also yield a threshold for the energy supply above which the excitation becomes coherent. Clearly the relevant model may be altered such that it does no longer exhibit coherence [23]; alternatively other models might be devised which also would provide coherence.

The consequences of both (i) and (ii) are very far reaching. Thus excitation of the metastable highly polar state in an enzyme might be identified with its activated state. The high internal field could then decrease activation energies and thus increase the rate of enzyme activity [24]. For proteins dissolved in membranes the highly polar metastable state through interaction with the membranes field may become the ground state thus giving a justification for the maintenance of these fields (cf. [21]). Excitation of metastable highly polar states may also act as energy storage required, e.g., in the transfer of energy from a source such as photon energy in photosynthesis to an energy rich molecule, e.g., ATP (cf. [25]).

The most important consequence of excitation of coherent electric vibrations is the establishment of long range selective interaction between systems with nearly equal frequencies.[1] Such interaction can lead to selective attraction [3,26]; to organisation as required in control of tissue growth (cancer); and to storage and transport of energy.

4. Experimental Evidence

A considerable literature on conformational changes and possible relations to electric field exists. It is much more difficult, however, to provide experimental evidence for the existence of coherent vibrations. The most direct way would be optical investigation of excitation of the relevant frequencies, e.g., by Raman effect. It is required to show that metabolic activity changes the Raman spectra in the required manner. This has indeed been verified [27] on bacteria (E coli and B megaterium) grown in two types of nutrients. Both resting cells and solutions of nutrients gave broad Raman bands in the relevant frequency region. Active cells, however, exhibit a number of Raman lines which disappear when the metabolic activity is intercepted by addition of KCN [28]. So far regrettably only Stokes lines have been investigated so that in principle the observed lines might arise from compounds produced as intermediate steps and sheltered from water whose friction gives rise to the broad bands. Evidence that the observed lines arise from excitation of the relevant frequencies requires in addition the measurement of anti Stokes lines which should then appear with nearly the same intensity as the Stokes lines.

[1] It should be mentioned that static electric sources (e.g., dipoles) are screened at short range through free ions.

Optical measurements of this and similar types are very difficult to perform: the relevant frequency regions are dominated by water always present in biological systems, the volume occupied by molecules or membrane regions active at certain frequencies may be very small and their influence might be drowned in the strong water background. A completely different approach consists in the impositions of electromagnetic radiation in the relevant frequency region on biological objects. If vibrations in such a frequency region are already excited during a particular biological activity then the outside vibration will enhance the activity and can thus be found biologically through quantitative measurements of the particular activity. A certain external frequency might on the other hand excite biological activities which interfere with the existing activity which thus would be decreased.

Investigation of the rate of growth of E coli bacteria has been investigated in the frequency region of 70-75 GHz (i.e., close to the proposed active region). It has been found [9] that near 71 and 73 GHz the rate of growth is substantially reduced while no significant effect is observed at other frequencies. Recently this observation has been confirmed by another group [10] who also point out that increase in temperature from the absorption of microwaves would lead to an increase rather than a decrease in the rate of growth. This experiment so far is the only one that has been confirmed by the work of two independent groups.

A great number of very startling experimental results along similar lines have been reported by Russian workers who use coherent radiation in the 5-8 mm region, i.e., again close to the proposed (Sec.2) frequency region [8]. A number of experiments have been started in order to confirm these findings. Though they deal with a very great number of biological objects and activities they find certain features common to all of them; (a) only in certain very narrow frequency regions are these biological effects observed (biological resonance) and (b) a threshold intensity of radiation exists below which no effect is observed and above which the observed effects are nearly independent of the intensity. These two features confirm the suggestion arising from the theoretical consideration, cf. Section 2 [6].

5. Further Consequences

The possibility of the existence of electric vibrations giving rise to selective long range interaction has been used in research in cancer to account for the communication between cells required to establish a control of growth [7]. In fact a formal theory of such control can be developed if, e.g., we assume that the nucleus of a differentiated cell is characterised by a well defined frequency - different for different types of differentiation [29]. On excitation the vibrations of these cells will organise themselves in the coherent pattern which yields the lowest free energy and which may exhibit size effects. Thus a collective mode of the whole system (order parameter) is excited such that individual cells which carry the relevant vibration also obey the instruction of the collective mode namely to vibrate with the appropriate phase.

Prior to cell division the DNA-protein complex of the nucleus must unfold. This may be expected to change its frequency and thus will require energy for removal from the collective mode. This energy may not be available, i.e., the collective mode controls cell division. Invasion of the nucleus by foreign materials will in general change the frequency and thus remove the particular cell from the control by the collective mode. It will be energetically favourable, however, for the particular cell to regain the resonance frequency. One may then expect a tendency for removal of the intruding material.

If, however, a considerable number of cells have been invaded and thus changed their frequency, then the collective mode itself will be weakened and lose its control over cell division. A phase transition into a state without control (cancer) will then take place.

It should be emphasised here that the proposed collective mode will in general have very small optical or Raman activity only based on surfaces and internal irregularities.

Finally it should be mentioned that the interaction of cells with the vibrations must influence the rate of mutations if these are considered as basically random processes. For then their rate must be governed by a Boltzmann factor depending on the energy change of the system which in our case will depend on the frequency of oscillation of the particular cell nucleus and on the magnitude of excitation of the controlling mode.

On a larger scale we might speak of the totality of vibrations of a biological entity as its vibrational field. This field will influence mutations and hence also evolution. Over relatively short periods this influence may appear to be minute, but being a systematic influence it might well be decisive over long periods. This type of influence may be illustrated on the example of a gas in the gravitational field. If we treat collisions of the gas molecules basically as random then the gravitational field will impose a certain bias which over a few collisions will be hardly noticeable but over long periods will exhibit a trend of motion towards the ground.

Note added on 1.5.77

F. Keilmann and W. Grundler have now repeated and confirmed the Russian experiment on the influence of coherent mm waves on the growth of yeast.

References

1. H. Fröhlich: in Theoretical Physics and Biology, ed. by M. Marois (North Holland, Amsterdam 1969) p.13
2. H. Fröhlich: Int.J.Quant.Chem. 2, 641 (1968)
3. H. Fröhlich: Phys.Lett. 39A, 153 (1972)
4. H. Fröhlich: in Synergetics, ed. by H. Haken (Teubner 1973) p.241
5. H. Fröhlich: Proc.Nat.Acad.Sci.USA 72, 4211 (1975)
6. H. Fröhlich: Phys.Lett. 51A, 21 (1975)
7. H. Fröhlich: Rivista del Nuovo Cimento (in print)
8. N.D. Devyatkov et al.: Sov.Phys. USPEKHI (Translation) 16, 568-579 (1974)
9. S.J. Webb, A.D. Booth: Nature 222, 1199 (1969)

10. A.J. Berteaud, M. Dardalhon, N. Rebeyrotte, D. Averbeck: C.R.Acad.Sci.Paris t281, Serie D, 843 (1975)
11. N. Kollias, W.R. Melander: Phys.Lett. 57A, 102 (1976)
12. S.J. Webb, M.E. Stoneham: Phys.Lett.
13. D. Bhaumik, K. Bhaumik, B. Dutta-Roy: Phys.Lett. 59A, 77 (1976)
14. D.E. Koshland, K.E. Neet: Ann.Rev.Biochem. 37, 359 (1968)
15. W.R. Adey, S.M. Bawin (eds.), Neurosciences Res.Prog.Bull. 15 No. 1 (1977)
16. R. Elul: Neurosciences Res.Prog.Bull. 12, 97-101 (1974)
17. B.M. Sayers, H.A. Beagley, W.R. Henshall: Nature 247, 481 (1974)
18. H. von Specht, Z.Sr. Kevanishvili: Nature 260, 461 (1976)
19. S. Takashima, A. Minakato: in Digest of Literature on Dielectrics, ed. by A. Vaughan (in print)
20. H.J. Fröhlich: Collective Phen. 1, 101 (1973)
21. H. Fröhlich: Nature 228, 1093 (1970)
22. H. Fröhlich: Biosystems (in print)
23. M.A. Livshits: Biofizika 17, 694 (1972). This author uses magnitudes which are not compelling and hence draws wrong conclusions
24. D.E. Green: Ann.N.Y.Acad.Sci. 227, 6 (1974)
25. H.T. Witt, E. Schlodder, P. Graeber: FEBS Letters 69, 272 (1976)
26. B.W. Holland: J.Theor.Biol. 35, 395 (1972)
27. S.J. Webb, M.E. Stoneham: Phys.Lett. A (in print)
28. S.J. Webb: personal communication
29. F. Fröhlich: in Cooperative Phenomena, ed. by H. Haken, M. Wagner (Springer, Berlin, Heidelberg, New York 1973) p.XL

General Systems

Sociocultural Systems and the Challenge of Sociobiology

W. Buckley

This paper represents an outline of some of the main points I am trying to develop in a monograph examining the possibilities of an evolutionary model on the truly sociocultural level of reality. It will contain a few general critical statements concerning theories of the genetic determination of sociocultural patterns of human actions and interactions. It will argue that much of the difficulty in the debate stems from the inability of many scientists to allow that sociocultural systems are structured entities that cannot easily be understood entirely in terms of psychology, biology, chemistry, or physics, and must therefore be studied also at their own level. Thus an attempt is made to distinguish three major evolved strategies of evolution: the phylogenetic, the ontogenetic, and the sociogenetic. On this basis, a model of sociocultural evolution on its own level, with its unique mechanisms of variation, selection, and adaptation to its relevant environment is outlined. Such a model is informed by modern system theoretic concepts from information theory and general systems theory, and is argued to be homomorphic to other levels of evolution. The most general concept here is that evolution, from the chemical to the sociocultural level, is the gradual, more and more refined and extensive mapping of the relevant environmental variety and its constraints into the structure of the evolving system in the form of organizing principles and coded information. Finally, an example of the application of the social evolutionary model to a vital contemporary public issue of societal adaptation is outlined, and the paper concludes with some cryptic remarks about the possibilities of developing a non-equilibrial "thermodynamics" of sociocultural systems.

E. O. Wilson's overly-confident conception of "sociobiology" challenges the sociologist to join the "grand Darwinian synthesis", which means, for him, to accept -- without any evidence -- the genetic determination of the central features of the social roles characteristic of human societies -- apparently in all their cultural diversity. Such speculation is not really worth taking seriously. The great bulk of anthropological and sociological evidence points to the enormous range of actual and potential patterned social activities, and attests to the inherent plasticity of such actions within the very broad limits set by man's biological and psychological make-up.

It is characteristic of the biological reductionist argument to limit discussion to static considerations, and forego treatment of developmental, morphogenetic, or epigenetic processes that account for the wide gulf between genetic codes and phenotypic expressions. But more than this, a rather narrow conservative view of the evolutionary process is typically assumed. What seems to be lost is the broader view that part of the grand Darwinian design has been the evolution of more effective evolutionary strategies themselves, with increasing plasticity of behavior an important facet. For example, the evolution of learning processes and mechanisms seems indisputable, but the very concept of learned behavior is contradictory to the notion of genetic determination. The more adaptive learning capabilities become, the more must the ties be cut between the genes and behavior. At the human extreme, there are no cultural patterns that are more than broadly circumscribed by biological considerations.

Speaking generally, it appears that we may see at least three major strategies that evolution has assumed as life has developed new forms of adaptation to the relevant environments. In discussing these, a very general conceptual framework may be helpful. Whatever else it may be, the evolutionary process can be seen as one in which, through variation and selection, information about the variety and constraints of the relevant environment is generated and incorporated in one way or another into the organization of the evolving system. More broadly still, evolving systems can be seen as mappings of their environments, though not necessarily simple one-to-one mappings. Such a formulation is in keeping with an information theoretic conception, since information is fundamentally a mapping between two or more sets of elements and their relations. It remains to suggest the sense in which three of the major strategies of evolution handle this information generating or mapping process. The phylogenetic strategy is, of course, the Darwinian process in which random mutations or genetic mixing are tested in their phenotypic forms in interaction with environmental pressures, with successful structures or processes that map the environment resulting in the blind natural selection of the corresponding genetic codes. But this strategy is enormously wasteful and slow, permitting rapid environmental changes to reduce the viability of species before the phylogenetic mapping can catch up. A second strategy, ontogenesis we call it, maps the relevant environment during the lifetime of the system by way of a neurophysiological code in a process we call learning. Instead of random genetic mutations, we have a plastic neural net responding selectively to trial and error probings of the environment. This permits a more rapid and refined process of generation of information about the environment which guides behavior with increasing degrees of inductive foresight and deductive certainty. Experimental studies of animals continue to push back the proportion of behavior seen to be determined by the genes and augment the proportion established by various levels of learning. This psychogenic level of mapping, though quite different from Darwinian evolution on the surface and only augmenting it, is modelled by the automata theorist in a way that is practically identical in principle. A change of structure in the system (morphogenesis) in the literal sense maps environmental changes or refinements through a variation and selection feedback process in which successful phenotypic characteristics reciprocally feed back to promote a coded ensemble or information reservoir (the gene pool or the engram pool).

But this strategy too has its weaknesses, and is unnecessarily wasteful, and provides the foundation for a higher level of biological organization: not merely the social, which had been tried earlier in insect societies without benefit of highly developed neural coding, but the sociocultural, which required also, of course, a permanent social organization along with an arbitrary symbolic coding. This we can refer to as the sociogenic strategy of evolution. At this new level of strategy, the genetic code and the neural code are augmented, and often dominated, by a now extra-somatic normative code. An obvious advantage here in storing the information about the environment outside the skin in a code available to all units of the system is that the information or mapping is not lost with the death of the unit, and furthermore can be passed on to the newborn units without their having to first directly experience the actual events. It is a little acknowledged but remarkable fact that most of what we "know" about the world today has never been directly experienced by us. (Of course, the notions of "directly" experiencing much of the scientific, aesthetic, or moral interpretations of the world becomes less and less meaningful.) The normative code, which is reproduced with variation in each generation via the socialization process, constitutes the basis of the role relationships defining sociocultural organization of social systems. It is this structure, not the biological structure of members of Homo sapiens, that is the focus of sociocultural evolution. It is noteworthy that most of the past and current discussions of social evolution do not concern themselves in fact with societal evolution at all. They are rather concerned with some social factors involved in the biological evolution of the elements of the higher level entity that the non-social scientist apparently has great intellectual difficulty in, or resistance to, acknowledging (partly due to our subtle but strong Western individualistic bias). For example, to the extent that the Social Darwinists, or their modern counterparts, attempted to demonstrate that certain social classes or ethnic groups that have attained high levels of material or intellectual or moral eminence represent the "fittest", they could be saying nothing at all about social evolution. Typical concerns of the latter include whether this type of normatively defined and regulated kinship system, or that type of economic environmental exploitation system, or this kind of political control structure, is more adaptive than another to given environmental conditions. Only recently have we come to appreciate that there is a real question as to whether modern sociocultural systems, and perhaps the species constituting them, can survive the coming environmental challenge. It is not a question of the biology of individuals but of the adaptive organization of groups, including nations.

Another advantage of the sociogenic strategy of evolution besides the extrasomatic storage of information is the potentially great rapidity with which the sociocultural system can restructure itself when the pressures become great enough, sometimes through destructive revolution and sometimes through constructive "evolution". In addition, the blind, opportunistic Darwinian selection process gives way to potential for continued development of non-blind processes -- that is, conscious, reasoned, and foresightful group decision processes that restructure aspects of the economic, political, or other social structures to better adapt to environmental exigencies (and to the internal social milieu). It is all too true that this potential is seldom utilized to its full extent, and much current societal action and interaction is as blind as the proverbial Darwinian struggle for survival. But the ongoing social evolutionary process involves the competition of ideas, norms, and role structure among social groups. Longer term adaptive success will depend not on the brute strength of the contenders, but on the continued viability and adaptability of their sociocultural systems in the face of environmental limitations and pressures, just as the adaptive success of scientific ideas depends not on the prestige or power of the scientist, but on their empirical relevance and verifiability.

To recapitulate the necessarily sketchy outline of a model of sociocultural evolution, it is important to note that its main principles are similar, if not isomorphic, to biological evolution, but its substantive mechanisms are quite different. Instead of a genetic code, we find a normative code performing pretty much the same general functions, i.e., guiding the social and psychological development of new generations to maturity, and shaping the patterned activities and interactions characterizing the sociocultural organization and its dynamic processes. Instead of a gene pool making up a population's stock of coded information or templates, we find a normative idea pool (if you wish), a reservoir of the culture's "templates" for the coordination and integration of unit activities and interactions. New ideas or ideologies are continually generated as "mutations" which are subject to various selection processes, some rather blind and often maladaptive in the long run (most sociocultural systems in history have not survived), and some consciously foresightful and adaptive over long periods. The counterpart of reproductive success as a criterion of adaptation would seem to be the intergenerational perpetuation of one normative system and social structure as against others. (But in both cases, a successful adaptation to today's environment holds no guarantee for tomorrow's. Hence systems with organized foresight would be a decided evolutionary advantage, a message gaining cogency every day.)

To be noted once again is the ease with which we can confuse the biological and social levels of analysis here. A good example of what can be called the fallacy of misplaced levels is the argument that the societal analogy with biological evolution breaks down because there are many biological species but for humans, there is only one: Homo sapiens. But if we follow out the logic of the parallel, we see that just as biological taxa and the distances between them are a function of the differences between genetic codes, sociocultural taxa and distances between them are a function of differences between normative codes (~~ideologies~~ or cultural patterns if you prefer). Hence, it would make perfectly logical sense to distinguish "species", "genera", or "families" of sociocultural systems, if it seemed helpful to do so.

Let me turn back now to elaborate just a little more on the notion of sociocultural systems as entities at a higher epistemological level than the organism or even the biological "population" or ecological "community". This has been a difficult concept to sell ever since Spencer introduced society as a "superorganism" and his followers elaborated it *ad absurdum*. Any entity acquires its status on the basis of a more or less stable bonding of components which give the system a continuity and wholeness with distinctive properties. Of great interest in sociology and social psychology is the attempt to trace the evolution of the social bond, which is fundamentally a symbol-based linkage and mapping between individual psyches making possible an empathic interpenetration of perspectives. The evolution of the bond from the pre-human primates is a complex, and still speculative, story that involves the interaction of bipedalism, the opposable thumb, tool use, permanent social organization and role differentiation, long infancy and rudimentary "family" attachments, a repertoire of gestures and vocalizations, and the rapid growth of the cortex. The precise way in which true language emerged will probably never be known, but social psychological theory suggests that it was a prerequisite to the development of uniquely human *self*-consciousness or reflexivity, the ability to take one's self as an object and hence compare self with others and make possible intersubjective standards, norms, and mutual role-taking.

We earlier suggested that biological systems evolve by mapping environmental structures into their own organizations in certain ways. In reading the descriptions of primate life in the field, it strikes one that a significant variation in this process begins: the individuals in a troop begin mapping the psychic states -- the intentions and expectations -- of other members into their own, a process made possible by close and continuous joint activity and by the development of often subtle gestures and sounds that presage or warn of responses that may ensue if a present course of action continues. The actions of one thus become contingent on those of another in more and more complex ways. This process is to be seen as fundamental to the emergence of the full sociocultural level of reality, in which some basic properties of the individual units stem -- not from their physical nature -- but from their membership and socialization in the group. True language then makes possible the internalization of a refined and extensive model or representation of the external world, as well as the higher thought processes that enable one to interact with, and manipulate, that world in an enormously effective way. This internal representation includes a mapping of one's own internal states as well as those of others, making possible the coordinated role structures that constitute social organizations and institutions. Thus the evolution of such macro structures must be understood in terms of the foregoing micro processes at the inter-psychic level.

The above model of sociocultural evolution is, of course, only one possible model, and its use in scientific analysis is not at all demonstrated. It will have to be refined and applied to concrete cases. As a brief example of a possible area of application, we might look at the historical development -- the evolution -- of societal regulatory structures, i.e., the dominant political structures. Without stopping to analyze earlier forms, we may argue that the appearance of substantially democratic forms of social regulation, emerging out of certain prerequisite micro and macro sociocultural processes, represents the evolution of a more adaptive procedure from the objective point of view of the science of cybernetics: for example, a more extensive idea pool available for decision-making, fuller information and feedback flow through the system, and more extensive mapping of the internal as well as external states of the system and environment.

Especially important is the balance between those institutional structures and processes designed to maintain the given structure and stability of the particular politico-economic order and those designed to change it to better adapt to environmental conditions. I have referred to the former as social morphostasis and the latter, social morphogenesis -- using those terms in their more literal sense. From this perspective, the cybernetics engineer could argue that most contemporary social systems have rather poor self-regulators and are poorly adapted control-wise to present and future environmental pressures. Their goal-seeking feedback and feed-forward systems are poorly designed or almost inoperative for a number of obvious cybernetic reasons, e.g., national goals are conflicting, poorly defined and operationalized; the feedback information-gathering and communicating loops contain gaps and inappropriate filters and exist in the most rudimentary institutional state; the comparators allow much too broad a range of error; and lag in the system is dangerously long compared to the frequency of world crises. But in addition, the systems can be expected to have great difficulty in evolving -- adapting their regulatory structures to meet the challenges -- because the morphostatic institutional structures are much more strongly built into the micro and macro organizations than the needed morphogenic arrangements. As a result, pressures tend to build up until the old structure can be changed only via potentially destructive revolutionary forces -- a poor strategy for evolution. An example of a small change in the right direction is the so-called "sunshine" and "sunset" legislation in the U.S. The former attempts to make available to the system more information about its own internal states -- quite essential for any self-regulating system. The latter is even more evolutionary, institutionalizing the dissolution of certain newly created social structures at the end of a specified period unless further debate confirms their continued adaptiveness.

Let me close by mentioning a long-standing pet intuition of mine which many of you here are more qualified to encourage or discourage. This is the development of a non-equilibrial or irreversible social "thermodynamics". The motive here is to bring the study of social phenomena a little closer to other fields of science, especially those concerned to study broad ecosystem relationships. The concept of entropy has been brought into economics, and social systems have been seen as "dissipative structures". The kind of sociological problem that turns me toward such an orientation is the question of whether certain widely desired social conditions are capable of attainment, given the current organization of the social system. For example, can the incidence of crime be reduced substantially given the present shaping of motives and drives under typically institutional structuring of inequality; or can inequality be substantially reduced and democracy increased under given economic and political structural arrangements? Individuals are inherently energetic, motivationally active beings whose drives and actions are shaped and channelled or dissipated by the sociocultural structures, as suggested earlier. Organized social life is essentially an interpsychic information flow system which sustains or changes these structures. Since many scholars see a close relation between information, entropy, free and bound energy, and structure or order, it would appear that the ingredients are present for the development of a theoretical framework that would allow us to understand better the dynamics of societies and answer questions like those above. The problem of defining and operationalizing the appropriate social and psychological variables, however, seems difficult, but worthy of further study. It is inherently an interdisciplinary task.

Applications of Synergy in Business Administration

M. Welge

With 4 Figures

1. Conceptualization

The term synergy is of Greek origin. "Syn" means as much as "with" or "together with" while "ergon" may be translated as "work" or "to work". Accordingly, the literal translation of synergy would be "working together". "Working together" in that sense can be observed in many different fields. It is for example the physical dimension of synergy that accounts for the functioning of a motor car, a fact that could not be explained with sole reference to its parts such as the engine, the tyres, the body of the car, etc. Neither does the total of the organs of living creatures explain their functioning. It is biological synergy that combines the organs to a viable coordinated entity. A company which succeeds in achieving an output exceeding the input to production in value has realized economic synergy. The precondition for the realization of these dimensions of synergy is, however, that there exists a psychological basis, a creative idea, a human spirit, i. e. psychological synergy [1]. In automobile manufacturing neither physical nor economic synergy is possible without the designer's ingenious idea and the know-how of the manager.

The necessity of psychological synergy as a prerequisite to the above-mentioned dimensions of synergy inevitably leads to a normative conceptualization of the phenomenon. Synergy-oriented strategies of design always aim at the optimal combination of individual functions (factors) [2] or an entity, i. e. no mere "and-combinations" (summations) but cooperating units are considered desirable [3], not additive combinations but synergetic combinations are the declared objectives of the design process [4]. For that reason mathematical laws are not applicable to the synergy concept and some widely accepted definitions of synergy such as "2 + 2 = 5-effect" or the "1 + 1 = 3-effect" are misleading rather than being of any help in understanding the synergy phenomenon.

Moreover, definitions of that kind suggest that synergy is always positive and that it occurs automatically. The analysis of concrete synergy-oriented decisions reveals, however, that more often than not synergy is negative . Neither is the release of positive synergy inevitable but highly dependent upon the creative work of management.

Synergy is here defined as the optimal integration of something that was previously differentiated [5]. A further specification is made in so far as the synergy phenomenon is reduced to economic synergy. This means that the relevant norms for the design process are predominantly economic categories like profit, sales, cost, etc. Thus, different functions of the enterprise usually serve as criteria for differentiating synergy into various types. A distinction is made between sales synergy, operations synergy, management synergy and financial synergy [6]. Since the optimal level of integration, i.e. the state in which the maximal synergy potential is released, is likely to change in the course of time, the time dimension is an additional specifying criterion.

Synergy is often confused with the by far more familiar phenomenon of specialization. Advantages of specialization are rendered possible by using a given capability potential for the processing of similar objects, which results in economies of scale. The advantages of synergy are brought about by using the given capability profile in skillfully integrating heterogeneous objects which results in economies of overhead [7].

2. Synergy as Criterion for the Differentiation of the Capability Potential

2.1. Differentiation by Diversification

Various reasons may motivate a company to diversify their production programme. To the positive result of a diversification decision it is essential that the company succeeds in finding a production programme matching the existing capability potential. To achieve this, a diversification strategy has to be selected that promises as high an integration effect as possible, in other words: that sets free as great a synergy potential as possible.

Thus the synergy criterion turns out to be an important heuristic in planning the diversification strategy [8]. Ideally, the objective function of the diversification planning system would aim at finding that diversification strategy which maximizes the synergy potential in all functional areas. This instruction, however, is not operational, since the synergy effects of alternative diversification strategies can neither be exactly quantified nor predicted. The only way out of this situation is that of subjective judgement on the basis of three dimensions (see Fig. 1).
The inclusion of these three dimensions - the time factor as a possible fourth dimension is neglected - results in a three-dimensional space. In its square segments synergy can be estimated as a function of the diversification strategy and the respective functional area. This may be done by means of adjectives(high, medium, unimportant), numerical values or symbols (+, ++, -, --) (see Fig. 2).

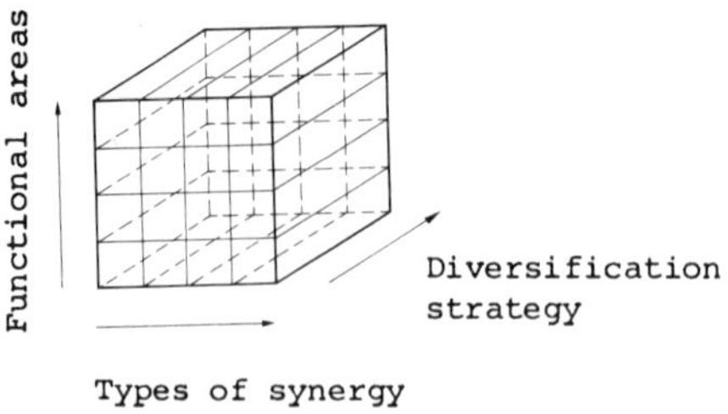

Fig. 1: Dimensions for subjective evaluation of synergetic effects

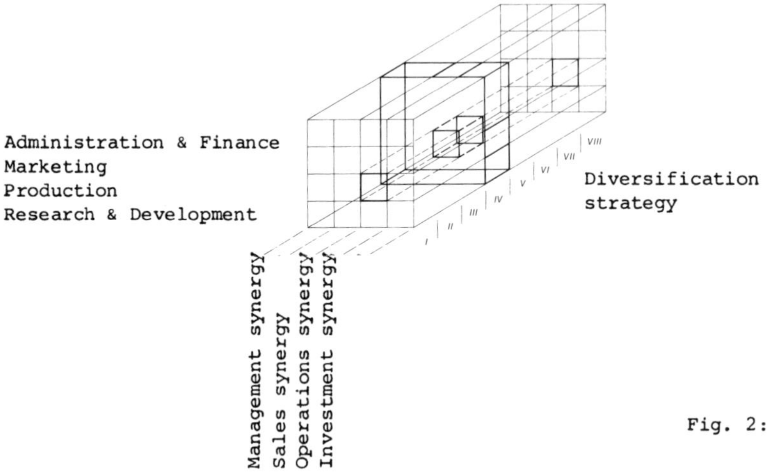

Fig. 2: Evaluation space for synergetic effects

The procedure is explained in some more detail for the case of <u>horizontal diversification</u>.

Synergy / Function	Management	Sales	Operation	Investment
Admin. & Finance	+		+	+
Marketing	++	++	++	++
Production	+		+	+
Research & Devel.			+	+

Fig. 3: Evaluation of Synergy Potential for the Case of Horizontal Diversification

As Fig. 3 shows, the highest synergy potential occurs in marketing. As the capabilities of the marketing managers are directly applicable to the new part of the programme, the risk of wrong decisions ranks low, high management synergy may be expected. Sales synergy is high as well, since there is room for rationalization in services and the new product can benefit from the good will of the old one which makes extra advertising when lancing the product unnecessary (positive spill-over effects). Operations synery is rendered possible under the condition that some distribution channels, existing transportation capacities, basically the whole distribution system can be used for the new product, too. Investment synergy may be expected if the new product requires no additional investment in repair shops or service cars, for example.

Because of the related technology used synergy may also be expected in production. If there are components that can be used for both products , the production of larger quantities will result in a better utilization of the aggregates. The capabilities of the production personnel are available to both products. Besides operations synergy, the use of multifunctional aggregates opens up the possibility of investment synergy.

In administration and finance a high potential of management synergy is likely to occur since experience in areas like planning and accounting is transferable to the new production line. Investment synergy could be realized by the common use of the office building or the central data processing unit whose capacity would otherwise not be fully utilized. Finally, there is the chance of synergy in the R & D area by the joint use of laboratory equipment, test apparatuses, test benches and personnel.

If according to the procedure suggested above a synergy profile is developed for each diversification strategy in question, a rank order of the different alternatives can be set up as to how great a synergy potential they will presumably bring about. Whether the synergy potential is actually realized highly depends on the management that in realizing synergy acts as a catalyst [9].

2. Differentiation by Merger

The synergy phenomenon is also discussed in connection with mergers [10]. Implying that the future value of the merged enterprise exceeds the total of the future values of the individual companies, some authors even speak of synergistic mergers [11]. The positive difference is often referred to as concentration rent [12].

In analogy to planning a diversification strategy the aim in selecting a merger strategy may be to find that strategy which presumably maximizes synergetic effects. The procedure for evaluating alternative merger strategies (horizontal, vertical, technology-concentric, marketing-concentric, pure conglomerate) on the basis of the synergy criterion equals that for evaluating alternative diversification strategies, with the exception that the dimension "diversification strategy" is replaced by "merger strategy".

It must, however, be taken into account that in actually evaluating the synergy potential as shown in Fig. 3 , problems are likely to arise from lack of information. A precondition to the evaluation would be detailed knowledge of the capability profile of all the prospective merger candidates, and this is not always available in an early planning state. Banks, consultants, business associations, etc. might be able to provide useful information.

The importance of synergy in line with decisions on mergers is disputed. An empirical study by KITCHING [13] about reasons for the failure of mergers revealed that in the opinion of the managers interviewed it was financial synergy that ranked highest and not production synergy as one might have expected from theoretical considerations (see also Fig. 4).

One explanation for this unexpected finding may be that when asked about the importance of different types of synergy in decisions on mergers, managers were predominantly thinking of how quickly the respective type of synergy could be realized. This assumption was confirmed by a corresponding additional question. Financial synergy can of course be realized much quicker than production synergy for instance, the latter often requiring time-consuming changes in the organization of the production process. Here the significance of the time dimension in evaluating synergy effects

Strategy \ Function	Production	Sales	Administr.[b)]	Finance
Horizontal	29	100	–	96
Vertical [a)]	–	–	–	–
Technol. concentric	27	72	–	100
Marketg. concentric	72	100	–	100
Pure conglomerate	32	58	–	100
All categor. (average)	36	74	–	100

a) The vertical strategy has been taken out, because the sample was too small
b) Administration was not included because of evaluation difficulties

Fig. 4: Synergy Potential in Line with Typical Merger Strategies
(According to a Study of KITCHING)

becomes evident. It seems therefore advisable to differentiate between potential short-term and long-term synergy in the actual planning process. Besides this, the period of investigation seems rather short (2 to 7 years), and the percentage of conglomerate mergers in the sample is relatively high (45 %), two more reasons for production synergy being considered of lesser importance.

3. Synergy as Criterion for Selecting the Organization Structure

In realizing synergy organization as one of the main instrumental functions of management becomes quite a significant element [14]. The relation between synergy and organization structure is characerized by a high degree of mutual interdependence [15]. The decision whether synergy and structure are to follow strategy or whether structure is to be followed by synergy and strategy represents an important management decision, i. e. irrespective of this decision being made in connection with diversification or merger projects.

In the first case strategy is the independent variable, synergy and structure are dependent variables. The decision problem consists in finding a structure that under the condition of maximizing potential synergy allows as effective a strategy as possible. In the second case structure is the independent variable while synergy and strategy are considered dependent variables. That strategy has to be selected that under the condition of releasing synergy can be realized best with the existing organization structure.

For practical decision problems it seems advisable to proceed from the first case [16]. If the selected diversification / merger strategy promises synergy potential in all functional areas, a close integration of the new tasks into the existing structure on the basis of the expected synergy effects is sensible, since the given capability potential of the enterprise may to a high degree be available to the new tasks as well. The consequent realization of this strategy implies that a large number of dissimilar tasks are integrated into the existing functional areas. The more dissimilar the tasks and the higher the cost of coordination the lesser are the chances of advantages of synergy [17]. If no synergy potential may be expected or only an insignificant one, negative synergy can be avoided by structuring independent units (divisions, legally independent enterprises). In that case holding companies are frequently established to function as linking elements. The selection of such structures often reduces the synergy potential to the area of finance.

If strategy is the independent variable and structure the dependent one, the design objectice is close integration of all functions and tasks promising a synergy potential, while those tasks of which no synergy effects or only insignificant ones can be expected should be decentralized. The spectrum of primarily decentralized structures is wide. It covers the product-oriented differentiation of functional segments and independent divisions as well as holding structures. To allow a balanced decision, possible advantages of synergy arising from specific structures have always to be contrasted with the inter- and intra-departmental costs of coordination they incur.

4. Problems in Measuring Synergy

The problems encountered in measuring synergy are probably the main reasons for the wide-spread scepticism towards the synergy phenomenon. Basically, economic synergy ought to be measured on the basis of its influence on economic criterion variables. Thus, there should be the possibility of measuring by what percentage sales could be increased, by what percentage cost would drop and how much investment would become unnecessary as a result of a strategy aiming at integration as compared to one that does not. In addition, it would have to be possible to specify how these data change as a function of time. Currently synergy effects cannot be quantified in such a direct manner as (1) there is a lack of information concerning the new products and markets, the candidates for prospective mergers, respectively, (2) the period of time between planning and implementing synergy-oriented strategies is rather long, (3) the relation between the criterion variables and synergy is confounded by a variety of uncontrollable factors.

As a possible way out of this dilemma the qualitative method of profile comparison is suggested [18]. The company that intends to diversify or to merge estimates its own capabilities by developing a capability profile. In the rows of the profile matrix the various functions are listed, the columns state qualifications and facilities related to personnel, organization, management, etc. Thus the individual cells contain qualitative and quantitative information about the capability of the respective unit. If then, for instance, the acquisition of a new company or the realization of a new project is considered, capability profiles have to be developed for these planned actions as well. The comparison of these two profile matrices reveals in what areas bottlenecks are likely to occur, i. e. where there is the danger of negative synergy and in what areas successful integration of the new tasks may be expected, i.e. where there is the chance of positive synergy. The result of such a profile comparison is the matrix of expected synergy effects.Again, its columns contain the different functions (general management, marketing, R & D, etc.) while the rows show dimensions like cost advantages in investment and operations in the beginning of the project, those in investment, operations, increase in sales, market position, etc. in the course of the project , The individual cells then contain information like "20% increase in sales", "20 employees in R & D redundant", etc. The total of this information serves as a basis for evaluating the synergy potential that may be expected in the respective areas.

Footnotes

1 Hampden-Turner,Ch.: Synergy as the Optimization of Differentiation and Integration by the Human Personality. In: Studies in Organization Design, J.W. Lorsch a. P.R. Lawrence (eds.), Homewood and Georgetown 1970, pp. 187-196

2 Heckmann, N.: Ein synergetisches Modell des Long-Range Planning, Düsseldorf 1965

3 Schlick, M.: Über den Begriff der Ganzheit. In: Logik der Sozialwissenschaften, E. Topitsch (ed.), Köln and Berlin 1965, pp. 213-224

4 Mee, J.F.: Ideational Items. A Collection. The Synergistic Effect. In: Business Horizons, March 1969, pp. 53-60

5 Hampden-Turner, Ch. op.cit.

6 Ansoff, H.I.: Corporate Strategy, New York 1965; Kitching, J.: Why do Mergers Miscarry? In: Harvard Business Review, Vol. 45,1967,pp. 84-101; Allen III, S.A.: Corporate-Divisional Relationships in Highly Diversified Firms. In: Studies in Organization Design, op. cit. pp. 16-35; Welge, M.K.: Profit-Center-Organisation, Wiesbaden 1975

7 Ansoff, H.I. and Brandenburg, R.G.: A Language for Organization Design. Part I and II. In: Management Science, Vol. 17,1971, pp. B-705-731

8 Ansoff, H.I. op.cit.; Lorsch, J.W. and S.A. Allen III: Managing Diversity and Interdependence. An Organizational Study of Multidivisional Firms, Boston 1973

9 Kitching,J.: op. cit.; Böckel, J.J.: Die Auswahl der Planungsmethode bei industriellen Diversifikationen durch Unternehmungserwerb. PhD Dissertation, University of Munich, 1971

10 Ansoff, H.I. and J.F. Weston: Merger Objectives and Organization Structure. In: Review of Economics and Business, 1963, pp. 49-58; Alberts, W.W.: The Profitability of Growth by Merger. In: The Corporate Merger, W.W. Alberts and J.E. Segall (eds.), Chicago and London 1967, pp. 235-287; Weston, J.F.: The Determination of Share Exchange Ratios in Mergers. In: The Corporate Merger, op.cit., pp. 117-138; Kitching, J. op. cit. ; Sigloch, J.: Unternehmenswachstum durch Fusion, Berlin 1974

11 Sigloch, J.: op. cit.

12 Böckel, J.J.: op.cit.

13 Kitching, J.: op.cit.

14 Grochla, E. (ed.): Management. Aufgaben und Instrumente, Düsseldorf and Wien 1974

15 Ansoff, H.I.: op.cit.; Ansoff, H.I. and R.G. Brandenburg: op.cit.; Fuchs-Wegner,G. and M.K. Welge: Kriterien für die Beurteilung und Auswahl von Organisationskonzeptionen. In: Zeitschrift für Organisation, Vol. 41, 1974, pp. 71-82 and 163-170; Welge, M.K.: op.cit.

16 Chandler,Jr.,A.D.: Strategy and Structure, Cambridge and London 1969; Thanheiser, H.T.: Strategy and Structure of German Industrial Enterprise. DBA Dissertation, Harvard University 1972; Channon, D.F.: The Strategy and Structure of British Enterprise, London 1973

17 Welge, M.K.: op.cit.

18 Ansoff, H.I.: op.cit.

The Linguistic Structure of the Chromosome Genetic Code and Language

F. Fröhlich

All the information that gives structure to and regulates living beings is communicated from one individual of the species to another through the genetic material in sequences of dinucleic acids - DNA. Such genomes constitute, so to speak, different languages for different species, but all species shart (perhaps surprisingly) the same code between the information carrying nucleic acids and the amino acids which go to make up the organism's proteins. Do all species also share other more complex structures which might be called the universal grammar and logic of the biological system? If we could find some universal linguistic structures, that might widen the viewpoint of genetics beyond that of individual molecular happenings, governed as Lucretius said by chance and shape, to that of the physical means of collective interaction.

This paper will begin with the already known universal genetic code and by extending the analogy from the code to larger units of discourse point to some universal linguistic structures which might be embodied in the chromosome.

The biologist who has decoded the genetic code - identifying the correspondences between nucleic acid triplets and amino acids - finds himself in the same position as a decoder who has succeeded in identifying the coded specification for the letters in an unknown language. In one sense he knows everything that can be contained in the language since all possible messages consist of these letters, and there are no more letters. In another sense he is only beginning to understand the language since he has not yet learned the meaning of the combination of the letters in words, nor the grammar nor logic nor usage. For the biologist, as for the linguist, there are two ways to go about learning more about the new language. The first method is to search for correlations between letter sequences and objects or happenings - to learn the meaning of more and more words. Biologically this method consists in identifying sequences of nucleic acids which code for the amino acids making up the proteins of the organism. Linguistically this method leads to the compilation of an enormous dictionary and biologically to the accumulation of correspondences between nucleic acid sequences and proteins. This alternative is the one employed almost exclusively by biologists today. It is the purpose of this paper to suggest that there may be a more sophisticated collective approach to such linguistic situations which might also be fruitfully followed by biologists. That is: one might look for certain systematic but not obvious limitations on the possible sequences of letters which determine those combinations which can make sense. Using this method one would biologically look for limitations on the vast number of mathematically possible combinations of nucleic acids. Although it may be objected that searching for linguistic counterparts to sections of the chromosome is blatantly anthropomorphic, it would seem not to add to the anthropomorphism already contained in the concepts of "code" and "message", but on the contrary to render it less rigid. Indeed by using some of the different perspectives on language worked out painstakingly during the last twenty-five centuries it may be hoped to liberate the "code" "message" model used currently from some of its inhibiting presuppositions. For instance once one has cracked a code, it would seem that the chief remaining problem is to identify the words. Then the grammar and logic will emerge automatically because they are embedded in the minds or linguistic competence of the users of the unknown tongue and of the translators. However, with a non-human language, such as the genetic communication system, these complex structures cannot be presupposed but require special investigation and examination because they may be essentially different. One may thus be misled by the simple model of a code between languages to expect that all that is necessary for complete understanding is an exhaustive dictionary. This may be the "philosophical" cause of the biologists' habit of considering that he possesses a complete explanation once he has identified the individual molecules without looking further for collective properties of the organism as a whole. Thus it might be illuminating to expand the narrow analogue of simple coding to embrace richer aspects of language.

The investigation here will proceed in a linear way from the shortest segments of the chromosome to progressively longer segments, finally including the chromosome as a while and indeed the whole group of an organism's chromosomes, the genome. In the following schematism these chromosome segments will be indicated under "I". The biological entities and functions controlled by these will be indicated under "II". Units of language suggested as analogues for these chromosome segments will be indicated under "III". Progressively longer stages are indicated by "A", "B", "C", etc.

A. I. Nucleic Acids II. None III, Dot-dash-dot of Morse code

B. I. Codons II. Amino Acids III. Letters

C. I. Genes II. Proteins III. Sentences

D. I. Operon or Super-gene II. Several proteins involved in one process such as the enzymes involved in the breakdown of a given substrate III. Several sentences connected by logical relations of implication or contradiction

E. I. Regions of chromosome longer than several genes II. Differentiation III. Longer groups of sentences, paragraphs or whole discourses.

F. I. Whole chromosome II. Self-regulation of function of chromosome as a whole such as asymmetry of protein production on only one strand of the double helix and switch between RNA production to DNA production.

G. I. Genome - all of the chromosomes in an organism or a species II. Avoidance of auto-immunity III. Consistency of language as a whole.

A. The smallest elements of the chromosome - the nucleic acids - have individually no product within the cell. Their linguistic analogue would be the dot-dash-dot of the Morse code.

B. The decoding of the genetic code discovered a (degenerate) correspondence between sequences of three nucleic acids, the condons, and the twenty amino acids which combine to make all of the cells' proteins just as a Morse code established correspondences between groups of dots and dashes and the letters of the alphabet. Thus the linguistic analogue of the condons would appear to be letters.

C. The next longer functionally distinguished sequence is the gene, which contains the information for a protein or for one of the protein's constituent polypeptide chains. The most obvious linguistic analogue for the gene would seem to be the word, specifically a noun. Taking such a linguistic analogue for the gene would reflect the biologist's method of searching for an isolateable enzyeme or protein corresponding to a particular gene. But a verb would reflect the method of identifying a gene by what it does - what processes are hindered by its absence. In a living organism these aspects coalesce. The protein combines functions of noun and verb in a complex subject-acting structure which corresponds more closely to an entire subject-predicate sentence [1].

In a human language once the words are established by convention, all combinations of words are no longer possible in a correct sentence. They are limited by the restrictions of grammatical correctness. According to Noam Chomsky [2] underlying the surface structure of sentences in various languages there is a universal deep grammatical structure common to all possible human languages. For example: The surface grammar sentence "I don't like John's cooking" could be transformed into two deep grammatical structures, "I don't like (the quality of) John's cooking" and "I don't like (the fact of) John's cooking (instead of his wife's cooking)." In the complex sentence "I don't like John's cooking any more than Richard's cooking", both parts could have to have the same deep grammatical structure, "I don't like (the quality of) John's cooking any more than (the quality of) Richard's cooking." The parts could not be expanded into alternative deep grammar structures. "I don't like the quality of John's cooking any more than the fact of Richard's cooking" cannot function within medium of language competence or mind of speaker. Now does an analogue for this universal grammar exist in the chromosome as some sort of general restriction common to the chromosome languages of all species? Its existence would mean that not every possible sequence of codons was possible within a gene - that given a certain sequence there would be limitations on those that could follow. But are there any such restrictions? Mathematical analysis of known sequences of codons has revealed no recognizable regularities in their occurrence [3]. Reflecting this lack of order, the newly assembled polypeptide chain also has no restrictions on its amino acid sequence. But it is not yet an active protein. In order to become active it must assume its correct three-dimensional shape. One might expect that at biological temperatures it could roll up in many ways, but in fact it (nearly) always assumes a very definite configuration. This then imposes severe physical conditions such that the free energy of this configuration must be sufficiently lower than that of all other configurations and possibly conditions on the pathways to reach it. Codon sequences in the chromosome must thus be restricted in an appropriate way. Such physical restriction on the possible codon sequences might be the biological analogue of universal grammar.

In linguistics the possibility of certain words combining in meaningful sentences is manifested within the wider context of the activities of a linguistically competent person (his mind or brain). This wider context might be said to correspond in biology to the cytoplasmic medium of the whole cell within which certain sequences of amino acids can assume a unique configuration and thus become functional proteins which have so to speak a "meaning" in the life of the organism, while others which do not fulfill the physical conditions cannot do so but remain as non-functioning nonsense. Such protein nonsense might constitute a disease in which there were amino acid chains floating inactively in the cytoplasm unable to take on their correct active configuration.

D. Thus far we have reached the situation in which a person would have the linguistic competence to make well-formed sentences but could not connect several sentences with each other in relations of implication, contradiction or even of relevance. Chomsky does not concern himself with relations between sentences. Biologically this would be the case if genes were separate, floating at random within the cytoplasm, but actually they are parts of chromosomes hundreds or even thousands of genes in length. One might thus be led to ask what kind of information beyond the single gene specification for protein manufacture could be contained in this long series. Within the bacterial chromatin (where the situation is simpler or perhaps even essentially different from that in differentiated organisms) the next longer operative segment of nucleic acids is the operon or supergene. This can be a group of genes which codes for several enzymes involved in the breakdown of a given substrate whose presence activates them all. Or it can be a self-regulatory feed-back inhibition group for which the presence of the end product switches off the whole group. Or it can be a feed-back excitation group. These short sequences linking the production of certain proteins which function together might parallel logical connections between several sentences. The feed-back circuits, being essentially logical complexes, would parallel the relations of entailment and contradiction between the protein sentences which they produce. The substrate-induced production of the group of enzymes capable of digesting it would provide an analogue for the mutual relevance of a group of sentences to each other and to the common occasion of their utterance. Thus these short gene sequences, the bacterial operon, suggest an analogue with the logical relations among sentences. The breakdown of operon functioning - linguistically the production of non sequitors or meaningless repetition - might correspond to diseases of lack of control in protein production.

E. I Beyond these short sequences of logically connected gene sentences, one might expect some complex information conveying function in this largest of all molecules, the chromosome. If this were not the case, then the very size of the chromosome, involving as it does complicated operations of unhelixing and separating during mitosis and thereafter being folded again without tangling into a space many time less than its own length, would work against its survival. By this time, having been filtered by countless efficiency tests according to the principle of survival of the fittest, the chromosome would be no longer than necessary - perhaps operon sized fragments. However, it has been found that beyond the sequence of several genes the gene order on the chromosome is not vital. Whole genes can be moved by translocation from one position to another and indeed from one chromosome to another without lethal result to the organism. So it appears that the great quantity of order in longer lengths of chromosome is not being used. But then what is its survival value? We might guess that this enormous molecule has several essentially different types of functions.

E. II Although the chromosomal material of single celled organism communicates normally only with wandering molecules - enzymes, substrates, etc. - and not with the cell as a whole (except during cell division), it may be that extrapolation from this is misleading in the case of a differentiated organism. There the chromosome must communicate at least during the embryonic stage with the surrounding cells in order that its cell becomes one of their type. During differentiation cells become gradually fixed from their originally labile condition into specialised protein production by their neighbouring cells. Does this fixing depend on some long range interaction and communication with surrounding cells through some collective property of the chromosome as a whole (or large sections of it)? It might be that the apparently excessive length of the chromosome is necessary for such a collective function. This could be its apparently absent survival value.

Two hypotheses about how the chromosome might function collectively in the control and maintenance of differentiation have been suggested. One by Cook [4] assumes a model with large control bands of nucleic acid and protein whose order is not significant and with small significant interbands. The large bands would function collectively in some sort of communication with the cell and fix the particular chromosomal configuration in that type of cell so that selected sections of the chromosome would be exposed for the production of relevant proteins. The second hypothesis by Herbert Fröhlich [5] suggests that long range coherent vibrations will lead to resonance between a differentiated cell (with its own characteristic frequency) and the chromosome such that its particular region responding to this characteristic frequency is activated or opened so it can produce appropriate RNA.

E. III Such a communication with the surrounding cells has analogies with more embracing aspects of language than do the grammatical rules or logical relations between several sentences. While grammar examines what sentences <u>could</u> be formed with the language, this aspect of language is concerned with what <u>would</u> be said in various situations - linguistic and extra-linguistic. It presupposes all the possibilities of the language, but it also involves the speaker's elaborate connection with other people - what they say and do and who they are and what is going on in his immediate environment. It considers how the given linguistic competence should be used in a concrete situation. Its Gedankenexperimental question is: "What would you say if...?" For example: the ambiguity of "I don't like John's cooking" is resolved by the context of usage. If someone has said previously "Women should do the cooking" or if someone has choked on the excessive pepper, the sentence would assume a unique alternative structure. Such investigation at the province of ordinary language philosophy, originated by Ludwig Wittgenstein but grown in England, whose motto might be "Meaning is usage". The precise specification of a small part of the chromosome's repetoire of possible utterances by the context of the surrounding cells displays analogies with this broader view of language as usage. Thus the chromosome itself determines what proteins can be made, but the cellular environment, and especially the membrane embedded with proteins, determines which of those possible will be actually made by a certain cell once it is differentiated.

F. I. Not only must the chromosome in some way communicate with its own and neighbouring cells, but under certain circumstances it must communicate also with itself. In linguistic terms it must function as its own meta-language.

F. II There are three cases in which the chromosome must regulate its own function and give directions for the way it itself is to be read. Firstly: the messenger RNA must be produced asymmetrically on only one strand of the unhelixed DNA although both strands would necessarily be simultaneously exposed. Mistakes in this regulation might be expected to lead to the production of mirror image proteins which could be disadvantageous for the cell. This might also sketch the general form of a possible disease. The experimental fact that this asymmetry is maintained only for relatively long sections of chromosome suggests that some collective property is giving instructions for asymmetrical transcription. The second case of reflexive self-regulation of chromosome function is the switch from the production of sections of messenger RNA during interphase to that of DNA before cell division begins. Thirdly: after cell division this must be reversed. These three different ways in which the chromosome's information is to be used must be contained within the chromosome itself. It would thus seem to be in the very centre of the classical situation in which self-referring paradoxes arise.

F. III The theory of meta-languages as developed by Tarski was intended to avoid such paradoxes by drawing a logical distinction between a language that is being talked about - the object language - and the language used to talk about it, the meta-language. Thus the meta-language - whether distinct from (as in mathematical logic) or embedded in (as in everyday language) the object language - can give instructions about object language statements without falling into self-referring paradoxes such as "This statement is false".

F. II The biological division between the production of messenger RNA and replication of DNA might have the logical function of a device for avoiding paradoxes in statements referring to themselves. This would be a control for the exclusive production of DNA or messenger RNA from the unhelixed chromosomes according to the state in the life cycle of the cell. It might involve some form of collective interaction between the chromosome as a whole and the stages of the life cycle of its cell, determining whether the chromosome copies itself as its own complete set of instructions or gives specific instructions for the production of independent proteins. Thus instructions to switch between these logically different functions corresponds to meta-language instructions about the use of the language. If the meta-language did not function properly, then the chromosome would continue reproducing itself with resulting cancerous proliferations. The intermediary step is now recognized - the production of DNA or RNA polymerase - but the question of the controlling interaction - why the right polymerase is produced at the right time in the cell's life - has scarcely been posed as a problem. One answer might be suggested by the hypothesis of long range coherence [5]. This could lead to an energy barrier against the unrolling of the chromosome into a different form to enter mitosis until the state of the cell warranted such a change. Cancer would thus be the disease arising from self-referring paradoxes.

G. I One might even look beyond the possible long range properties of the single chromosome and ask whether any chromosomes whatever could combine in some viable organism (or species) or whether there are limitations on those which can be so combined - whether there are conditions of consistency for the whole genome.

What would "inconsistency" in the genome mean? The genome of a species must somehow contain the information to make antibodies, not only against the currently circulating diseases or possible new diseases which may attack it, but also against any foreign proteins, which entering the organism (with no harmful intent so to speak) act as antigens provoking the production of specific antibodies. There is a sliding scale of species similarity within which foreign proteins are accepted, beyond which they are rejected. Now what prevents an organism from producing on one part of its chromosome a protein which is antigenic to proteins produced on other parts? The presence of these would be combatted by the production of antibodies. This indeed happens in auto-immune diseases, but not normally. That these auto-immune diseases are the rare exception rather than the rule indicates that normally the chromosome set contains no genes whose relation to the rest of its genes is anti-genic. This could give a meaning to the concept of mutual self-consistency among all of the genes in the genome. How this self-consistency is established and maintained involves one of the most vexed questions in immunology, that of self-recognition. Although in immunology the question is posed most acutely when the mechanism for self-recognition is not acting - either in the production of antibodies to combat invading foreign bodies or when self-recognition breaks down and anti-self antibodies are produced - a more fundamental question is how it normally functions. When the genome is functioning normally, in logical terms one might say that inconsistent statements are not being produced from one and the same body of information. This is a meta-mathematical problem and not one that has a solution in everyday language wherein the generation of inconsistent statements is all too possible. However, the question has been posed for the system of real numbers for which it was hoped to prove that inconsistent statements could not be generated out of the laws and operations of arithmetic. Gödel showed that within these rules one could not prove that an inconsistency would not arise from their application. Perhaps the genome has solved this meta-mathematical problem which has remained insoluble for us. Auto-immune diseases could be the rare inconsistencies that arise when the genome s consistency mechanism fails to function. Thus once again disease might appear as a logical mistake.

This paper has suggested one among many possible systems of analogies between chromosome information and human language, distinguishing individual from collective operations. In biology this distinction falls between molecular interactions and long range physical forces and stabilities. In linguistics the distinction is between letter by letter coding and the collective complexity of language. These alternative approaches should not be mutually exclusive but cooperative. It is hoped thus to extend the range of concepts available in biology toward collective phenomena.

References

1. F. Fröhlich: in Cooperative Phenomena, ed. H. Haken, W. Wagner (Springer, Berlin, Heidelberg, New York 1973) VII-XII
2. N. Chomsky: Language and Mind (MIT Press)
3. J. Monod: Chance and Necessity
4. R.R. Cook: Nature 245, 23 (1973)
5. H. Fröhlich: Rivista de Nuovo Cimento 7 (in print)

Cooperative Phenomena

Editors: H. Haken, M. Wagner
86 figures, 13 tables, XIII, 458 pages. 1973
ISBN 3-540-06203-3

Contents

Springer-Verlag
Berlin Heidelberg New York